Chemistry 1 Class Notes

Mr. Bigler
Lynn Public Schools
September 2018

http://www.mrbigler.com/Chemistry-1

Copyright © 2006–2018 Mr. Bigler.

This document is licensed under a Creative Commons Attribution-NonCommercial-ShareAlike 4.0 International (CC BY-NC-SA 4.0) License.

You are free to:

Share — copy and redistribute the material in any medium or format

Adapt — remix, transform, and build upon the material

The licensor cannot revoke these freedoms as long as you follow the license terms.

Under the following terms:

Attribution — You must give appropriate credit, provide a link to the license, and indicate if changes were made. You may do so in any reasonable manner, but not in any way that suggests the licensor endorses you or your use.

NonCommercial — You may not use the material for commercial purposes.

ShareAlike — If you remix, transform, or build upon the material, you must distribute your contributions under the same license as the original.

No additional restrictions — You may not apply legal terms or technological measures that legally restrict others from doing anything the license permits.

Notices:

You do not have to comply with the license for elements of the material in the public domain or where your use is permitted by an applicable exception or limitation.

No warranties are given. The license may not give you all of the permissions necessary for your intended use. For example, other rights such as publicity, privacy, or moral rights may limit how you use the material.

If you have specific questions about your desired use of this material, or if you would like to request a specific exemption to these license terms, please contact the author at mrbigler@mrbigler.com

ISBN-13: 978-1517557300

ISBN-10: 1517557305

This is a set of class notes for a first-year high school chemistry course. These notes can be used for any honors or CP1 chemistry course by omitting information that is specific to the higher-level course.

This hardcopy is provided so that you can fully participate in class discussions without having to worry about writing everything down.

While a significant amount of detail is included in these notes, they are intended to supplement the textbook, classroom discussions, experiments and activities. These class notes and the textbook discussion of the same topics are intended to be complementary. In some cases, the notes and the textbook differ in method or presentation, but the chemistry is the same. There may be errors and/or omissions in any textbook. There are almost certainly errors and omissions in these notes, despite my best efforts to make them clear, correct, and complete.

About the Homework Problems

The homework problems include a mixture of easy and challenging problems. (Remember that these notes are intended for use in both regular-level and honors-level chemistry classes.) *The process of making yourself smarter involves challenging yourself, even if you are not sure how to proceed.* By spending at least 10 minutes attempting each problem, you build neural connections between what you have learned and what you are trying to do. Even if you are not able to get the answer, when we go over those problems in class, you will reinforce the neural connections that led in the correct direction.

Answers to the problems are often provided so you can check your work and see if you are on the right track. Do not simply write those answers down, in order to receive credit for work you did not do. This will give you a false sense of confidence, and will actively prevent you from using the problems to make yourself smarter. *You have been warned.*

Using These Notes

As we discuss topics in class, you will want to add your own notes to these. If you have purchased this copy, you are encouraged to write directly in it, just as you would write in your own notebook. If this copy was issued to you by the school and you intend to return it at the end of the year, you will need to write your supplemental notes on separate paper. If you do this, be sure to write down page numbers in your notes, to make cross-referencing easier.

You should bring these notes to class every day, because lectures and discussions will follow these notes, which will also be projected onto the SMART board.

These notes, and the course they accompany, are designed to follow both the 2016 Massachusetts Curriculum Frameworks, and the 2016 Massachusetts Curriculum Frameworks, which are based on the Next Generation Science Standards (NGSS). The notes also utilize strategies from the following popular teaching methods:

- Each topic includes Mastery Objectives and Success Criteria, as presented in the *Studying Skillful Teaching* course, from Research for Better Teaching (RBT).

- Each topic includes Tier 2 vocabulary words and language objectives, from the Rethinking Equity and Teaching for English Language Learners (RETELL) course.

- Notes are organized in two-column notes format, with a top-down web at the beginning of each unit and a page for students to summarize the unit at the end, from Keys To Literacy.

The order of topics in a chemistry course is a hotly-debated subject. These notes and the course they accompany are organized as follows:

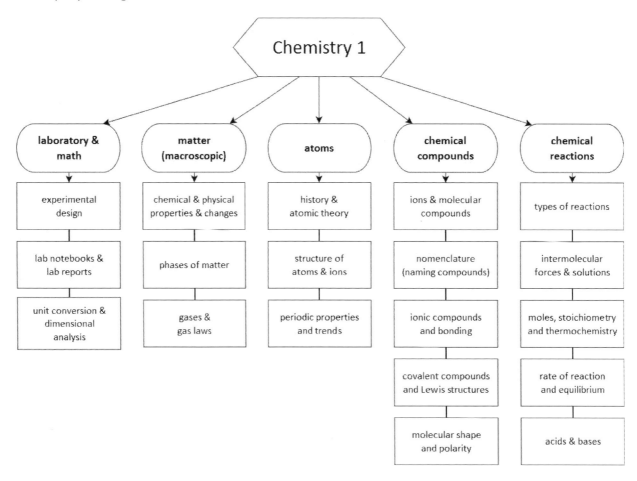

The most controversial decision was to place gases & gas laws at the beginning of the course, right after matter. The rationale is to keep the macroscopic study of matter together, and to have a topic that is rich in lab experiments and demonstrations early in the course. One downside to this approach is that moles need to be introduced in a rudimentary fashion early in the course, and re-introduced and expanded on later.

Acknowledgements

These notes would not have been possible without the assistance of many other people. I cannot mention everyone, but I would particularly like to thank:

- Every student I have ever taught, for helping me learn how to teach, and how to explain and convey challenging concepts.

- Every teacher I have worked with, for their kind words, sympathetic listening, helpful advice and suggestions, and other contributions great and small that have helped me to enjoy and become competent at the profession of teaching.

- Many of the department heads and principals I have worked with, for mentoring and encouraging me and allowing me to develop my own teaching style. In particular, Mark Greenman, Marilyn Hurwitz, Scott Gordon, Barbara Osterfield, Wendell Cerne, John Graceffa, Maura Walsh, Lauren Mezzetti, and Tom Strangie.

- The chemistry teachers I have worked with over the years who have generously shared their time, expertise, and materials. In particular, Kathy McGrane, who shared her teaching notes with me, which ultimately inspired me to create these; Kirstin Bunker, who gave significant help with the Keys to Literacy® top-down webs; Beth Hocking, with whom I have had more discussions than I can count about creative ideas for teaching chemistry; and Harvey Gendreau, whose contributions to the AP Chemistry email discussion forums and ChemEd conferences have taught me much about the chemistry that high school students need to understand in order to be successful.

- Everyone else who has shared insights, stories, and experiences in chemistry, many of which are reflected in some way in these notes.

I am reminded of Sir Isaac Newton's famous quote, *"If I have seen further it is by standing on the shoulders of giants."*

Table of Contents

MA Curriculum Frameworks for Chemistry 7
Two-Column Notes 9
1. What Is Chemistry? 15
2. Laboratory 25
3. Math & Measurement 71
4. Matter 107
5. Gases 127
6. Atomic Structure 161
7. Nuclear Chemistry 189
8. Electronic Structure 211
9. Periodicity 251
10. Nomenclature & Formulas 275
11. Bonding & Molecular Geometry 303
12. Intermolecular Forces 327
13. The Mole 347
14. Solutions 369
15. Chemical Reactions 399
16. Oxidation & Reduction 431
17. Stoichiometry 443
18. Thermochemistry (Heat) 469
19. Kinetics & Equilibrium 511
20. Acids & Bases 531
    Appendix: Chemistry Reference Tables 549
    Periodic Table of the Elements 563
    Index 564

MA Curriculum Frameworks for Chemistry

Standard	Topics	Chapter(s)
HS-PS1-1	Periodic table & periodic trends: ionization energy, atomic & ionic radius	9
HS-PS1-2	Types of chemical reactions. Predicting products. Intermolecular Forces (IMF) *vs.* physical state at room temperature	12, 14, 15
HS-PS1-3	IMF *vs.* bulk properties (melting point/boiling point, density, vapor pressure, *etc.*)	11, 12
HS-PS1-4	Energy of reaction, heat of formation	18
HS-PS1-5	Reaction kinetics as related to Kinetic Molecular Theory (KMT) & collision theory	19
HS-PS1-6	Equilibrium, Le Châtelier's principle, as related to KMT	19
HS-PS1-7	Conservation of mass & atoms, balancing equations, law of constant composition, moles, stoichiometry, percent yield	4, 13, 15, 17
HS-PS1-9(MA)	pH as measure of acid/base strength, Arrhenius & Brønsted-Lowry theories as related to bases & monoprotic acids	20
HS-PS1-10(MA)	Oxidation/reduction (REDOX) reactions, oxidation numbers, predicting products of REDOX reactions, conceptual electrochemistry	16
HS-PS1-11(MA)	Mixture separation based on chemical & physical properties (*e.g.*, chromatography, distillation, centrifuging, precipitation reactions)	2, 4, 14
HS-PS2-6	Molecular structures of ionic compounds, acids, bases, metals & polymers	10, 11
HS-PS2-7(MA)	Solvent polarity & why ions dissolve in polar solvents	11, 12, 14
HS-PS2-8(MA)	KMT & gases (electrostatic forces, interactions between molecules in solids, liquids & gases), combined gas law	5
HS-PS3-4b	Conservation of energy with respect to enthalpy, entropy, and free energy (conceptual)	18

MA Science Practices

Practice	Description
SP1	Asking questions.
SP2	Developing & using models.
SP3	Planning & carrying out investigations.
SP4	Analyzing & interpreting data.
SP5	Using mathematics & computational thinking.
SP6	Constructing explanations.
SP7	Engaging in argument from evidence.
SP8	Obtaining, evaluating and communicating information.

Two-Column Notes

Unit: Introduction

MA Curriculum Frameworks (2016): N/A

MA Curriculum Frameworks (2006): N/A

Mastery Objective(s): (Students will be able to…)

- Use the two-column note-taking system to take effective notes, or add to existing notes.

Success Criteria:

- Notes are in two columns with appropriate main ideas on the left and details on the right.

Language Objectives:

- Understand and describe how two-column notes are different from other forms of note-taking.

Notes:

The two-column note-taking system is based on the Cornell note-taking system, which was developed in the 1950s at Cornell University. Besides being a useful system for note-taking in general, it is an especially useful system for interacting with someone else's notes (such as these) in order to get more out of them.

The main features of the two-column note-taking system are:

1. The main section of the page is for the details of what actually gets covered in class.

2. The left section (officially 2½ inches, though I have shrunk it slightly to 2¼ inches for these notes) is for "big ideas"—the organizational headings that help you organize these notes and find details that you are looking for. These have been left blank for you to add throughout the year, because the process of deciding what is important is a key element of understanding and remembering.

3. In the Cornell Notes system, the bottom section (2 inches) is officially for you to add a 1–2 sentence summary of the page in your own words. This is always a good idea, but you may also choose to use that space for other things you want to remember that aren't in these notes.

Use this space for summary and/or additional notes:

How to Get Nothing Worthwhile Out Of These Notes

Because this book serves as a combination of your textbook and a set of notes, you may be tempted to sleep through class because "it's all in the book," and then use these notes look up how to do the homework problems when you get confused. If you do this, you will learn very little chemistry, and you will find this class to be both frustrating and boring.

How to Get the Most Out Of These Notes

These notes are provided so you can pay attention and participate in class without having to worry about writing everything down. However, because active listening, participation and note-taking improve your ability to understand and remember, it is important that you interact with these notes and the discussion.

The "Big Ideas" column on the left of each page has been deliberately left blank. This is to give you the opportunity to go through your notes and categorize each section according to the big ideas it contains. Doing this throughout the year will help you keep the information organized in your brain—it's a lot easier to remember things when your brain has a place to put them!

If we discuss something in class that you want to remember, mark or highlight it in the notes! If we discuss an alternative way to think about something that works well for you, write it in! You paid for these notes—don't be afraid to use them!

There is a summary section at the bottom of each page, and a Keys to Literacy® style summary template at the end of each chapter. Utilize both of these. If you can summarize something, you understand it; if you understand something, it is much easier to remember.

Use this space for summary and/or additional notes:

Reading & Taking Notes from a Textbook

Unit: Introduction

MA Curriculum Frameworks (2016): N/A

MA Curriculum Frameworks (2006): N/A

Mastery Objective(s): (Students will be able to...)
- Use information from the organization of a textbook to take well-organized notes.

Success Criteria:
- Section headings from text are represented as main ideas.
- All information in section summary is represented in notes.
- Notes include page numbers.

Language Objectives:
- Understand and be able to describe the strategies presented in this section.

Notes:

If you read a textbook the way you would read a novel, you probably won't remember much of what you read. Before you can understand anything, your brain needs enough context to know how to file the information. This is what Albert Einstein was talking about when he said, "It is the theory which decides what we are able to observe."

When you read a section of a textbook, you need to create some context in your brain, and then add a few observations to solidify the context before reading in detail.

René Descartes described this process in 1644 in the preface to his *Principles of Philosophy*:

> "I should also have added a word of advice regarding the manner of reading this work, which is, that I should wish the reader at first go over the whole of it, as he would a romance, without greatly straining his attention, or tarrying at the difficulties he may perhaps meet with, and that afterwards, if they seem to him to merit a more careful examination, and he feels a desire to know their causes, he may read it a second time, in order to observe the connection of my reasonings; but that he must not then give it up in despair, although he may not everywhere sufficiently discover the connection of the proof, or understand all the reasonings—it being only necessary to mark with a pen the places where the difficulties occur, and continue reading without interruption to the end; then, if he does not grudge to take up the book a third time, I am confident that he will find in a fresh perusal the solution of most of the difficulties he will have marked before; and that, if any remain, their solution will in the end be found in another reading."

Use this space for summary and/or additional notes:

Reading & Taking Notes from a Textbook

Big Ideas	Details
	The following 4-step system takes about the same amount of time you're used to spending on reading and taking notes, but it will probably make a tremendous difference in how much you understand and remember.

1. Make a two-column notes template. Copy the title/heading of each section as a big idea in the left column. (If the author has taken the trouble to organize the textbook, you should take advantage of it!) Leave about ¼ to ½ page of space for the details for each big idea. (Don't do anything else yet.) This should take about 1–2 minutes.

2. Do not write anything yet! Look through the section for pictures, graphs, and tables. Take a minute to look at these—the author must have thought they were important. Also read over (but don't try to answer) the homework questions/problems at the end of the section. (For the visuals, the author must think these things illustrate something that is important enough to dedicate a significant amount of page real estate to it. For the homework problems, these illustrate what the author thinks you should be able to do once you know the content.) This process should take about 10–15 minutes.

3. Actually read the text, one section at a time. For each section, jot down keywords and sentence fragments that remind you of the key ideas. You are not allowed to write more than the ¼ to ½ page allotted. (You don't need to write out the details—those are in the book, which you already have!) This process is time consuming, but shorter than what you're probably used to doing for your other teachers.

4. Read the summary at the end of the chapter or section—this is what the author thinks you should know now that you've finished the reading. If there's anything you don't recognize, go back and look it up. This process should take about 5–10 minutes.

You shouldn't need to use more than about one sheet of paper (both sides) per 10 pages of reading!

Use this space for summary and/or additional notes:

Taking Notes on Math Problems

Unit: Introduction

MA Curriculum Frameworks (2016): SP5

MA Curriculum Frameworks (2006): N/A

Mastery Objective(s): (Students will be able to…)

- Take notes on math problems that both show and explain the steps.

Success Criteria:

- Notes show the order of the steps, from start to finish.
- A reason or explanation is indicated for each step.

Language Objectives:

- Be able to describe and explain the process of taking notes on a math problem.

Notes:

If you were to copy down a math problem and look at it a few days or weeks later, chances are you'll recognize the problem, but you won't remember how you solved it.

Solving a math problem is a process. For notes to be useful, they need to describe the process as it happens, not just the final result.

If you want to take good notes on how to solve a problem, you need your notes to show what you did at each step.

For example, consider the following problem:

> How much heat is needed to increase the temperature of a 25 g sample of a metal with a specific heat capacity of $0.375 \frac{J}{g \cdot °C}$ by 40 °C?

The process of solving this problem involves applying the equation $Q = mC\Delta T$, where Q is the amount of heat, m is the mass of the metal, C is the specific heat capacity of the metal, and ΔT is the temperature change. (Note that ΔT is only one quantity, even though it uses two symbols.)

Use this space for summary and/or additional notes:

Taking Notes on Math Problems

A good way to document the process is to use a T-chart, in which you show the steps of the solution on the left side, and you write an explanation of what you did and why for each step on the right side.

For this problem, your T-chart might look like the following:

Step	Description/Explanation
$m = 25$ g $C = 0.375 \frac{J}{g \cdot °C}$ $\Delta T = 40 °C$ $Q =$ quantity desired	Declare variables.
$\underline{Q} = \underline{m} \, \underline{C} \, \underline{\Delta T}$	Choose a formula that gives the desired quantity. Make sure we have values for the other variables.
$Q = m \quad C \quad \Delta T$ $Q = (25)(0.375)(40)$ $Q = 375$	Look up the values of any constants needed to solve the problem. Substitute for the variables and solve. (Show as much of the algebra as you think you'll need later.)
$Q = \boxed{380 \text{ J}}$	Round to the appropriate number of significant figures, include the units, and box the final answer.

You will notice that the answers are provided for many of the homework problems in these notes. This is because students are often unsure of whether they are doing a problem correctly until they see whether or not they got the correct answer. This means that if your teacher assigns these problems for homework, <u>it is not sufficient to just write down the answer</u>.

When a teacher says "show work," this does not necessarily mean you should show what <u>you</u> did to obtain the answer. Rather, it means:

1. Declare variables and assign them to values (with units).
2. Write down the relevant equation.
3. Substitute numbers for variables in the equation.
4. Solve for the missing variable.
5. Round to the appropriate number of significant figures and tack on the correct units.

This process would correspond to the left column of the above T-chart.

Use this space for summary and/or additional notes:

Introduction: What Is Chemistry?

Unit: What is Chemistry?

Topics covered in this chapter:

Chemistry 16

The Scientific Method 19

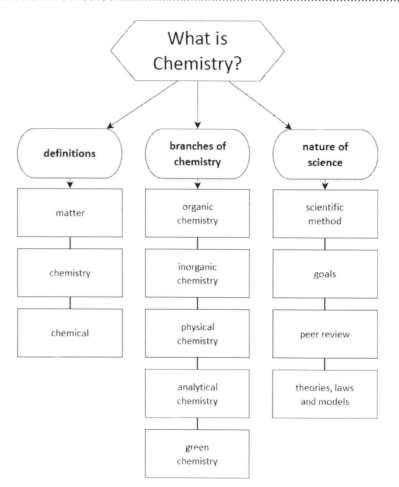

Standards addressed in this chapter:

Massachusetts Curriculum Frameworks & Science Practices (2016):

No 2016 MA curriculum frameworks are specifically addressed in this chapter.

Massachusetts Curriculum Frameworks (2006):

No 2006 MA curriculum frameworks are specifically addressed in this chapter.

Use this space for summary and/or additional notes:

Chemistry

Unit: What Is Chemistry?

MA Curriculum Frameworks (2016): N/A

MA Curriculum Frameworks (2006): N/A

Mastery Objective(s): (Students will be able to...)
- Explain what chemistry is and what is studied in different branches of chemistry.

Success Criteria:
- Explanation describes what is studied in each of the branches of study described in this section.

Tier 2 Vocabulary: matter

Language Objectives:
- Understand and correctly use terms relating to each branch of chemistry.

Notes:

matter: the "stuff" that everything is made of. Matter is anything that has mass and takes up space (has volume).

chemistry: the study of matter, its properties, how it behaves, how it's put together, and how it can be changed or rearranged.

chemical: a specific substance (regardless of size or shape) that has a specific arrangement of the atoms that it's made of, and has specific properties because of that arrangement.

Use this space for summary and/or additional notes:

Chemistry

Page: 17
Unit: What Is Chemistry?

The major units we will study this year include:

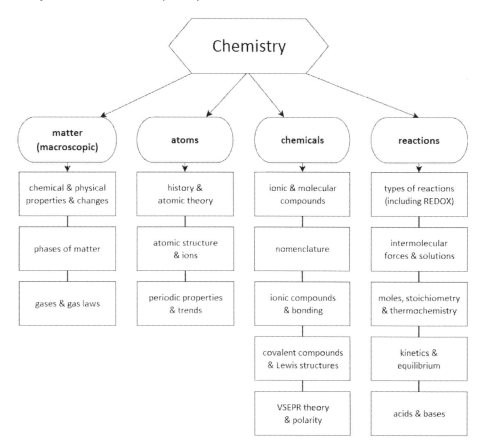

- **macroscopic properties of matter**
    - solids & liquids
    - gases
    - when a change is or is not caused by a chemical reaction
- **atoms**
    - what they're made of (protons, neutrons & electrons)
    - what properties they have (periodic table and periodic properties)
- **chemicals**
    - how atoms combine
    - how the names tell us what's they're made of
    - the shapes of the molecules or crystals (bonding and molecular geometry)
    - how the shapes affect the properties they have
    - dissolving in water (solutions) and forces between molecules

Use this space for summary and/or additional notes:

- **chemical reactions**
  - different ways atoms can rearrange (chemical reactions & equations)
  - calculating how much of the reactants you use and products you make (stoichiometry)
  - heat produced (or consumed) by chemical reactions
  - how fast chemicals react (kinetics)
  - how much chemicals react (equilibrium)
  - acids & bases

Branches of Chemistry

The study of chemistry is divided into different branches, including:

organic chemistry: the study of chemicals and reactions involving molecules that contain carbon and hydrogen.

inorganic chemistry: the study of chemicals and reactions involving molecules that do not contain both carbon and hydrogen.

biochemistry: the study of chemicals that play important roles in biological processes, such as amino acids, lipids, and sugars.

physical chemistry: the study of energy changes in chemistry. Some sub-fields include thermodynamics (the study of heat energy), statistical mechanics (the study of molecular collisions and momentum), and quantum mechanics (the study of discrete energy changes at the sub-atomic level).

analytical chemistry: quantitative aspects of chemistry, such as determining what a chemical is made of, how much of it reacts under certain conditions, *etc.*

green chemistry: the study of making decisions about how chemicals are made or used in order to reduce the impact on the environment.

Use this space for summary and/or additional notes:

Big Ideas	Details	The Scientific Method	Page: 19
			Unit: What Is Chemistry?

The Scientific Method

Unit: What Is Chemistry?

MA Curriculum Frameworks (2016): SP1, SP2, SP6, SP7

MA Curriculum Frameworks (2006): N/A

Mastery Objective(s): (Students will be able to...)

- Explain how the scientific method can be applied to a problem or question.

Success Criteria:

- Steps in a specific process are connected in consistent and logical ways.
- Explanation correctly uses appropriate vocabulary.

Tier 2 Vocabulary: theory, model, claim, law, peer

Language Objectives:

- Understand and correctly use terms relating to the scientific method, such as "peer review".

Notes:

The scientific method is a fancy name for "figure out what happens by trying it."

In the middle ages, "scientists" were called "philosophers." These were church scholars who decided what was "correct" by a combination of observing the world around them and then arguing and debating with each other about the mechanisms and causes.

During the Renaissance, scientists like Galileo Galilei and Leonardo da Vinci started using experiments instead of argument to decide what really happens in the world.

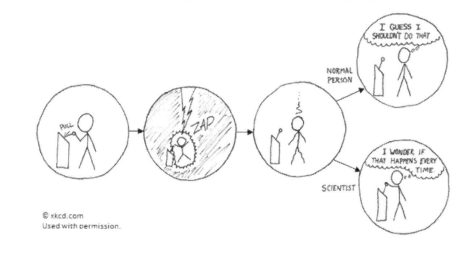

Use this space for summary and/or additional notes:

A Mindset, Not a Recipe

The scientific method is a mindset, which basically amounts to "let nature speak". Despite what you may have been taught previously, the scientific method does not have specific "steps," and does not necessarily require a hypothesis.

The scientific method looks more like a web, with testing ideas (experimentation) at the center:

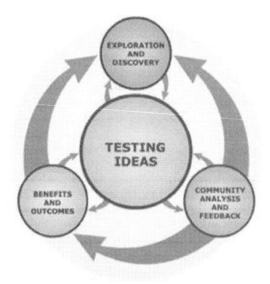

from the *Understanding Science* website[*]

---

[*] Understanding Science. 2018. University of California Museum of Paleontology. 1 July 2018 <http://www.understandingscience.org>. Reprinted with permission.

Use this space for summary and/or additional notes:

The Scientific Method

Page: 21
Unit: What Is Chemistry?

Big Ideas	Details

Each of the circles in the above diagram is a broad area that contains many processes:

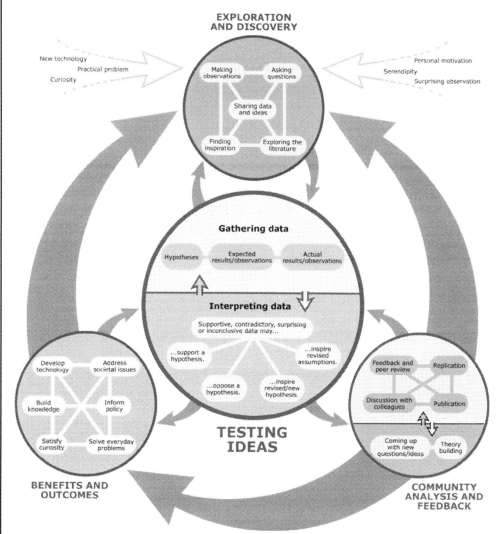

from the *Understanding Science* website

When scientists conclude something interesting that they think is important and want to share, they state it in the form of a *claim*, which states that something happens, under what conditions it happens, and in some cases gives a possible explanation.

Use this space for summary and/or additional notes:

The Scientific Method

Big Ideas | Details | Unit: What Is Chemistry?

Before a claim is taken seriously, the original scientist and any others who are interested try everything they can think of to disprove the claim. If the claim holds up despite many attempts to disprove it, the claim gains support.

peer review: the process by which scientists scrutinize, evaluate and attempt to disprove each other's claims.

If a claim has gained widespread support among the scientific community and can be used to predict the outcomes of experiments (and it has *never* been disproven), it might eventually become a theory or a law.

theory: a claim that has never been disproven, that gives an explanation for a set of observations, and that can be used to predict the outcomes of experiments.

model: a way of viewing a set of concepts and their relationships to one another. A model is one type of theory.

law: a claim that has never been disproven and that can be used to predict the outcomes of experiments, but that does not attempt to model or explain the observations.

Note that the word "theory" in science has a different meaning from the word "theory" in everyday language. In science, a theory is a model that:

- *has never failed* to explain a collection of related observations
- *has never failed* to successfully predict the outcomes of related experiments

For example, the theory of evolution *has never failed* to explain the process of changes in organisms caused by factors that affect the survivability of the species.

If a repeatable experiment contradicts a theory, and the experiment passes the peer review process, the theory is deemed to be wrong. If the theory is wrong, it must either be modified to explain the new results, or discarded completely.

Use this space for summary and/or additional notes:

Theories *vs.* Natural Laws

The terms "theory" and "law" developed organically over many centuries, so any definition of either term must acknowledge that common usage, both within and outside of the scientific community, will not always be consistent with the definitions.

Nevertheless, the following rules of thumb may be useful:

A *theory* is a model that attempts to explain <u>why</u> or <u>how</u> something happens. A *law* simply describes or quantifies what happens without attempting to provide an explanation. Theories and laws can both be used to predict the outcomes of related experiments.

> For example, the *Law of Gravity* states that objects attract other objects based on their masses and distances from each other. It is a law and not a theory because the Law of Gravity does not explain *why* masses attract each other.

> *Atomic Theory* states that matter is made of atoms, and that those atoms are themselves made up of smaller particles. The interactions between these particles are used to explain certain properties of the substances. This is a theory because we cannot see atoms or prove that they exist. However, the model gives an explanation for *why* substances have the properties that they do.

A theory cannot become a law for the same reasons that a definition cannot become a measurement, and a postulate cannot become a theorem.

Use this space for summary and/or additional notes:

Introduction: Laboratory

Unit: Laboratory

Topics covered in this chapter:

Designing & Performing Experiments 27
Laboratory Equipment 33
Accuracy & Precision 36
Uncertainty & Error Analysis 38
Significant Figures 47
Keeping a Laboratory Notebook 55
Internal Laboratory Reports 59
Formal Laboratory Reports 64

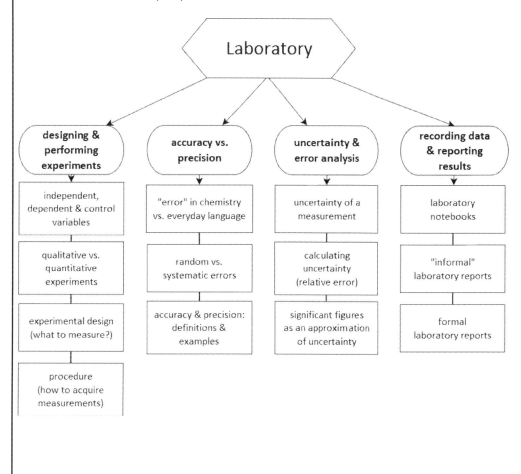

Use this space for summary and/or additional notes:

Introduction: Laboratory

Big Ideas | Details

Standards addressed in this chapter:

Massachusetts Curriculum Frameworks & Science Practices (2016):

SP3: Planning & carrying out investigations

SP4: Analyzing & interpreting data

SP6: Constructing explanations

SP8: Obtaining, evaluating & communicating information

Massachusetts Curriculum Frameworks (2006):

No MA curriculum frameworks are specifically addressed in this chapter.

Use this space for summary and/or additional notes:

Designing & Performing Experiments

Unit: Laboratory

MA Curriculum Frameworks (2016): SP1, SP3, SP8

MA Curriculum Frameworks (2006): N/A

Mastery Objective(s): (Students will be able to...)

- Create a plan and procedure to answer a question through experimentation.

Success Criteria:

- Experimental Design utilizes backward design.
- Experimental Design uses logical steps to connect the desired answer or quantity to quantities that can be observed or measured.
- Procedure gives enough detail to set up experiment.
- Procedure establishes values of control and independent variables.
- Procedure explains how to measure dependent variables.

Tier 2 Vocabulary: inquiry, independent, dependent, control

Language Objectives:

- Understand and correctly use the terms "dependent variable" and "independent variable."
- Understand and be able to describe the strategies presented in this section.

Notes:

Unlike many biology experiments in which the purpose is to observe something, most chemistry experiments involve observing changes to find out *what* happens, *under what conditions* it happens, and measuring *to what extent* it happens. Part of the purpose of chemistry experiments is to get practice making things happen, but another part of the purpose is for you to figure out *how* to get them to happen and how to measure them. This makes chemistry well-suited for teaching you how to design experiments.

The education "buzzword" for this is **inquiry-based experiments**, which means you (or your lab group) will need to figure out what to do to perform an experiment that answers a question about some aspect of chemistry. In this course, you will usually be given only an objective or goal and a general idea of how to go about achieving it. You and your lab group (with help) will decide the specifics of what to do, what to measure (and how to measure it), and how to make sure you are getting good results. This is a form of *guided* inquiry.

Use this space for summary and/or additional notes:

Designing & Performing Experiments

Framing Your Experiment

Experiments are motivated by something you want to find out, observe, or calculate.

Independent, Dependent, and Control Variables

In an experiment, there is usually something you are doing, and something you are measuring or observing.

independent variable: the conditions you are setting up. These are the predetermined values (the ones you pick). Because you choose the values, they are *independent* of what happens in the experiment. For example, if you are trying to figure out how much sugar you can dissolve in water at different temperatures, you are choosing the temperatures to test, so temperature is the *independent* variable.

dependent variable: the things that happen in the experiment. These are the numbers you measure as a result of the experiment, which means they are *dependent* on what happens in the experiment. For example, if you are trying to figure out how much sugar you can dissolve in water, you don't know how much sugar dissolves until you measure it. This means the amount of sugar that dissolves is the *dependent* variable.

control variable: things that you need to keep constant in the experiment. If there are a lot of factors that could affect your dependent variable, and you want to test the effect of one of them, you need to keep the other ones the same. Otherwise, you don't know what caused the changes that you are seeing.

Control variables are usually parameters that could become independent variables in other experiments. For example, if you are trying to figure out how much sugar you can dissolve in water at different temperatures, you need to use the same volume (amount) of water each time. This means the volume of water that you use is a *control* variable.

If someone asks what your independent, dependent and control variables are, the question simply means:

- "What did you vary on purpose (independent variables)?"
- "What did you measure (dependent variables)?"
- "What did you keep the same for each trial (control variables)?"

Use this space for summary and/or additional notes:

Designing & Performing Experiments

Qualitative Experiments

If the goal of your experiment is to find out ***whether or not*** something happens at all, you need to set up a situation in which the phenomenon you want to observe can either happen or not, and then observe whether or not it does. The only hard part is making sure the conditions of your experiment don't bias whether the phenomenon happens or not.

If you want to find out ***under what conditions*** something happens, what you're really testing is whether or not it happens under different sets of conditions that you can test. In this case, you need to test three situations:

1. A situation in which you are sure the thing will happen, to make sure you can observe it. This is your **positive control**.

2. A situation in which you sure the thing cannot happen, to make sure your experiment can produce a situation in which it doesn't happen and you can observe its absence. This is your **negative control**.

3. A condition or situation that you want to test to see whether or not the thing happens. The condition is your independent variable, and whether or not the thing happens is your dependent variable.

Quantitative Experiments

If the goal of your experiment is to quantify (find a numerical relationship for) the extent to which something happens (the dependent variable), you need to figure out a set of conditions under which you can measure the thing that happens. Once you know that, you need to figure out how much you can change the parameter you want to test (the independent variable) and still be able to measure the result. This gives you the highest and lowest values of your independent variable. Then perform the experiment using a range of values for the independent value that cover the range from the lowest to the highest (or *vice-versa*).

For quantitative experiments, a good rule of thumb is the **8 & 10 rule**: you should have at least 8 data points, and the range from the highest to the lowest values tested should span at least a factor of 10.

Use this space for summary and/or additional notes:

Letting the Chemistry Design the Experiment

Determining what to measure usually means determining what you need to know and working backwards to figure out how to get there from *quantities that you can measure*.

Especially in chemistry, most of the complexity of experiments comes from the fact that there are few quantities that we can measure directly. When we have to measure quantities indirectly, we need to be clever, and we need to think about all of the possible ways we could end up with an invalid result, so we can design the appropriate safeguards into the experiment.

For a quantitative experiment, it's often best to start with a mathematical formula that includes the quantity you want to determine. Then, you need to find the values of the other quantities in the equation, either by measuring them directly or by performing an experiment in which you can determine them by measuring other things.

As an illustrative example, suppose you want to calculate the heat released per mole of reactant in a specific chemical reaction. (Don't worry if you don't understand what this means—we will get there later in the course.) You might go through the following thought process:

1. We can't measure heat directly, so we need an equation to calculate it.

    We can use the equation:

    $$Q = mC\Delta T$$

    In order to use this equation to calculate heat (Q), we need to find m (mass), C (specific heat capacity), and ΔT (temperature change), which means we need to use the heat to *change the temperature* of a *mass* of something.

    An experimental device that does this is a bomb calorimeter[*], so we will use one of those.

2. Determine our independent variables in the above equation.

    a. We predetermine m (the mass of the water in the calorimeter) by putting a known mass of water into the calorimeter. We can measure the mass of the water using a balance.

    b. C (the specific heat capacity of the water in the calorimeter) is a constant that we can look up.

---

[*] A "bomb calorimeter" is a device that uses the heat of a chemical reaction to heat up a known amount of water so you can measure how much heat is produced. Despite the cool-sounding name, a bomb calorimeter doesn't actually involve blowing anything up.

Use this space for summary and/or additional notes:

3. Determine our dependent variables in the equation.

    - We need to calculate ΔT (the temperature change of the water). This means we need to measure the temperature at the beginning and at the end, and subtract. We can measure the temperatures with a thermometer.

4. Determine our control variables.

    a. We want the heat per mole of reactant. If we're going to calculate the heat, we need to control the number of moles of reactant (*i.e.,* how much of the chemical we use), which means we need to measure the mass of the chemical and convert it to moles.

    b. We need to make sure all of the chemical reacts, so we need use excess amounts of everything else.

    c. We need to make sure no heat is lost to the environment. This means we need to set the experiment up so we can start the reaction after the chemicals are already inside the calorimeter.

Notice that our entire experiment is ultimately determined by starting with what we wanted to know at the end of it, and figuring out what makes it happen, and what problems we need to avoid.

Use this space for summary and/or additional notes:

Designing & Performing Experiments

The Experimental Design Process

1. **Figure out what you want to know.** Decide on an experiment that you could use to find it. If what you want to know is a term in an equation, your experiment will involve finding out values for the other variables in the equation.

2. **Determine your independent variables.** Based on your experiment and the equation that goes with it if there is one, determine your independent variables. Figure out what they are, how you are going to determine/measure them, and which values you are going to choose for them. (You may want to wait to choose the values until you know what your dependent and control variables are, and use the expected values to backward calculate your choices for your independent variables.)

3. **Determine your dependent variables.** Figure out what they are, and how *and when* you are going to measure them.

4. **Determine your control variables.** Think about the things that you already know that you need to keep constant. Then start thinking about what could go wrong, which will lead you to other things you need to keep constant.

5. **Set up your experiment and do a test run.** Use your test run to make sure you can actually measure what you think you can measure and make sure you are getting results that make sense. *This means you need to perform the calculations for your test run before doing the rest of the experiment,* in case you need to modify your procedure. You will be sad if you finish your experiment and go home, only to find out at 2:00 am the night before the write-up is due that it didn't work.

More complex experiments use this same process, except that each step might have several sub-steps, and/or that each step or sub-step might be its own completely separate experiment.

Use this space for summary and/or additional notes:

Laboratory Equipment

Page: 33
Unit: Laboratory

Big Ideas | Details

Laboratory Equipment

Unit: Laboratory

MA Curriculum Frameworks (2016): SP3

MA Curriculum Frameworks (2006): N/A

Mastery Objective(s): (Students will be able to…)

- Recognize and identify common laboratory equipment used in chemistry.

Success Criteria:

- Be able to give the name and describe the use for each of the pieces of equipment in this section.

Tier 2 Vocabulary: graduated, ring, stand

Language Objectives:

- Identify laboratory equipment by name and explain what it is used for.

Notes:

The following are some of the common pieces of laboratory apparatus.

Description	Picture	Description	Picture
beaker — Used as a cup/container. May be heated. Volume markings are approximate.		**graduated cylinder** — Used for accurate measurement of liquid volume.	
Erlenmeyer flask — Used as a container. Contents may be swirled to mix. May be heated.		**test tube** — Used for mixing and reacting small quantities of liquids.	
Bunsen burner — Used to heat chemicals in beakers, flasks or test tubes.		**ring stand** — Used to support lab apparatus.	

Use this space for summary and/or additional notes:

Laboratory Equipment

Big Ideas | Details

Page: 34
Unit: Laboratory

Description	Picture	Description	Picture
tongs Used for picking up and holding hot things.		**test tube holder** Used for holding a test tube while heating.	
clamp Used to attach a piece of equipment to a ring stand.		**pinch clamp** Used to clamp rubber tubing so no liquid comes out.	
wire gauze Used to spread the heat of a flame onto the bottom of a beaker or flask.		**iron ring** Fastens to ring stand. Used to hold lab apparatus.	
mortar & pestle Used to grind chemicals to a fine powder.		**evaporating dish** Used to hold solutions over a Bunsen burner to evaporate the liquid.	
stirring rod Used to stir mixtures and to help pour liquids without spilling.		**pipette** Used to transfer small amounts of liquid.	
watch glass Used to cover liquids being heated in a beaker to avoid spatter. Should not be heated.		**crucible & cover** Used to heat small amounts of solids to high temperatures.	
funnel Used to help pour liquids without spilling and/or to hold filter paper.		**pipestem triangle** Used to hold a crucible while heating.	

Use this space for summary and/or additional notes:

Homework Problem

In the following setup, label the piece of lab equipment that each of the arrows is pointing to.

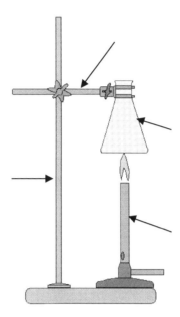

Use this space for summary and/or additional notes:

Accuracy & Precision

Unit: Laboratory
MA Curriculum Frameworks (2016): SP3
MA Curriculum Frameworks (2006): N/A
Mastery Objective(s): (Students will be able to...)

- Correctly use the terms "accuracy" and "precision" in a scientific context.

Success Criteria:

- Be able to give definitions for "accuracy" and "precision."
- Be able to recognize situations as accurate/inaccurate and/or precise/imprecise.

Tier 2 Vocabulary: accurate, precise
Language Objectives:

- Be able to describe the difference between accuracy and precision.

Notes:

Science relies on making and interpreting measurements, and the accuracy and precision of these measurements affect what you can conclude from them.

Random vs. Systematic Errors

Random errors are natural uncertainties in measurements because of the limits of precision of the equipment used. Random errors are assumed to be distributed around the actual value, without bias in either direction. Systematic errors occur from specific problems in your equipment or your procedure. Systematic errors are often biased in one direction more than another, and can be difficult to identify.

Accuracy vs. Precision

The words "accuracy" and "precision" have specific meanings in science.

<u>accuracy:</u> for a single measurement, how close the measurement is to the "correct" or accepted value. For a group of measurements, how close the <u>average</u> is to the accepted value.

<u>precision:</u> for a single measurement, how finely the measurement was made. (How many decimal places it was measured to.) For a group of measurements, how close the measurements are to each other.

Use this space for summary and/or additional notes:

Accuracy & Precision

Examples:

Suppose the following drawings represent arrows shot at a target.

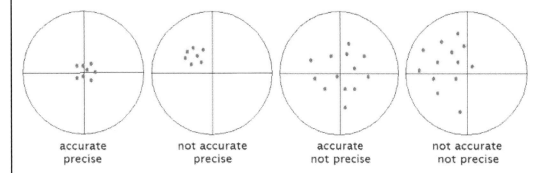

The first set is both accurate (the average is close to the center) and precise (the data points are all close to each other.)

The second set is precise (close to each other), but not accurate (the average is not close to the correct value). This is an example of *systematic* error—some problem with the experiment caused all of the measurements to be off in the same direction.

The third set is accurate (the average is close to the correct value), but not precise (the data points are not close to each other). This is an example of *random* error—the measurements are not biased in any particular direction, but there is a lot of scatter.

The fourth set is neither accurate nor precise, which means that there are significant errors present, both random and systematic.

For another example, suppose two classes estimate Mr. Bigler's age. The first class's estimates are 73, 72, 77, and 74 years old. These measurements are fairly precise (close together), but not accurate. (Mr. Bigler is actually 53 years old at the time of publication.) The second class's estimates are 10, 11, 97 and 98. This <u>set</u> of data is accurate (because the average is 54, which is close to correct), but the <u>set</u> is not precise because the individual values are not close to each other.

Use this space for summary and/or additional notes:

Uncertainty & Error Analysis

Unit: Laboratory

MA Curriculum Frameworks (2016): SP4

MA Curriculum Frameworks (2006): N/A

Mastery Objective(s): (Students will be able to...)

- Determine the uncertainty of a measured or calculated value.

Success Criteria:

- Take analog measurements to one extra digit of precision.
- Correctly estimate measurement uncertainty.
- Correctly read and interpret stated uncertainty values.
- Correctly propagate uncertainty through calculations involving addition/subtraction and multiplication/division.

Tier 2 Vocabulary: uncertainty, error

Language Objectives:

- Understand and correctly use the terms "uncertainty" and "relative error."
- Correctly explain the process of estimating and propagating uncertainty.

Notes:

In science, unlike mathematics, there is no such thing as an exact answer. Ultimately, every quantity is limited by the precision and accuracy of the measurements that it came from. If you can only measure a quantity to within 10 %, that means any calculation that is derived from that measurement can't be any better than ±10 %.

Error analysis is the practice of determining and communicating the causes and extents of uncertainty in your results. Error analysis involves understanding and following the uncertainty in your data, from the initial measurements to the final calculated and reported results.

Note that the word "error" in science has a different meaning from the word "error" in everyday language. In science, **"error" means "uncertainty."** If you report that you drive (2.4 ± 0.1) miles to school every day, you would say that this distance has an error of ±0.1 mile. This does not mean your car's odometer is wrong; it means that the actual distance *could be* 0.1 mile more or 0.1 mile less— *i.e.*, somewhere between 2.3 and 2.5 miles. **When you are analyzing your results, *never* use the word "error" to mean mistakes that you might have made!**

Use this space for summary and/or additional notes:

Uncertainty & Error Analysis

Uncertainty

The uncertainty or error of a measurement describes how close the actual value is likely to be to the measured value. For example, if a length was measured to be 22.34 cm, and the uncertainty was 0.31 cm (meaning that the measurement is only known to within ±0.31 cm), we could represent this measurement in either of two ways:

$$(22.34 \pm 0.31) \text{ cm} \qquad 22.34(31) \text{ cm}$$

The first of these states the variation (±) explicitly in cm (the actual unit). The second is shows the variation in the last digits shown.

What it means is that the true length is approximately 22.34 cm, and is statistically likely* to be somewhere between 22.03 cm and 22.65 cm.

Absolute Error

Absolute error (or absolute uncertainty) refers to the uncertainty in the actual measurement. For the measurement (22.34 ± 0.31) cm, the absolute error is ± 0.31 cm.

Relative Error

Relative error shows the error or uncertainty as a fraction of the total.

The formula for relative error is $\text{R.E.} = \dfrac{\text{uncertainty}}{\text{measured value}}$

For the measurement (22.34 ± 0.31) cm, the relative error would be 0.31 out of 22.34. Mathematically, we express this as:

$$\text{R.E.} = \frac{0.31}{22.34} = 0.0139$$

Note that relative error is dimensionless (does not have any units). This is because the numerator and denominator have the same units, so the units cancel.

Percent Error

Percent error is relative error expressed as a percentage. You can turn relative error into percent error by multiplying by 100.

In the example above, the relative error of 0.0139 would be 1.39 % error.

---

* Statistically, the uncertainty is one standard deviaition. *I.e.,* if multiple measurements are taken, approximately two-thirds of those measurements will lie within the uncertainty (plus or minus) of the stated value.

Use this space for summary and/or additional notes:

Big Ideas | Details | Unit: Laboratory

Uncertainty & Error Analysis

Uncertainty of Measurements

If you have the ability to measure a quantity that is not changing (such as the mass or length of an object), you will get the same value every time you measure it. This means you have only one data point.

When you have only one data point, the uncertainty is the limit of how well you can measure it. This will be your best educated guess, based on how closely you think you actually measured the quantity. This means you need to take measurements as carefully and precisely as possible, because *every careless measurement needlessly increases the uncertainty of the result.*

Digital Measurements

For digital equipment, if the reading is *stable* (not changing), look up the published precision of the instrument in its user's manual. (For example, many balances used in high schools have a readability of 0.01 g but are only precise to within ± 0.02 g.) If there is no published value (or the manual is not available), assume the uncertainty is ± 1 in the last digit.

Analog Measurements

When making analog measurements, always estimate one extra digit beyond the finest markings on the equipment. For example, in the diagram below, the graduated cylinder is marked in 1 mL increments. When measuring volume in this graduated cylinder, you would estimate and write down the volume to the nearest 0.1 mL, as shown:

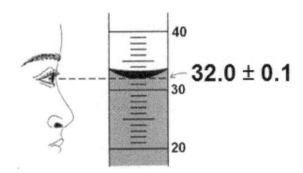

In the above experiment, you should record the volume as 32.0 ± 0.1 mL. It would be inadequate to write the volume as 32 mL; you *must* write 32.0 mL, or better yet, (32.0 ± 0.1) mL

The zero at the end of 32.0 mL is not extra. It is necessary to show that *you measured the volume to the nearest tenth, not to the nearest one.*

Use this space for summary and/or additional notes:

Uncertainty & Error Analysis

When estimating, the uncertainty depends on how well you can see the markings, but you can usually assume that the estimated digit has an uncertainty of $\pm \frac{1}{10}$ of the finest markings on the equipment. Here are some examples:

Equipment	Markings	Estimate To	Assumed Uncertainty
ruler	1 mm	0.1 mm	± 0.1 mm
25 mL graduated cylinder	0.2 mL	0.02 mL	± 0.02 mL
thermometer	1 °C	0.1 °C	± 0.1 °C

Homework Problems

Write the readings that you would record (estimated to one extra decimal place) and the assumed uncertainty for each of the following thermometers and graduated cylinders.

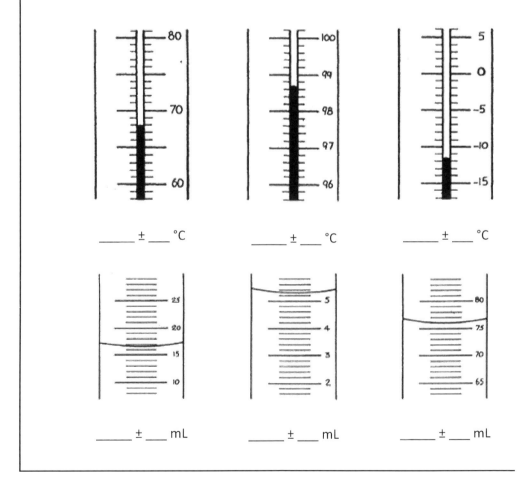

_____ ± _____ °C   _____ ± _____ °C   _____ ± _____ °C

_____ ± _____ mL   _____ ± _____ mL   _____ ± _____ mL

Use this space for summary and/or additional notes:

Uncertainty & Error Analysis

Propagating Uncertainty in Calculations

When you perform calculations using numbers that have uncertainty, you need to propagate the uncertainty through the calculation.

Addition & Subtraction

When quantities with uncertainties are added or subtracted, add the quantities to get the answer, then add the uncertainties to get the total uncertainty.

Sample Problem:

Q: A substance is being heated. You record the initial temperature as (23 ± 0.2) °C, and the final temperature as (84± 0.2) °C. You need to calculate the temperature change (ΔT) with its uncertainty to use in a later calculation. What is the temperature change?

A: To calculate ΔT, simply subtract:

$$\Delta T = T_{final} - T_{initial} = 84 - 23 = 61°C$$

To calculate the uncertainty, add the individual uncertainties (even though the quantities were subtracted):

$$u = 0.2 + 0.2 = 0.4 \ °C$$

Report the value as: $\Delta T = (61 \pm 0.4)°C$

Multiplication & Division

Because most calculations that we will perform in chemistry involve multiplication and/or division, you can

For calculations involving multiplication and division, estimate the uncertainty of your calculated answer by adding the relative errors and applying the total relative error to your result.

1. Perform the calculation for the desired quantity.

2. Divide the uncertainty (the ±) for each quantity by its measured value to determine its relative error.

$$R.E. = \frac{\text{uncertainty}}{\text{measured value}}$$

3. Add up all of the relative errors to get the total relative error.

4. Multiply your calculated result by the total relative error to get its uncertainty (the ± amount).

Use this space for summary and/or additional notes:

Uncertainty & Error Analysis

Unit: Laboratory

Note: *Most of the calculations that you will perform in chemistry involve multiplication and/or division, so almost all of your uncertainty calculations throughout the course will use relative error.*

Sample Problem #1:

Q: You want to determine the amount of heat released by a chemical reaction. You use the heat from the reaction to heat up some water in an insulated container called a "bomb calorimeter". You will calculate the heat using the equation: $Q = mC\Delta T$.

Suppose you recorded the following data (including uncertainties):

- The mass of the water in the calorimeter is (24.8 ± 0.1) g.
- The temperature change of the water was (12.4 ± 0.2) °C.
- The specific heat capacity of water is $4.181 \frac{J}{g \cdot °C}$. (This is a published value. The uncertainty of this value is so small that we can leave it out of our calculations.)

A: The heat released by the reaction is given by the equation:

$$Q = mC\Delta T$$
$$Q = (24.8)(4.181)(12.4)$$
$$Q = 1\,285.74 \text{ J}$$

The relative errors for the two quantities that we measured are:

- mass: $\dfrac{0.1}{24.8} = 0.004\,03$

- temperature change: $\dfrac{0.2}{12.4} = 0.016\,13$

The total relative error is $0.004\,03 + 0.016\,13 = 0.020\,16$

The uncertainty is therefore $(0.020\,16)(1\,285.74) = \pm 25.92$ J

(Note that the absolute uncertainty has the same units as the measurement.)

We would report the measurement as $(1\,285.74 \pm 25.92)$ J.

Use this space for summary and/or additional notes:

Rounding

In the example above, the uncertainty tells us that our actual result could be different from our calculated value by as much as 25.92 J.

However, we only estimated one digit (which happened to be the tenths place) when we took our measurements. This means we have only one digit of uncertainty. Because we can't report more precision than we actually have, we need to round the calculated uncertainty off, so that we have only one unrounded digit. This means we should report our uncertainty as ± 30 J.

It wouldn't make sense to report our answer as (1 285.74 ± 30) J. Think about that—if the *tens* digit could be different from our calculated value, there is no point in reporting the ones or tenths digits. So we need to round our calculated answer to the same place value as the uncertainty—the tens place.

This means our final, rounded answer should be (1 290 ± 30) J.

Use this space for summary and/or additional notes:

Uncertainty & Error Analysis

Page: 45
Unit: Laboratory

Sample Problem #2:

Q: You need to find the density of a piece of metal. We measure its mass on a balance to be (24.75 ± 0.02) g. You measure its volume in a graduated cylinder using water displacement, and you find the volume to be (7.2 ± 0.1) mL. Calculate the density, including its uncertainty.

A: 1. Calculate the density.

$$\rho = \frac{m}{V} = \frac{24.75 \text{ g}}{7.2 \text{ mL}} = 3.4375 \tfrac{g}{mL}$$

2. Calculate the relative errors of your two measurements:

$$R.E._{mass} = \frac{\text{uncertainty}}{\text{measured value}} = \frac{0.02}{24.75} = 0.000808$$

$$R.E._{volume} = \frac{0.1}{7.2} = 0.013889$$

3. Add the individual relative errors together to get the total R.E.:

$$0.000808 + 0.013889 = 0.014697$$

4. Multiply the total R.E. by the density to get the uncertainty:

$$3.4375 \times 0.014697 = 0.050521$$

Because you only estimated one decimal place of uncertainty, you need to round the uncertainty off to ± 0.05.

Because uncertainty is rounded to the hundredths place, you need to also round your answer to the hundredths place:

$$\rho = (3.44 \pm 0.05) \tfrac{g}{mL}$$

Use this space for summary and/or additional notes:

Uncertainty & Error Analysis

Homework Problems

Because the answers are provided, you must show sufficient work in order to receive credit.

1. You need to combine three liquids and calculate the total volume. You measured the individual volumes as (12.36 ± 0.02) mL, (37.4 ± 0.2) mL, and (61.0 ± 0.1) mL. What is the total volume, including the uncertainty?

    Answer: (110.8 ± 0.3) mL

2. A sample of (0.517 ± 0.008) moles of a chemical is dissolved to make (1.362 ± 0.005) liters of solution.

    a. What is the concentration in moles per liter? (Divide the amount in moles by the volume in liters.)

    Answer: $0.3796 \frac{mol}{L}$ (Don't worry about rounding yet.)

    b. What are the relative errors of the number of moles and the number of liters? What is the total relative error?

    Answers: moles: R.E. = 0.015
    volume: R.E. = 0.0037
    total R.E. = 0.019

    c. Calculate the uncertainty of the concentration of the solution and express your answer as the concentration (from part a above) plus or minus the uncertainty that you just calculated, with correct rounding.

    Answer: $(0.380 \pm 0.007) \frac{mol}{L}$

Use this space for summary and/or additional notes:

Big Ideas	Details	Unit: Laboratory

Significant Figures

Unit: Laboratory

MA Curriculum Frameworks (2016): N/A

MA Curriculum Frameworks (2006): N/A

Mastery Objective(s): (Students will be able to…)

- Identify the significant figures in a number.
- Perform calculations and round the answer to the appropriate number of significant figures

Success Criteria:

- Be able to identify which digits in a number are significant.
- Be able to count the number of significant figures in a number.
- Be able to determine which places values will be significant in the answer when adding or subtracting.
- Be able to determine which digits will be significant in the answer when multiplying or dividing.
- Be able to round a calculated answer to the appropriate number of significant figures.

Tier 2 Vocabulary: significant, round

Language Objectives:

- Explain the concepts of significant figures and rounding.

Notes:

Because it would be tedious to calculate the uncertainty for error for every calculation in chemistry, we often use significant figures (or significant digits) as a simple way to estimate and represent the uncertainty.

Significant figures are based on the following approximations:

- All stated values are rounded off so that the uncertainty is only in the last unrounded digit.
- Assume that the uncertainty in the last unrounded digit is ±1.
- The results of calculations are rounded so that the uncertainty of the result is only in the last unrounded digit and is assumed to be ±1.

Use this space for summary and/or additional notes:

Significant Figures

Note that using significant figures gives less information than stating the measurement with its uncertainty. This is why, when you take measurements and perform calculations in the laboratory, you will estimate the actual uncertainty of each measurement and calculate the uncertainty of your results. However, for homework problems and written tests, you will use significant figures as a simple way to keep track of the approximate effects of uncertainty on your answers.

In the example **Error! Bookmark not defined.**, we rounded the number 1 285.74 off to the tens place, resulting in the value of 1 290, because we couldn't show more precision than we actually had.

In the number 1 290, we would say that the first three digits are "significant", meaning that they are the part of the number that is not rounded off. The zero in the ones place is "insignificant," because the digit that was there was lost when we rounded.

significant figures (significant digits): the digits in a measured value or calculated result that are not rounded off. (Note that the terms "significant figures" and "significant digits" are used interchangeably.)

insignificant figures: the digits in a measured value or calculated result that were "lost" (became zeroes before a decimal point or were cut off after a decimal point) due to rounding.

Use this space for summary and/or additional notes:

Significant Figures

Identifying the Significant Digits in a Number

The first significant digit is where the "measured" part of the number begins—the first digit that is not zero.

The last significant digit is the last "measured" digit—the last digit whose true value is known.

- If the number doesn't have a decimal point, the last significant digit will be the last digit that is not zero. (Anything after that has been rounded off.)

  Example: If we round the number 234 567 to the thousands place, we would get 235 000. (Note that because the digit after the "4" in the thousands place was 5 or greater, so we had to "round up".) In the rounded-off number, the first three digits (the 2, 3, and 5) are the significant digits, and the last three digits (the zeroes at the end) are the insignificant digits.

- If the number has a decimal point, the last significant digit will be the last digit shown. (Anything rounded after the decimal point gets chopped off.)

  Example: If we round the number 11.223 344 to the hundredths place, it would become 11.22. When we rounded the number off, we "chopped off" the extra digits.

- If the number is in scientific notation, it has a decimal point. Therefore, the above rules tell us (correctly) that all of the digits before the "times" sign are significant.

In the following numbers, the significant figures have been underlined:

- <u>13</u> 000
- 0.0<u>275</u>
- 0.0<u>150</u>
- <u>6 804.305 00</u>
- <u>6.0</u> × 10^{23}
- <u>3400.</u> (note the decimal point at the end)

Use this space for summary and/or additional notes:

Significant Figures

Mathematical Operations with Significant Figures

Addition & Subtraction

When adding or subtracting, calculate the total normally. Then identify the smallest place value where nothing is rounded. Round your answer to that place.

For example, consider the following problem.

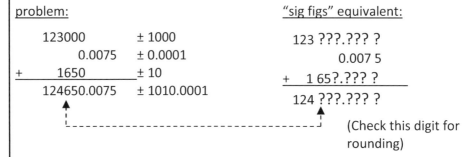

problem:

```
   123000       ± 1000
     0.0075     ± 0.0001
+    1650       ± 10
   124650.0075  ± 1010.0001
```

"sig figs" equivalent:

```
   123 ???.??? ?
     0.007 5
+  1 65?.??? ?
   124 ???.??? ?
```
(Check this digit for rounding)

In the first number (123 000), the hundreds, tens, and ones digit are zeros, presumably because the number was rounded to the nearest 1000. The second number (0.0075) is presumably rounded to the ten-thousandths place, and the number 1650 is presumably rounded to the tens place.

The first number has the largest uncertainty, so we need to round our answer to the thousands place to match, giving 125 000 ± 1 000.

A silly (but correct) example of addition with significant digits is:

$$100 + 37 = 100$$

Multiplication and Division

When multiplying or dividing, calculate the result normally. Then count the total *number* of significant digits in the values that you used in the calculation. Round your answer so that it has the same number of significant digits as the value that had the *fewest*.

Consider the problem:

$$34.52 \times 1.4$$

The answer (without taking significant digits into account) is $34.52 \times 1.4 = 48.328$

The number 1.4 has only two significant digits, so we need to round our answer so that it also has only two significant digits. This means we should round our answer to 48.

A silly (but correct) example of addition with significant digits is:

$$234 \times 1 = 200$$

Use this space for summary and/or additional notes:

Significant Figures

Mixed Operations

For mixed operations, keep all of the digits until you're finished (so round-off errors don't accumulate), but keep track of the last significant digit in each step by putting a line over it (even if it's not a zero). Once you have your final answer, round it to the correct number of significant digits. Don't forget to use the correct order of operations (PEMDAS)!

For example:

$$137.4 \times 52 + 120 \times 1.77$$
$$(137.4 \times 52) + (120 \times 1.77)$$
$$7,\overline{1}44.8 + 2\overline{1}2.4 = 7,\overline{3}57.2 = 7,400$$

Note that in the above example, we kept all of the digits until the end. This is to avoid introducing small rounding errors at each step, which can add up to enough to change the final answer. Notice how, if we had rounded off the numbers at each step, we would have gotten the wrong answer:

$$137.4 \times 52 + 120 \times 1.77$$
$$(137.4 \times 52) + (120 \times 1.77)$$
$$7,\overline{1}00 + 2\overline{1}0 = 7,\overline{3}10 = 7,300 \quad \leftarrow \; \; \text{☹}$$

Use this space for summary and/or additional notes:

Significant Figures

What to Do When Rounding Doesn't Give the Correct Number of Significant Figures

If you have a different number of significant digits from what the rounding shows, you can place a line over the last significant digit, or you can place the whole number in scientific notation. Both of the following have <u>four</u> significant digits, and both are equivalent to writing 13,000 ± 10

- 13 0$\bar{0}$0
- 1.300×10^4

When Not to Use Significant Figures

Significant figure rules only apply in situations where the numbers you are working with have a limited precision. This is usually the case when the numbers represent measurements. <u>Exact</u> numbers have infinite precision, and therefore have an infinite number of significant figures. Some examples of exact numbers are:

- Pure numbers, such as the ones you encounter in math class.
- Anything you can count. (*E.g.,* there are 24 people in the room. That means <u>exactly</u> 24 people, not 24.0 ± 0.1 people.)
- Whole-number exponents in formulas. (*E.g.,* the area of a circle is πr^2. The exponent "2" is a pure number.)

You should also avoid significant figures any time the uncertainty is likely to be substantially different from what would be implied by the rules for significant figures, or any time you need to quantify the uncertainty more exactly.

Summary

Significant figures are a source of ongoing stress among chemistry students. To make matters simple, realize that few formulas in chemistry involve addition or subtraction, so you can usually just apply the rules for multiplication and division: look at each of the numbers you were given in the problem. Find the one that has the fewest significant figures, and round your final answer to the same number of significant figures.

If you have absolutely no clue what else to do, round to three significant figures. You would have to measure quite carefully to have more than three significant figures in your original data, and three is usually enough significant figures to avoid unintended loss of precision, at least in a high school chemistry course. ☺

Use this space for summary and/or additional notes:

Significant Figures

Homework Problems

1. For each of the following, Underline the significant figures in the number and Write the assumed uncertainty as ± the appropriate quantity.

    <u>57 3</u>00 ± 100 ← Sample problem with correct answer.

    a. 13 500

    b. 26.0012

    c. 01902

    d. 0.000 000 025

    e. 320.

    f. 6.0×10^{-7}

    g. 150.00

    h. 10

    i. 0.005 310 0

2. Round off each of the following numbers as indicated and indicate the last significant digit if necessary.

    a. 13 500 to the nearest 1000

    b. 26.0012 to the nearest 0.1

    c. 1902 to the nearest 10

    d. 0.000 025 to the nearest 0.000 01

    e. 320. to the nearest 10

    f. 6.0×10^{-7} to the nearest 10^{-6}

    g. 150.00 to the nearest 100

    h. 10 to the nearest 100

Use this space for summary and/or additional notes:

Significant Figures

Big Ideas | Details

3. Solve the following math problems and round your answer to the appropriate number of significant figures.

   a. 3521×220

   b. $13\,580.160 \div 113$

   c. $2.71828 + 22.4 - 8.31 - 62.4$

   d. $23.5 + 0.87 \times 6.02 - 105$ (Remember PEMDAS!)

Use this space for summary and/or additional notes:

Keeping a Laboratory Notebook

Unit: Laboratory
MA Curriculum Frameworks (2016): SP3, SP8
MA Curriculum Frameworks (2006): N/A
Mastery Objective(s): (Students will be able to...)

- Determine which information to record in a laboratory notebook.
- Record information in a laboratory notebook according to practices used in industry.

Success Criteria:

- Record data accurately and correctly, with units and including estimated digits.
- Use the correct protocol for correcting mistakes.

Language Objectives:

- Understand and be able to describe the process for recording lab procedures and data.

Notes:

A laboratory notebook serves two important purposes:

1. It is a diary of what you did in case you want to look up the details later.
2. It is a legal record of what you did and when you did it.

In a research laboratory, you would normally do a write-up in your lab notebook whenever you do a significant experiment that you believe you might want to refer back to sometime in the future.

Your Notebook as an Official Record

Laboratory notebooks are kept by scientists in research laboratories and high tech companies. If a company or research institution needs to prove that you did a particular experiment on a particular date and got a particular set of results (perhaps to apply for a patent), your lab notebook is the primary evidence. This means you need to maintain your lab notebook in a way that gives it the best chance of being able to prove beyond a reasonable doubt exactly what you did and exactly when you did it.

Use this space for summary and/or additional notes:

Keeping a Laboratory Notebook

For companies that use laboratory notebooks in this way, there are a set of guidelines that exist to prevent mistakes that could compromise the integrity of the notebook. Details may vary somewhat from one company to another, but are probably similar to these, and the spirit of the rules is the same.

- All entries in a lab notebook must be hand-written in ink.
- Your actual procedure and all data must be recorded directly into the notebook, not recorded elsewhere and copied in.
- All pages must be numbered consecutively, to show that no pages have been removed. If your notebook did not come with pre-numbered pages, you need to write the page number on each page before using it.
- Start each experiment on a new page.
- Sign and date the bottom of the each page when you finish recording information on it. (In industry, each page needs to be witnessed by someone else, usually your supervisor. The date that an entry is considered to have happened is the date it was witnessed, even if that is much later than the date when it was originally recorded.)
- When crossing out an incorrect entry in a lab notebook, never obliterate it. Always cross it out with a single line through it, so that it is still possible to read the original mistake. (This is to prove that it was a mistake, and you didn't change your data or observations.) Any time you cross something out, write your initials and the date next to the change.
- Never remove pages from a lab notebook for any reason. If you need to cross out an entire page, you may do so with a single large "X". If you do this, write a brief explanation of why you crossed out the page, and sign and date the cross-out.
- Never, ever change data after the experiment is completed. Really. Your data, right or wrong, is what you actually observed. Changing your data constitutes fraud, which is a form of cheating that is every bit as bad as plagiarism.
- Never change anything on a page you have already signed and dated. If you realize that an experiment was flawed, leave the bad data where it is and add a note that says "See page ____." with your initials and date next to the addendum. On the new page, refer back to the page number of the bad data and describe briefly what was wrong with it. Then, give the correct information and sign and date it as you would an experiment.
- Never, ever erase or cover with white-out anything in a lab notebook. Erased or covered-up data is considered the same as faked or changed data in the scientific community.

Use this space for summary and/or additional notes:

Keeping a Laboratory Notebook

Recording Data

Here are some general rules for working with data. (Most of these are courtesy of Dr. John Denker, at http://www.av8n.com/physics/uncertainty.htm):

- Write something about what you did on the same page as the data, even if it is a very rough outline. Your procedure notes should not get in the way of actually performing the experiment, but there should be enough information to corroborate the detailed summary of the procedure that you will write afterwards. (Also, for evidence's sake, the sooner after the experiment that you write the detailed summary, the more weight it will carry in court.)
- Keep *all* of the raw data, whether you will use it or not.
- Don't discard a measurement, even if you think it is wrong. Record it anyway and put a "?" next to it. You can always choose not to use the data point in your calculations (as long as you give an explanation).
- Never erase or delete a measurement. The only time you should ever cross out recorded data is if you accidentally wrote down the wrong number.
- Record all digits. Never round off original data measurements. If the last digit is a zero, you must record it anyway!
- For analog readings (*e.g.*, ruler, graduated cylinder, thermometer), always estimate and record one extra digit.
- Always write down the units with each measurement!
- Record *every* quantity that will be used in a calculation, whether it is changing or not.
- Don't convert in your head before writing down a measurement. Record the original data in the units you actually measured it in, and convert in a separate step.

Calculations

- Use enough digits to avoid unintended loss of significance. (Don't introduce round-off errors in the middle of a calculation.) This usually means use at least two more digits than the number of "significant figures" you expect your answer to have.
- Use few enough digits to be reasonably convenient.
- Record uncertainty separately from the measurement. (Don't rely on "sig figs" to express uncertainty.)
- Leave digits in the calculator between steps. (Don't round until the end.)
- When in doubt, keep plenty of "guard digits" (digits after the place where you think you will end up rounding).

Use this space for summary and/or additional notes:

Integrity of Data

Your data are your data. In classroom settings, people often get the idea that the goal is to report an uncertainty that reflects the difference between the measured value and the "correct" value. That idea certainly doesn't work in real life—if you knew the "correct" value you wouldn't need to make measurements!

In all cases—in the classroom and in real life—you need to determine the uncertainty of your own measurement by scrutinizing your own measurement procedures and your own analysis. Then you judge how well they agree.

For example, we would say that the quantities 10 ± 2 and 11 ± 2 agree reasonably well, because there is considerable overlap between their probability distributions. However, 10 ± 0.2 does not agree with 11 ± 0.2, because there is no overlap.

If you get an impossible result or if your results disagree with well-established results, you should look for and comment on possible problems with your procedure and/or measurements that could have caused the differences you observed. You must *never* fudge your data to improve the agreement.

Your Laboratory Notebook is *Not* a Report

Many high school students are taught that a laboratory notebook should be a journal-style book in which they must write perfect after-the-fact reports, but they are not allowed to change anything if they make a mistake. This is not at all what laboratory notebooks were ever meant to be. A laboratory notebook does not need to be anything more than an official signed and dated record of your procedure (what you did) and your data (what happened) at the exact instant that you took it and wrote it down.

Of course, because it is your journal, your laboratory notebook *may* contain anything else that you think is relevant. You may choose to include an explanation of the motivations for one or more experiments, the reasons you chose the procedure that you used, alternative procedures or experiments you may have considered, ideas for future experiments, *etc.* Or you may choose to record these things separately and cross-reference them to specific pages in your lab notebook.

Use this space for summary and/or additional notes:

Internal Laboratory Reports

Unit: Laboratory
MA Curriculum Frameworks (2016): SP3, SP8
MA Curriculum Frameworks (2006): N/A
Mastery Objective(s): (Students will be able to...)

- Write an internal laboratory report that appropriately communicates all of the necessary information.

Success Criteria:

- The report has the correct sections in the correct order.
- Each section contains the appropriate information.

Language Objectives:

- Understand and be able to describe the sections of an internal laboratory report, and which information goes in each section.
- Write an internal laboratory report with the correct information in each section.

Notes:

An internal laboratory report is written for co-workers, your boss, and other people in the company or research facility that you work for. It is usually a company confidential document that is shared internally, but not shared outside the company or facility. Every lab you work in, whether in high school, college, research, or industry, will have its own preferred internal report format.

It is much more important to understand what *kinds* of information you need to report and what you will use it for than it is to get attached to any one format. The format we will use in this class is based on the outline of the actual experiment.

Title & Date

Each experiment should have the title and date the experiment was performed written at the top. The title should be a descriptive sentence fragment (usually without a verb) that gives some information about the purpose of the experiment.

Objective

This should be a one or two-sentence description of what you are trying to determine or calculate by performing the experiment.

Use this space for summary and/or additional notes:

Internal Laboratory Reports

Experimental Design

Your background or experimental plan needs to convey your plan for carrying out the experiment. This section should follow the design process as described in the Experimental Design section on page 32, and should include:

- an overview of the experiment, including any relevant equations that will be used to calculate the desired quantity
- a description of the independent variables
- a description of the dependent variables
- a description of the control variables
- a brief description of how you will calculate the desired quantity/quantities once you have performed the experiment.

Procedure

This is a detailed description of exactly what you did to set/measure the values of each of the variables. You need to include:

- A *labeled* sketch or photograph of your experimental set-up, even if the experiment is simple. The sketch will serve to answer many questions about how you set up the experiment and most of the key equipment you used.
- A list of any significant equipment that you used other than what you labeled in your sketch. (You do not need to mention generic items like pencils and paper. Basic lab safety equipment is assumed, but mention any unusual precautions that you need to take.)
- A description of how you set up the experiment, including the values of your independent variables and how you set them.
- A description of your control variables, including their values and how you are ensuring that they remain constant.
- A description of your dependent variables and how you are measuring their values. (Do not include the values of the dependent variables here—you will present those in your Data & Observations section.)
- Any significant things you did as part of the experiment besides the ones mentioned above.

Use this space for summary and/or additional notes:

Data & Observations

This is a section in which you present all of your data.

For a high school lab, it is usually sufficient to present a single data table that includes the values of your independent, control, and dependent variables for each trial. However, if you have other data or observations that you recorded during the lab, they must be listed here.

You must also include estimates of the uncertainty for each measured quantity, and your calculated uncertainty for the final quantity that your experiment is intended to determine.

Analysis

The analysis section is where you interpret your data. (Note that calculated values in the table in the Data & Observations section are actually part of your analysis, even though they appear in the Data & Observations section.) Your analysis should mirror your Experimental Design section (possibly in the same order, possibly in reverse), with the goal of guiding the reader from your data to the quantity you ultimately want to calculate or determine.

Your analysis needs to include:

- A narrative description (one or more paragraphs) of the outcome of the experiment that guides the reader from your data through your calculations to the quantity you set out to determine.

- One (and only one) sample calculation for each separate equation that you used. For example, if you calculated acceleration for each of five data points, you would write down the formula, and then choose one set of data to plug in and show how you got the answer.

- Any calculated values that did not appear in the data table in your Data & Observations section

- For some experiments, a carefully-plotted graph showing the data points you took for your dependent *vs.* independent variables. Note that **any graphs you include in your write-up must be drawn accurately to scale, using graph paper, and using a ruler/straightedge wherever a straight line is needed**. (When an accurate graph is required, you will lose points if you include a freehand sketch instead.)

Use this space for summary and/or additional notes:

- Quantitative error analysis. In general, most quantities in a high school physics class are calculated from equations that use multiplication and division. Therefore, you need to use relative error:

    1. Determine the uncertainty of each your measurements.

    2. Calculate the relative error for each measurement.

    3. Combine your relative errors to get the total relative error for your calculated value(s).

    4. Multiply the total relative error by your calculated values to get the absolute uncertainties (±).

- Sources of uncertainty: this is a list of factors **inherent in your procedure** that limit how precise your answer can be.

    You need to list *one source of human-derived uncertainty* (*e.g.,* "It was unclear exactly when the reaction was finished. We declared it to have finished when nothing appeared to be changing."), and *two sources of non-human uncertainty* (*e.g.,* "The graduated cylinder was marked in 1 mL increments, so the volume was estimated to ± 0.1 mL.")

    Never include mistakes, especially mistakes you aren't sure whether or not you made! A statement like "We might have written down the wrong number." or "We might have done the calculations incorrectly." is really saying, "We might be stupid and you shouldn't believe anything else in this report." (Any "we might be stupid" statements will not count toward your required number of sources of uncertainty.)

Note, however, that if a problem *actually occurred*, and if you *used that data point in your calculations anyway*, you need to explain what happened and give an estimate of the effects on your results.

Conclusion

Your conclusion should be worded similarly to your objective, but this time including your final calculated result(s) and uncertainty. You do not need to restate sources of uncertainty in your conclusions unless you believe they were significant enough to create some doubt about your results.

Your conclusion should also include 1–2 sentences describing ways the experiment could be improved. These should specifically address the sources of uncertainty that you listed in the analysis section above.

Use this space for summary and/or additional notes:

Internal Laboratory Reports

Page: 63
Unit: Laboratory

Big Ideas | Details

Summary

You can think of the sections of the report in pairs. For each pair, the first part describes the intent of the experiment, and the corresponding second part describes the result.

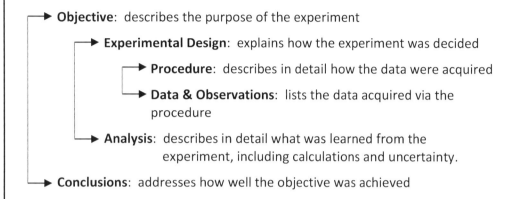

- **Objective**: describes the purpose of the experiment
  - **Experimental Design**: explains how the experiment was decided
    - **Procedure**: describes in detail how the data were acquired
    - **Data & Observations**: lists the data acquired via the procedure
  - **Analysis**: describes in detail what was learned from the experiment, including calculations and uncertainty.
- **Conclusions**: addresses how well the objective was achieved

Use this space for summary and/or additional notes:

Chemistry 1 — Mr. Bigler

Formal Laboratory Reports

Unit: Laboratory

MA Curriculum Frameworks (2016): SP3, SP8

MA Curriculum Frameworks (2006): N/A

Mastery Objective(s): (Students will be able to...)

- Write a formal (journal article-style) laboratory report that appropriately communicates all of the necessary information.

Success Criteria:

- The report has the correct sections in the correct order.
- Each section contains the appropriate information.
- The report contains an abstract that conveys the appropriate amount of information.

Tier 2 Vocabulary: abstract

Language Objectives:

- Understand and be able to describe the sections of a formal laboratory report, and which information goes in each section.
- Write a formal laboratory report with the correct information in each section.

Notes:

A formal laboratory report serves one important purpose: to communicate the results of your experiment to other scientists outside of your laboratory or institution.

A formal report is a significant undertaking. In a research laboratory, you might submit as many as one or two articles to a scientific journal in a year. Some college professors require students to submit lab reports in journal article format.

Use this space for summary and/or additional notes:

Formal Laboratory Reports

The format of a formal journal article-style report is as follows:

Abstract

This is the most important part of your report. It is a (maximum) 200-word executive summary of everything about your experiment—the procedure, results, analysis, and conclusions. In most scientific journals, the abstracts are searchable via the internet, so it needs to contain enough information to enable someone to find your abstract, and after reading it, to know enough about your experiment to determine whether or not to purchase a copy of the full article (which can sometimes cost $100 or more). It also needs to be short enough that the person doing the search won't just say "TL; DR" ("Too Long; Didn't Read") and move on to the next abstract.

Because the abstract is a complete summary, it is always best to write it last, after you have already written the rest of your report.

Introduction

Your introduction is actually an mini research paper on its own, including citations. (For a high school lab report, it should be 1–3 pages; for scientific journals, 5–10 pages is not uncommon.) Your introduction needs to describe any general background information that another scientist might not know, plus all of the background information that specifically led up to your experiment. Assume that your reader has a similar knowledge of physics as you, but does not know anything about this experiment. The introduction is usually the most time-consuming part of the report to write.

Materials and Methods

This section combines both the experimental design and procedure sections of an informal lab write-up. Unlike an informal write-up, the Materials and Methods section of a formal report is written in paragraph form, in the past tense, using the passive voice, and avoiding pronouns. As with the informal write-up, a labeled photograph or drawing of your apparatus is a necessary part of this section, but you need to *also* describe the set-up in the text.

Also unlike the informal write-up, your Materials and Methods section needs to give some *explanation* of your choices of the values used for your control and independent variables.

Use this space for summary and/or additional notes:

Data and Observations

This section is similar to the same section in the lab notebook write-up, except that:

1. You should present only data you actually recorded/measured in this section. (Calculated values are presented in the Discussion section.)

2. You need to *introduce* the data table. (This means you need to describe the important things someone should notice in the table first, and then say something like "Data are shown in Table 1.")

Note that all figures and tables in the report need to be numbered separately and consecutively.

Discussion

This section is similar to the Analysis section in the internal report, but with some important differences.

As with the rest of the formal report, your discussion must be in paragraph form. Your discussion is essentially a long essay discussing your results and what they mean. You need to introduce and present a table with your calculated values and your uncertainty. After presenting the table, you should discuss the results, uncertainties, and sources of uncertainty in detail. If your results relate to other experiments, you need to discuss the relationship and include citations for those other experiments.

Your discussion needs to include each of the formulas that you used as part of your discussion and give the results of the calculations, but you do not need to show the intermediate step of substituting the numbers into the equation.

Conclusions

Your conclusions are written much like in the internal write-up. You need at least two paragraphs. In the first, restate your findings and summarize the significant sources of uncertainty. In the second paragraph, list and explain improvements and/or follow-up experiments that you suggest.

Works Cited

As with a research paper, you need to include a complete list of bibliography entries for the references you cited in your introduction and/or discussion sections.

Your ELA teachers probably require MLA-style citations; scientific papers typically use APA style. However, in a high school chemistry class, while it is important that you know which information needs to be cited and *what* information needs to go into each citation, you may use any format you like as long as you use it consistently.

Use this space for summary and/or additional notes:

Big Ideas	Details

Typesetting Superscripts and Subscripts

Because formal laboratory reports need to be typed, and because chemistry uses superscripts and subscripts extensively, it is important to know how to typeset superscripts and subscripts.

You can make use of the following shortcuts:

<u>superscript</u>: text that is raised above the line, such as the "3+" in Al^{3+}.

In most Microsoft programs, select the text, then hold down "Ctrl" and "Shift" and press the "+" key.

On a Macintosh, select the text, then hold down "Command" and "Control" and press the "+" key.

<u>subscript</u>: text that is lowered below the line, such as the "2" in $CaCl_2$.

In most Microsoft programs, select the text, then hold down "Ctrl" and press the "–" key.

On a Macintosh, select the text, then hold down "Command" and "Control" and press the "–" key.

Note that you will lose credit in your laboratory reports if you don't use superscripts and subscripts correctly. For example, you will lose credit if you type NO3– instead of NO_3^-.

Use this space for summary and/or additional notes:

Big Ideas	Details
	# Summary: Laboratory
Unit: Laboratory
List the main ideas of this chapter in phrase form:

Write an introductory sentence that categorizes these main ideas.

Turn the main ideas into sentences, using your own words. You may combine multiple main ideas into one sentence.

Add transition words to make your writing clearer and rewrite your summary below.

Use this space for summary and/or additional notes: |

Introduction: Math & Measurement

Unit: Math & Measurement

Topics covered in this chapter:

The International System of Units 73
Scientific Notation 80
Using Math in Calculations 84
Conversions (Factor-Label Method) 91
Dimensional Analysis 96
Logarithms 102

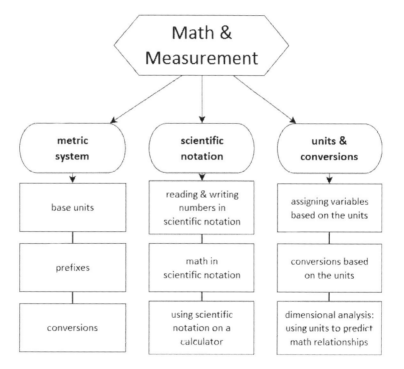

Standards addressed in this chapter:

Massachusetts Curriculum Frameworks & Science Practices (2016):

This chapter addresses the following MA science and engineering practices:

Practice 4: Analyzing and Interpreting Data

Practice 5: Using Mathematics and Computational Thinking

Practice 8: Obtaining, Evaluating, and Communicating Information

Use this space for summary and/or additional notes:

Introduction: Math & Measurement

Massachusetts Curriculum Frameworks (2006):

No MA curriculum frameworks are specifically addressed in this chapter. However, this chapter addresses the following skills described in the Massachusetts Curriculum Frameworks for High School Chemistry either as "specific skills from the Mathematics Framework that students in this course should have the opportunity to apply" or as "skills [that] are not detailed in the Mathematics Framework, but [which] are necessary for a solid understanding in this course":

- Measure with accuracy and precision (*e.g.,* length, volume, mass, temperature, time)
- Convert within a unit (*e.g.,* centimeters to meters).
- Use common prefixes such as milli-, centi-, and kilo-.
- Use scientific notation, where appropriate.
- Determine the appropriate number of significant figures.
- Determine percent error from experimental and accepted values.
- Use appropriate metric/standard international (SI) units of measurement for mass (g); length (cm); and time (s).

Use this space for summary and/or additional notes:

Big Ideas	Details	Unit: Math & Measurement

The International System of Units

Unit: Math & Measurement

MA Curriculum Frameworks (2016): SP5

MA Curriculum Frameworks (2006): N/A

Mastery Objective(s): (Students will be able to...)

- Use and convert between metric prefixes attached to units.

Success Criteria:

- Conversions between prefixes move the decimal point the correct number of places.
- Conversions between prefixes move the decimal point in the correct direction.
- The results of conversions have the correct answers with the correct units, including the prefixes.

Tier 2 Vocabulary: prefix

Language Objectives:

- Set up and solve problems relating to the concepts described in this section.

Notes:

This section is intended to be a brief review. You learned to convert between metric prefixes in elementary or middle school. **You are expected to be able to fluently perform calculations that involve converting between metric prefixes.**

A unit is a specifically defined measurement. Units describe both the type of measurement, and a base amount.

For example, 1 cm and 1 inch are both lengths. They are used to measure the same dimension, but the specific amounts are different. (In fact, 1 inch is exactly 2.54 cm.)

Every measurement is a number multiplied by its units. In algebra, the term "3x" means "3 times x". Similarly, the distance "75 m" means "75 times the distance 1 meter".

The number and the units are both necessary to describe any measurement. You *always* need to write the units. Saying that "12 is the same as 12 g" would be as ridiculous as saying "12 is the same as 12 × 3".

Use this space for summary and/or additional notes:

The International System of Units

Big Ideas | **Details** | Unit: Math & Measurement

The International System (often called the metric system) is a set of units of measurement that is based on natural quantities (on Earth) and powers of 10.

The metric system has 7 fundamental "base" units:

Unit	Quantity	Currently Based On
meter (m)	length	the distance light travels in a specific time
kilogram (kg)	mass	the mass of the official prototype kilogram
second (s)	time	the time it takes for a particular type of radiation from a cesium-133 atom
Kelvin (K)	temperature	the temperature of the triple point of water
mole (mol)	amount of substance	the number of atoms in a specific mass of carbon-12
ampere (A)	electric current	the amount of current that produces a specific force under specific conditions
candela (cd)	intensity of light	the amount of light per unit of area at a specific distance

All other S.I. units are combinations of one or more of these seven base units.

For example:

Velocity (speed) is a change in distance over a period of time, which would have units of distance/time (m/s).

Force is a mass subjected to an acceleration. Acceleration has units of distance/time^2 (m/s^2), and force has units of mass × acceleration. In the metric system this combination of units (kg·m/s^2) is called a Newton, which means:
$1 \text{ N} \equiv 1 \text{ kg·m/s}^2$

As of 2018, Each of these base units is defined in some way that could be duplicated in a laboratory anywhere on Earth (except for the kilogram, which is defined by a physical object in a safe in France).

Use this space for summary and/or additional notes:

The International System of Units

Big Ideas | **Details** | Unit: Math & Measurement

In May 2019, all of the above S.I. units will be defined based on specifying exact values for certain fundamental constants:

- The Planck constant h is exactly $6.626\,070\,15 \times 10^{-34}$ J·s
- The elementary charge e is exactly $1.602\,176\,634 \times 10^{-19}$ C
- The Boltzmann constant k is exactly $1.380\,649 \times 10^{-23}$ J·K^{-1}
- The Avogadro constant N_A is exactly $6.022\,140\,76 \times 10^{23}$ mol^{-1}
- The speed of light c is exactly $299\,792\,458$ m·s^{-1}
- The ground state hyperfine splitting frequency of the caesium-133 atom $\Delta\nu(^{133}\text{Cs})_{hfs}$ is exactly $9\,192\,631\,770$ Hz
- The luminous efficacy K_{cd} of monochromatic radiation of frequency 540×10^{12} Hz is exactly 683 lm·W^{-1}

The S.I. base units are calculated from these seven definitions, after converting the derived units (joule, coulomb, hertz, lumen and watt) into the seven base units (second, meter, kilogram, ampere, kelvin, mole and candela).

Use this space for summary and/or additional notes:

The International System of Units

Rules for Writing S.I. Numbers and their Units

- The value of a quantity is written as a number followed by a space (representing a multiplication sign) and a unit symbol; *e.g.,* 2.21 kg, 7.3×10^2 m², or 22 K. This rule explicitly includes the percent sign (10 %, not 10%) and the symbol for degrees of temperature (37 °C, not 37°C). (However, note that angle measurements in degrees are written next to the number without a space.)

- Units do not have a period at the end, except at the end of a sentence.

- A prefix is part of the unit and is attached to the beginning of a unit symbol without a space. Compound prefixes are not allowed.

- Symbols for derived units formed by multiplication are joined with a center dot (·) or a non-breaking space; *e.g.,* N·m or N m.

- Symbols for derived units formed by division are joined with a solidus (fraction line), or given as a negative exponent. E.g., "meter per second" can be written m/s, m s^{-1}, m·s^{-1}, or $\frac{m}{s}$.

- The first letter of symbols for units derived from the name of a person is written in upper case; otherwise, they are written in lower case. *E.g.,* the unit of pressure is named after Blaise Pascal, so its symbol is written "Pa" (note that "Pa" is a two-letter symbol), but the symbol for mole is written "mol". However, the symbol for liter is "L" rather than "l", because a lower case "l" can be confused with the number "1".

- A plural of a symbol must not be used; *e.g.,* 25 kg, not 25 kgs.

- Units and prefixes are case-sensitive. *E.g.,* the quantities 1 mW and 1 MW represent two different quantities (milliwatt and megawatt, respectively).

- The symbol for the decimal marker is either a point or comma on the line. In practice, the decimal point is used in most English-speaking countries and most of Asia, and the comma in most of Latin America and in continental European countries.

- Spaces should be used as a thousands separator (1 000 000) in contrast to commas (1,000,000) or periods (1.000.000), to reduce confusion resulting from the variation between these forms in different countries.

- Any line-break inside a number, inside a compound unit, or between number and unit should be avoided.

Use this space for summary and/or additional notes:

The International System of Units

Unit: Math & Measurement

Prefixes

The metric system uses prefixes to indicate multiplying a unit by a power of ten. There are prefixes for powers of ten from 10^{-24} to 10^{24} but in chemistry, only the following four are commonly used:

- kilo (k) = 10^3 = 1000
- milli (m) = 10^{-3} = $\frac{1}{1\,000}$ = 0.001
- centi (c) = 10^{-2} = $\frac{1}{100}$ = 0.01
- micro (μ) = 10^{-6} = $\frac{1}{1\,000\,000}$ = 0.000 001

Any metric prefix is allowed with any metric unit. For example, if a mole (mol) is 6.02×10^{23} objects, then a millimole (mmol) would be

$(6.02 \times 10^{23}) \times \frac{1}{1\,000} = 6.02 \times 10^{20}$ objects.

An easier way to convert is to use the powers of ten that correspond with the prefixes to determine how many places to move the decimal point.

Metric Prefixes

Factor		Prefix	Symbol
1 000 000 000 000 000 000 000 000	10^{24}	yotta	Y
1 000 000 000 000 000 000 000	10^{21}	zeta	Z
1 000 000 000 000 000 000	10^{18}	exa	E
1 000 000 000 000 000	10^{15}	peta	P
1 000 000 000 000	10^{12}	tera	T
1 000 000 000	10^{9}	giga	G
1 000 000	10^{6}	mega	M
1 000	10^{3}	kilo	k
100	10^{2}	hecto	h
10	10^{1}	deca	da
1	10^{0}	—	—
0.1	10^{-1}	deci	d
0.01	10^{-2}	centi	c
0.001	10^{-3}	milli	m
0.000 001	10^{-6}	micro	μ
0.000 000 001	10^{-9}	nano	n
0.000 000 000 001	10^{-12}	pico	p
0.000 000 000 000 001	10^{-15}	femto	f
0.000 000 000 000 000 001	10^{-18}	atto	a
0.000 000 000 000 000 000 001	10^{-21}	zepto	z
0.000 000 000 000 000 000 000 001	10^{-24}	yocto	y

Use this space for summary and/or additional notes:

The International System of Units

Unit: Math & Measurement

Note that some of the prefixes skip by a factor of 10 and others skip by a factor of 10^3. This means you can't just count the steps—you have to actually look at the exponents.

Sample Problem:

15 Tm = _____ nm

- You need to move the decimal point 12 places to get to 10^0, and 9 more places to get to 10^{-9}, for a total of 21 places.
- Terameters are huge, and nanometers are much smaller. That means we're going to have a lot more nanometers than terameters, so we have to move the decimal point in the direction that makes the number larger (to the right).

Therefore, we need to move the decimal point 21 places to the right, which means we need to multiply by 10^{21}.

You could simply write your answer as 15×10^{21} m, and it would be correct. (And you can enter it into your calculator that way and the right thing will happen.)

However, to be proper scientific notation, you need to make the part before the multiplication sign between 1 and 10, which means you need to make it 1.5. If the number before the × sign gets smaller, then the number after the × sign needs to get larger so the end result stays the same. Therefore, 15×10^{21} m is the same as 1.5×10^{22} m, which is our final answer.

There is a popular joke based on the ancient Greek heroine Helen of Troy. She was said to have been the most beautiful woman in the world, and when she was kidnapped, the Trojan War was fought to bring her back to Sparta. Her beauty was described as "the face that launched a thousand ships." Therefore a milliHelen must be the amount of beauty needed to launch one ship.

Use this space for summary and/or additional notes:

The International System of Units

Big Ideas	Details
	Homework Problems

Perform the following conversions.

1. 2.5 m = _____ cm

2. 18 mL = _____ L

3. 68 kJ = _____ J

4. 6 500 mg = _____ kg

5. 101 kPa = _____ Pa

6. 325 ms = _____ s

Use this space for summary and/or additional notes:

Scientific Notation

Unit: Math & Measurement
MA Curriculum Frameworks (2016): SP5
MA Curriculum Frameworks (2006): N/A
Mastery Objective(s): (Students will be able to…)

- Correctly use numbers in scientific notation in mathematical problems.

Success Criteria:

- Numbers are converted correctly to and from scientific notation.
- Numbers in scientific notation are correctly entered into a calculator.
- Math problems that include numbers in scientific notation are set up and solved correctly.

Language Objectives:

- Explain how numbers are represented in scientific notation, and what each part of the number represents.

Notes:

This section is intended to be a brief review. You learned to use scientific notation in elementary or middle school. **You are expected to be able to fluently perform calculations that involve numbers in scientific notation, and to express the answer correctly in scientific notation when appropriate.**

Scientific notation is a way of writing a very large or very small number in compact form. The value is always written as a number between 1 and 10 multiplied by a power of ten.

For example, the number 1 000 would be written as 1×10^3. The number 0.000 075 would be written as 7.5×10^{-5}. The number 602 000 000 000 000 000 000 000 would be written as 6.02×10^{23}. The number 0.000 000 000 000 000 000 000 000 000 000 000 663 would be written as 6.63×10^{-34}.

Scientific notation is really just math with exponents, as shown by the following examples:

$$5.6 \times 10^3 = 5.6 \times 1000 = 5600$$

$$2.17 \times 10^{-2} = 2.17 \times \frac{1}{10^2} = 2.17 \times \frac{1}{100} = \frac{2.17}{100} = 0.0217$$

Use this space for summary and/or additional notes:

Scientific Notation

Notice that if 10 is raised to a positive exponent means you're multiplying by a power of 10. This makes the number larger, and the decimal point moves to the right. If 10 is raised to a negative exponent, you're actually dividing by a power of 10. This makes the number smaller, and the decimal point moves to the left.

Significant figures are easy to use with scientific notation: all of the digits before the "×" sign are significant. The power of ten after the "×" sign represents the (insignificant) zeroes, which would be the rounded-off portion of the number. In fact, the mathematical term for the part of the number before the "×" sign is the *significand*.

Math with Scientific Notation

Because scientific notation is just a way of rewriting a number as a mathematical expression, all of the rules about how exponents work apply to scientific notation.

Adding & Subtracting: adjust one or both numbers so that the power of ten is the same, then add or subtract the significands.

$$(3.50 \times 10^{-6}) + (2.7 \times 10^{-7}) = (3.50 \times 10^{-6}) + (0.27 \times 10^{-6})$$
$$= (3.50 + 0.27) \times 10^{-6} = 3.77 \times 10^{-6}$$

Multiplying & dividing: multiply or divide the significands. If multiplying, add the exponents. If dividing, subtract the exponents.

$$\frac{6.2 \times 10^8}{3.1 \times 10^{10}} = \frac{6.2}{3.1} \times 10^{8-10} = 2.0 \times 10^{-2}$$

Exponents: raise the significand to the exponent. Multiply the exponent of the power of ten by the exponent to which the number is raised.

$$(3.00 \times 10^8)^2 = (3.00)^2 \times (10^8)^2 = 9.00 \times 10^{(8 \times 2)} = 9.00 \times 10^{16}$$

Use this space for summary and/or additional notes:

Scientific Notation

Unit: Math & Measurement

Using Scientific Notation on Your Calculator

Scientific calculators are designed to work with numbers in scientific notation. It's possible to can enter the number as a math problem (always use parentheses if you do this!) but math operations can introduce mistakes that are hard to catch.

Scientific calculators all have either an "EE" or "EXP" button. The entire purpose of this button is to enter numbers in scientific notation and make sure the calculator stores them properly. On Texas Instruments calculators, such as the TI-30 or TI-89, you would do the following:

What you type	What the calculator shows	What it means
6.6 EE –34	6.6E–34	6.6×10^{-34}
1.52 EE 12	1.52E12	1.52×10^{12}
–4.81 EE –7	–4.81E–7	-4.81×10^{-7}

On some calculators, the scientific notation button is labeled EXP instead of EE.

Important notes:

- *Many high school students are afraid of the EE button because it is unfamiliar. If you are afraid of your EE button, you need to get over it and start using it anyway. However, if you insist on clinging to your phobia, you need to at least use parentheses around all numbers in scientific notation, in order to minimize the likelihood of PEMDAS errors in your calculations.*

- *Regardless of how you enter numbers in scientific notation into your calculator, always place parentheses around the denominator of fractions.*

$$\frac{2.75 \times 10^3}{5.00 \times 10^{-2}} \text{ becomes } \frac{2.75 \times 10^3}{(5.00 \times 10^{-2})}$$

- *You need to **write** answers using correct scientific notation. For example, if your calculator displays the number 1.52E12, you need to write 1.52×10^{12} (plus the appropriate unit, of course) in order to receive credit.*

Use this space for summary and/or additional notes:

Scientific Notation

Unit: Math & Measurement

Homework Problems

Convert each of the following between scientific and algebraic notation.

1. $2.65 \times 10^9 =$

2. $387\,000\,000 =$

3. $1.06 \times 10^{-7} =$

4. $0.000\,000\,065 =$

Solve each of the following on a calculator that can do scientific notation.

5. $(2.8 \times 10^6)(1.4 \times 10^{-2}) =$

   Answer: 3.9×10^4

6. $\dfrac{3.75 \times 10^8}{1.25 \times 10^4} =$

   Answer: 3.00×10^4

7. $\dfrac{1.2 \times 10^{-3}}{5.0 \times 10^{-1}} =$

   Answer: 2.4×10^{-3}

Use this space for summary and/or additional notes:

Using Math in Calculations

Unit: Math & Measurement

MA Curriculum Frameworks (2016): SP5

MA Curriculum Frameworks (2006): N/A

Mastery Objective(s): (Students will be able to…)

- Substitute values for variables in equations and solve them.

Success Criteria:

- Values are substituted for the correct variables.
- Equations are correctly solved for the missing variable using basic algebra.
- Answers have the correct units and are rounded to the appropriate number of significant figures.

Language Objectives:

- Set up and solve word problems relating to the concepts described in this section.

Notes:

Unlike biology, chemistry is a physical science. Among other things, this means chemistry involves calculations, which means you need to be comfortable with algebraic expressions.

Variables and Units

Unlike expressions in math class, which make a clear distinction between constants (the numbers you know the value of) and the variables,

- Equations in chemistry are written as all variables, because each equation works the same way no matter which quantity (or quantities) you are looking for.

- Each of the variables is a letter that relates to the quantity that it represents. For example, volume is V, mass is m, temperature is T, and the number of moles of substance is n. In chemistry, the same quantity *always* uses the same variable.

Use this space for summary and/or additional notes:

Using Math in Calculations

Page: 85
Unit: Math & Measurement

Big Ideas | Details

- Almost all quantities are measured and have units. . These units are your key to what kind of quantity the numbers describe. For example, in the quantity 12.5 mL, the mL means "milliliters." In the quantity 37.21 g, the g means "grams." In the quantity 21.5 °C, the °C means "degrees Celsius."
  - The unit is part of the quantity. For example, you can't say your height is 1.62. (1.62 what?) You would need to say that your height is 1.62 m (1.62 meters).
  - The unit tells you which type of quantity, and the type of quantity tells you the variable. For example, in the quantity 12.5 mL, the mL (milliliters) is a volume, which uses the variable "V." This means that if you have an equation in which the letter V represents volume, you would *replace* the letter V with the quantity 12.5 mL.
  - Be careful! In many cases, the same letter can be a unit or a number. For example, the letter "m" next to a number means the unit "meters" (a distance), but the variable "m" in an equation means "mass".

Some Quantities Used in Chemistry

Quantity	Unit	Variable	Quantity	Unit	Variable
mass	g	m	temperature	°C, K	T
length	m, cm	ℓ	velocity	$\frac{m}{s}$	v
area	m^2	A	heat	J	q^*
volume	mL	V	energy	J	E
number of moles	mol	n	pressure	bar, atm, kPa	P
density	$\frac{g}{mL}$	$\rho^\dagger$	time	s	t
concentration	$\frac{mol}{L}$	c	equilibrium constant	—	K
distance	m	d, ℓ	charge	C	q^*

---

[*] Notice that q is used for both heat and electrical charge. You need to figure out which quantity is meant from context.

[†] Some chemistry books use the Roman letter "D" for density, but the Greek letter "ρ" ("rho") is preferred. Be careful not to confuse it with the letter "P" (pressure).

Use this space for summary and/or additional notes:

Chemistry 1   Mr. Bigler

Big Ideas	Details
	## Variable Substitution

Variable substitution simply means taking the numbers you have from the problem and substituting those numbers for the corresponding variable in an equation. A simple version of this is a density problem:

If you have the formula:

$$\rho = \frac{m}{V}$$ and you're given: $m = 12.3 \text{ g}$ and $V = 2.8 \text{ cm}^3$

simply substitute 12.3 g for m, and 2.8 cm³ for V, giving:

$$\rho = \frac{12.3 \text{ g}}{2.8 \text{ cm}^3} = 4.4 \tfrac{g}{cm^3}$$

Use this space for summary and/or additional notes:

Using Math in Calculations

Unit: Math & Measurement

Equations

Math is a language. Like other languages, it has nouns (numbers), pronouns (variables), verbs (operations), and sentences (equations), all of which must follow certain rules of syntax and grammar.

This means that turning a word problem into an equation is translation from English to math.

Mathematical Operations

You have probably been taught translations for most of the common math operations:

word	meaning	word	meaning
and, more than (but not "is more than")	+	is	=
less than (but not "is less than")	−	is at least	≥
of	×	is more than	>
per	÷	is at most	≤
percent	÷ 100	is less than	<
change in x, difference in x	Δx*		

Suppose you were given the equation:

$$\rho = \frac{m}{V}$$

Using the table on page 85, we can see that m is mass and V is volume, which means the equation says "density is mass divided by volume". This means that if we knew that the mass of an object was 10.5 g and its volume was 23.7 mL, we could substitute those numbers into the equation to find the density:

$$\rho = \frac{m}{V} = \frac{10.5\ g}{23.7\ mL} = \frac{10.5}{23.7}\ \frac{g}{mL} = 0.443\ \frac{g}{mL}$$

We can use the same approach no matter which variable we are looking for.

---

* Note: The Greek letter Δ (delta) is attached to a variable to indicate the change in that variable. For example, ΔT represents a change in temperature. ΔT is one variable in the equation, even though it uses two symbols.

Use this space for summary and/or additional notes:

Using Math in Calculations

Unit: Math & Measurement

Sample Problems:

Q: An object has a volume of 17.7 mL and a density of $2.35 \frac{g}{mL}$. What is its mass?

A: Start with the equation and substitute:

$$\rho = \frac{m}{V}$$

$$2.35 \tfrac{g}{mL} = \frac{m}{17.7 \text{ mL}}$$

Now we have to do algebra. We want to get m by itself, which means we need to move 17.7 mL to the other side. Because it's in the denominator (on the bottom), we have to multiply both sides by it.

$$(17.7 \text{ mL})(2.35 \tfrac{g}{mL}) = \frac{m}{17.7 \text{ mL}}(17.7 \text{ mL})$$

$$41.6 \text{ g} = m$$

(Notice that the 17.7 mL cancels on the right because it's in both the numerator and the denominator. Notice also that the mL cancels on the left for the same reason, leaving g, which happens to be the correct unit. This is called "dimensional analysis," and we will study it in more depth in a future section.)

Q: An object has a mass of 44.7 g and a density of $1.68 \frac{g}{mL}$. What is its volume?

A: Again, start with the equation and substitute:

$$\rho = \frac{m}{V}$$

$$1.68 \tfrac{g}{mL} = \frac{44.7 \text{ g}}{V}$$

The variable we want is on the bottom. Again, following the rules of algebra, in order to get it off the bottom, we first have to multiply both sides by it to clear the fraction:

$$V \cdot 1.68 \tfrac{g}{mL} = \frac{44.7 \text{ g}}{V} \cdot V$$

Then, we can solve for V in a subsequent step:

$$\frac{V \cdot 1.68 \tfrac{g}{mL}}{1.68 \tfrac{g}{mL}} = \frac{44.7 \text{ g}}{1.68 \tfrac{g}{mL}}$$

$$V = 26.6 \text{ mL}$$

Use this space for summary and/or additional notes:

Using Math in Calculations

Big Ideas | **Details**

Unit: Math & Measurement
Page: 89

Note: Whenever you have to solve an equation for a quantity in the denominator, always do it in two steps: clear the fraction first, then divide. If you try to cleverly rearrange the quantities without doing this, you are almost certain to get the wrong answer!

Q: Find the volume taken up by 3.10 mol of a gas at 298 K and 1.25 atm.

A: When you see a problem like this, the first thing you should do is use the units to figure out what quantities you have in the problem, and label them with their variables:

Find the volume taken up by 3.10 mol of a gas at 298 K and 1.25 atm. (For this problem, use $0.0821 \frac{\ell \cdot atm}{mol \cdot K}$ for the gas constant.)
$$\underset{V}{} \quad \underset{n}{} \quad \underset{T}{} \quad \underset{P}{} \quad \underset{R}{}$$

To solve this problem, we need an equation that relates V, n, T, and P. This turns out to be the ideal gas law:

$$PV = nRT$$

Now substitute the numbers in place of the variables in the equation:

$$\underset{P}{(1.25\,atm)}\,\underset{V}{V} = \underset{n}{(3.10\,mol)}\,\underset{R}{(0.0821 \tfrac{\ell \cdot atm}{mol \cdot K})}\,\underset{T}{(298\,K)}$$

Then solve, using algebra. This means we need to divide both sides by 1.25 atm to get the answer.

$$V = \frac{(3.10\,mol)(0.0821 \tfrac{\ell \cdot atm}{mol \cdot K})(298\,K)}{1.25\,atm} = 60.7\,\ell$$

The Problem-Solving Process

1. Identify the quantities in the problem, using the units.
2. Assign variables to those quantities.
3. Make a list of all of your variables (including the one you're looking for), and what they're equal to.
4. Write down an equation that relates all of those variables.
5. Substitute the values of the variables into the equation. You should have only one variable left, which is the one you're looking for.
6. Solve the equation, using algebra.
7. Don't forget to round your answer correctly and include the units!

Use this space for summary and/or additional notes:

Chemistry 1 — Mr. Bigler

Using Math in Calculations

Page: 90
Unit: Math & Measurement

Homework Problems

For these problems, use the table of units and variables on page 85 to determine which quantities represent which variables. Then substitute the variables into the equation given. *You do not have to solve the equations.*

1. 375 J of heat is added to a 75 g block of metal that has a specific heat capacity of $C = 0.450 \frac{J}{g \cdot °C}$. What is the temperature change of the metal?

   $q = mC\Delta T$

   Answer: $(375\,J) = (75\,g)(0.450 \frac{J}{g \cdot °C})\Delta T$

2. A rock has a density of $6.4 \frac{g}{cm^3}$ and a mass of 1 500 g. What is its volume.

   $\rho = \frac{m}{V}$

3. 2.5 mol of an ideal gas has a pressure of 1.5 bar and a temperature of 325 K. The gas constant is $0.081 \frac{L \cdot bar}{mol \cdot K}$. What is its volume?

   $PV = nRT$

Use this space for summary and/or additional notes:

Conversions (Factor-Label Method)

Unit: Math & Measurement
MA Curriculum Frameworks (2016): N/A
MA Curriculum Frameworks (2006): N/A
Mastery Objective(s): (Students will be able to...)

- Use algebra and units to create a strategy for problem-solving.

Success Criteria:

- Conversion factors are arranged so that the numerator and denominator are equal.
- Units in conversion factors are arranged so they cancel unwanted units and provide desired units.
- Answers are correct with the correct units.

Tier 2 Vocabulary: unit, convert, conversion

Language Objectives:

- Understand and explain that a conversion factor is two quantities (including their units) that are equal.

Notes:

A conversion is based on the idea that you can express the same quantity using different numbers and units.

For example, Mr. Bigler is 5 feet 4 inches tall. We could express this as 64 inches, $5\frac{1}{3}$ feet, $1.\overline{7}$ yards, 0.001 mile, 163 cm, 1.63 m, or 5.3×10^{-13} parsecs.[*]

The process of getting from one of these numbers to another is called a unit conversion.

Conversions are based on two strategies:

1. Canceling units you don't want and replacing them with units you do want.
2. Repeatedly multiplying by fractions that equal 1 (*i.e.*, the numerator equals the denominator), so the actual quantity doesn't change.

---

[*] A parsec is a distance of about 3.26 light years, or about 3×10^{13} km.

Use this space for summary and/or additional notes:

Conversions (Factor-Label Method)

Unit: Math & Measurement

To show how this works, consider the following math problem:

$$\frac{1}{2} \times \frac{2}{3} \times \frac{3}{4} \times \frac{4}{5} \times \cdots \times \frac{99}{100} = ?$$

The answer is $\frac{1}{100}$, because everything else cancels:

$$\frac{1}{\cancel{2}} \times \frac{\cancel{2}}{\cancel{3}} \times \frac{\cancel{3}}{\cancel{4}} \times \frac{\cancel{4}}{\cancel{5}} \times \cdots \times \frac{\cancel{99}}{100} = \frac{1}{100}$$

As you may know from algebra, this also works with numbers and variables:

$$\left(\frac{4w}{1}\right)\left(\frac{3x}{w}\right)\left(\frac{y}{2x}\right)\left(\frac{5z}{2y}\right) = \frac{\cancel{4}\cdot 3 \cdot 5 \cdot z}{\cancel{2}\cdot\cancel{2}} = 15z$$

Units work just like variables, so the algebra that you can do with a variable also works with a unit:

$$\frac{2\,\text{yd.}}{1} \times \frac{3\,\text{ft.}}{1\,\text{yd.}} \times \frac{12\,\text{in.}}{1\,\text{ft.}} = \frac{2 \times 3 \times 12\,\text{in.}}{1} = 72\,\text{in.}$$

Notice also that each time we multiplied by a fraction, the numerator was equal to the denominator. (3 ft. = 1 yd. and 12 in. = 1 ft.) This means we were multiplying by 1 each time. That's why 72 in. *must be* the <u>same</u> distance as 2 yd. We converted by multiplying:

$$2\,\text{yd.} \times \text{"1"} \times \text{"1"} = 72\,\text{in.}$$

(The "1"s are in quotes because the fractions derived from the conversion factors are all equal to one, even if they don't look like it.)

Some chemistry teachers prefer to use a table with lines to keep the conversion factors neat. The following are two equivalent ways to represent the same calculation. Note that conversion factors (fractions that equal 1) are in vertical columns:

$$\frac{2\,\text{yd.} \;|\; 3\,\text{ft.} \;|\; 12\,\text{in.}}{1 \;\;|\;\; 1\,\text{yd.} \;|\; 1\,\text{ft.}} = \frac{2 \times 3 \times 12\,\text{in.}}{1 \times 1} = 72\,\text{in.}$$

$$\frac{2\,\text{yd.}}{1} \times \frac{3\,\text{ft.}}{1\,\text{yd.}} \times \frac{12\,\text{in.}}{1\,\text{ft.}} = \frac{2 \times 3 \times 12\,\text{in.}}{1} = 72\,\text{in.}$$

Use this space for summary and/or additional notes:

Conversions (Factor-Label Method)

Page: 93

Big Ideas | Details | Unit: Math & Measurement

To convert a quantity from one unit to another:

1. Write down the number *and* *units* that you're starting with
2. Find a conversion factor that contains the unit you want to get rid of.
3. Turn the conversion into a fraction and arrange it to cancel the unit you want to get rid of. (If the unit you want to cancel is in the numerator, the same unit needs to be in the denominator in the fraction.)
4. Repeat steps 2 & 3 until you end up with the unit you want.
5. After canceling units, multiply and divide the numbers in the numerator & denominator and simplify the expression.

Working Example:

What is the mass (in grams) of 2.75 moles of sodium chloride (NaCl)?

Conversion factor for NaCl:
$$1 \text{ mol} = 58.44 \text{ g}$$

Solution:

1. We are starting with 2.75 moles (2.75 mol) of chlorine. This means we need to write 2.75 mol in fraction form, as $\dfrac{2.75 \text{ mol}}{1}$.

2. We want to multiply $\dfrac{2.75 \text{ mol}}{1} \times 1$ (so we don't change the actual amount).

   This will become $\dfrac{2.75 \text{ mol}}{1} \times \dfrac{}{}$ and the unknown fraction needs to equal 1.

3. We know we need to cancel moles, so moles will end up on the bottom of the next fraction. This gives us:

   $$\dfrac{2.75 \text{ mol}}{1} \times \dfrac{}{\text{mol}}$$

Use this space for summary and/or additional notes:

Conversions (Factor-Label Method)

Unit: Math & Measurement

4. Our conversion factor was: 1 mole NaCl = 58.44 g NaCl which means our conversion factor is:

$$\frac{1\,\text{mol}}{58.44\,\text{g}} = \frac{58.44\,\text{g}}{1\,\text{mol}} = 1$$

We need moles on the bottom, which means we need to use the second fraction. Multiplying our original 2.75 mol by this fraction gives us:

$$\frac{2.75\,\cancel{\text{mol}}}{1} \times \frac{58.44\,\text{g}}{1\,\cancel{\text{mol}}}$$

5. Now, the only unit left is the one we want (grams), so we're ready to solve the problem:

$$\frac{2.75\,\cancel{\text{mol}}}{1} \times \frac{58.44\,\text{g}}{1\,\cancel{\text{mol}}} = \frac{2.75 \times 58.44\,\text{g}}{1 \times 1} = \frac{160.71\,\text{g}}{1} = 160.71\,\text{g}$$

6. Because we had only 3 significant figures in the original number, we need to round our answer to 3 "sig figs". This gives us our final, rounded-off answer of $\boxed{161\,\text{g}}$.

Use this space for summary and/or additional notes:

Conversions (Factor-Label Method)

Homework Problems

Perform each of the following conversions.

1. 23.6 cm = __.24__ m

2. 15.9 L = __15,900__ mL

3. 0.89 km = __890,00__ mm

4. 7.31×10^{24} mmol = _____ mol

5. 15.0 gallons = _____ L

6. 3.65 miles = __5.88__ km

7. 64 inches = __162.5__ cm

    $\dfrac{64\ i}{1}\ \dfrac{cm}{i}$

8. 183 pounds = __83.01__ kg

    $\dfrac{183\ P}{1}\ \dfrac{kg}{1\ P}$

9. $65\,\dfrac{\text{miles}}{\text{hour}}$ = _____ $\dfrac{m}{s}$

    $\dfrac{65\ m}{h} = \dfrac{m}{s}$

10. $13.2\,\dfrac{g}{cm^3}$ = _____ $\dfrac{pounds}{foot^3}$

    $\dfrac{13.2\ g}{cm^3}\ \dfrac{\ }{\ }$

11. $3.65\,\dfrac{\text{dollars}}{\text{gallon}}$ = _____ $\dfrac{\text{cents}}{L}$

    $\dfrac{3.65\ d}{g}$

12. $32\,\dfrac{\text{miles}}{\text{gallon}}$ = _____ $\dfrac{km}{L}$

    $\dfrac{32\ m}{g}\ \dfrac{km}{L}$

Conversion Factors

1 gallon = 3.785 L	1 mile = 1.61 km	(1 inch)³ ≡ (2.54 cm)³	1 hour ≡ 60 min.
1 pound = 454 g	1 inch ≡ 2.54 cm	(1 foot)³ ≡ (12 inch)³	1 min. ≡ 60 s
	1 foot ≡ 12 inches		

Use this space for summary and/or additional notes:

Dimensional Analysis

Unit: Math & Measurement
MA Curriculum Frameworks (2016): SP5
MA Curriculum Frameworks (2006): N/A
Mastery Objective(s): (Students will be able to…)

- Use compound units to infer conversion factors.
- Infer equations from compound units.

Success Criteria:

- Equations relate quantities in the same way that units do.
- Word problems involving conversions that use inferred quantities are correct, including the correct units, and are rounded to the appropriate number of significant figures.

Tier 2 Vocabulary: dimension, unit, conversion

Language Objectives:

- Explain the process of using a compound unit as a conversion factor using appropriate vocabulary.

Notes:

<u>dimensions</u>: the units that something is measured in.

<u>dimensional analysis</u>: using the relationship between units (dimensions) and equations in order to analyze (and solve) problems. This can involve either:

- Using units to predict the relationships between quantities (and sometimes the equations that relate them)
- Using an equation to determine what the dimensions (units) should be.

Remember that units are like variables. They can be multiplied and divided by their coëfficients, and by each other. If you divide meters (a unit of distance) by seconds (a unit of time) you end up with the units $\frac{m}{s}$ (a unit of velocity[*]). Because this works for one set of units, it must also work for other units that measure the same dimensions. Because we divided the units of distance by the units of time and got units of velocity, this means we can divide <u>*any*</u> distance (regardless of the units) by <u>*any*</u> time (regardless of the units) to get velocity.

---

[*] There is a subtle difference between velocity and speed, which you will study if you take physics.

Use this space for summary and/or additional notes:

Dimensional Analysis

| Big Ideas | Details | Unit: Math & Measurement |

For example, density measures how heavy something is for its size. The density of iron is $7.87 \frac{g}{cm^3}$. Because grams measure mass and cm^3 measure volume, it must be true that:

$$\text{density units} = \frac{\text{mass units}}{\text{volume units}} \quad \text{which means} \quad \text{density} = \frac{\text{mass}}{\text{volume}}$$

This must mean that *any* mass unit divided by *any* volume unit gives a density unit (which is, in fact, true).

Some of the units that density can be measured in (mass units divided by volume units) include:

$$\frac{g}{mL} \quad \frac{g}{L} \quad \frac{g}{cm^3} \quad \frac{kg}{L} \quad \frac{kg}{m^3} \quad \frac{tonne}{m^3} \quad \frac{lb.}{ft.^3}$$

You can use dimensional analysis in the same manner to help you figure out how to solve many problems that include units of measure. In fact, dimensional analysis has led to new scientific discoveries. Scientists can sometimes determine a relationship between two quantities based on the units, and are then able to prove that relationship in the laboratory!

However, there are pitfalls. For example, mechanical work and torque have the same units (N·m), but they describe different kinds of quantities and cannot be used interchangeably. (In fact, in a physics problem involving tightening a bolt with a wrench, 36 N·m of torque ended up doing 19 N·m of work on the bolt!)

Also, dimensional analysis cannot predict constants that might appear in an equation. For example, the unit for energy is a joule, which equals one $\frac{kg \cdot m^2}{s^2}$.

Because mass is measured in kg and velocity in $\frac{m}{s}$, this suggests that the equation for kinetic energy should be mv^2. However, the equation is actually $\frac{1}{2}mv^2$. The factor of $\frac{1}{2}$ is a necessary part of the equation, but cannot be discovered by dimensional analysis.

Use this space for summary and/or additional notes:

Dimensional Analysis

Page: 98
Unit: Math & Measurement

Big Ideas | Details

Making Compound Units Into Conversion Factors

Any time you have a number with a compound unit, you can make a "temporary" conversion factor out of it. For example, if a car is driving $60\,\frac{mi.}{hr.}$, then:

$$60\,\tfrac{mi.}{hr.} = 60 \times \frac{mi.}{hr.} = \frac{7.87}{1} \times \frac{mi.}{hr.} = \frac{60\,mi.}{1\,hr.}$$

In other words, the quantity $60\,\frac{mi.}{hr.}$ gave us the conversion factor 60 mi. = 1 hr. (Notice that the coëfficient of 60 ended up with miles, which is the unit that was on top.) This means that for this problem—as long as the car keeps going the same speed—60 miles takes 1 hour. That means you can use this conversion factor *in this problem* to convert miles to hours, or hours to miles.

Examples:

- The density of iron (Fe) is $7.87\,\frac{g}{cm^3}$. What is the volume of 15.0 g of iron?

    To solve this problem, we recognize that the compound unit $7.87\,\frac{g}{cm^3}$ can be written as follows:

    $$7.87\,\tfrac{g}{cm^3} = 7.87 \times \frac{g}{cm^3} = \frac{7.87}{1} \times \frac{g}{cm^3} = \frac{7.87\,g}{1\,cm^3}$$

    This means 7.87 g = 1 cm³. We can use this as the conversion factor between grams of iron and cubic centimeters of iron. We will use in this problem to convert grams of iron into cm³ of iron:

    $$\frac{15.0\,g}{1} \times \frac{1\,cm^3}{7.87\,g} = \frac{15.0\,cm^3}{7.87} = 1.91\,cm^3$$

Use this space for summary and/or additional notes:

Chemistry 1 — Mr. Bigler

Dimensional Analysis

Unit: Math & Measurement

- The concentration of sodium chloride (NaCl) in sea water is about $0.48 \frac{mol}{L}$. How many moles of NaCl are in 55,000 L of sea water?

  To solve this problem, we recognize that the compound unit $0.48 \frac{mol}{L}$ is going to be the conversion factor 0.48 mol = 1 L. This is the temporary conversion factor that we will use in this problem to convert liters of sea water into moles of NaCl:

  $$\frac{55\,000\,\cancel{L}}{1} \times \frac{0.48\,mol}{1\,\cancel{L}} = 55\,000 \times 0.48\,mol = 26\,400\,mol = 26\,000\,mol$$

  (Note that we had to round off the final answer to 2 significant figures.)

- The molar mass of NaOH is $40.00 \frac{g}{mol}$. What is the mass of 2.85 mol of NaOH?

  To solve this problem, we recognize that the compound unit $40.00 \frac{g}{mol}$ is going to be the conversion factor 40.00 g = 1 mol. This is the temporary conversion factor that we will use in this problem to convert moles of NaOH into grams:

  $$\frac{2.85\,\cancel{mol}}{1} \times \frac{40.00\,g}{1\,\cancel{mol}} = 2.85 \times 40.00\,g = 114\,g$$

Use this space for summary and/or additional notes:

Dimensional Analysis

Unit: Math & Measurement

Big Ideas | Details

Homework Problems

1. An object has a density of $3.65 \frac{g}{cm^3}$. If the volume of the object is 12.5 cm^3, what is its mass?

   Answer: 45.6 g

2. A liquid solution has a salt concentration of 2.5 mol/l. How many moles of salt are in 0.50 ℓ of the solution?

   Answer: 1.3 mol

3. A car is travelling at a speed of 65 mi./hr. How many hours would it take for this car to travel 250 mi.?

   Answer: 3.8 hr

4. Suppose the average temperature of the Earth is rising at a rate of 2.0 °C/100 years. When Mr. Bigler gives this same homework assignment to one of your children 25 years from now, how much warmer will the average temperature of the Earth be?

   Answer: 0.5 °C

Use this space for summary and/or additional notes:

Dimensional Analysis

Unit: Math & Measurement

5. If a gas at "standard temperature and pressure" has a molar volume of $22.7 \frac{L}{mol}$, how many moles of this gas would there be in a 5.5 L balloon?

   Answer: 0.24 mol

6. Suppose you have a job at which you earn $8.00 per hour (which you can write as 8.00 $/hr). How many hours would you have to work to have enough money for a $1200 car?

   Answer: 150 h

7. The continent of South America is drifting away from Africa at a rate of about $2.5 \frac{cm}{year}$. If South America was once touching Africa, and the speed of the plates was constant, how many years did it take for South America to get to its present location, which is about 5000 km away from Africa?

   (*Hint: don't forget that you will need to convert* km *to* mm.)

   Answer: ?

Use this space for summary and/or additional notes:

Logarithms

Unit: Math & Measurement

MA Curriculum Frameworks (2016): SP5

MA Curriculum Frameworks (2006): N/A

Mastery Objective(s): (Students will be able to...)

- Use logarithms to solve for a variable in an exponent.

Success Criteria:

- Equations use logarithms to turn equations of the form $a^x = b$ into equations of the form $x \log(a) = \log(b)$.

Tier 2 Vocabulary: logarithm

Language Objectives:

- Explain what the logarithm function is used for.

Notes:

The logarithm may well be the least well-understood function encountered in high school mathematics. In high school chemistry, logarithms are used for the pH function for measuring the strength of acids & bases (which we will cover at the end of the year).

The simplest way to understand logarithms is to start with the base ten logarithm. You can think of the (base ten) logarithm of a number as the number of zeroes after a number.

x		$\log_{10}(x)$
100 000	10^5	5
10 000	10^4	4
1 000	10^3	3
100	10^2	2
10	10^1	1
1	10^0	0
0.1	10^{-1}	−1
0.01	10^{-2}	−2
0.001	10^{-3}	−3
0.000 1	10^{-4}	−4
0.000 01	10^{-5}	−5

As you can see from the above table, the logarithm of a number turns a set of numbers that vary exponentially (powers of ten) into a set that vary linearly.

Use this space for summary and/or additional notes:

Logarithms

You can get a visual sense of the logarithm function from the logarithmic number line below:

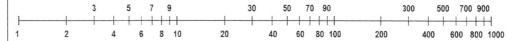

Notice that the *distance* from 1 to 10 is the same as the *distance* from 10 to 100 and from 100 to 1000. In fact, the relative distance to every number on this number line is the logarithm of the number.

x	$\log_{10}(x)$	distance from beginning of number line
10^0	0	0
$10^{0.5} \approx 3.16$	0.5	½ cycle
$10^1 = 10$	1	1 cycle
$10^2 = 100$	2	2 cycles
$10^3 = 1000$	3	3 cycles

By inspection, you can see that the same is true for numbers that are not exact powers of ten. The logarithm function compresses correspondingly more as the numbers get larger.

The most useful mathematical property of logarithms is that they move an exponent into the linear part of the equation:

$$\log_{10}(10^3) = 3\log_{10}(10) = (3)(1) = 3$$

In fact, the logarithm function works the same way for any base, not just 10:

$$\log_2(2^7) = 7\log_2(2) = (7)(1) = 7$$

(In this case, the word "base" means the base of the exponent.) The general equation is:

$$\log_x(a^b) = b\log_x(a)$$

This is a powerful tool in solving for the exponent in an equation. This is, in fact, precisely the purpose of using logarithms in most mathematical equations.

Use this space for summary and/or additional notes:

Logarithms

Unit: Math & Measurement

Sample problem:

Q: Solve $3^x = 15$ for x.

A: Take the logarithm of both sides. (Note that writing "log" without supplying a base implies that the base is 10.)

$$\log(3^x) = \log(15)$$
$$x \log(3) = \log(15)$$
$$(x)(0.477) = 1.176$$
$$x = \frac{1.176}{0.477} = 2.465$$

This is the correct answer, because $3^{2.465} = 15$

A powerful tool that follows from this is using logarithmic graph paper to solve equations. If you plot an exponential function on semi-logarithmic ("semi-log") graph paper (meaning graph paper that has a logarithmic scale on one axis but not the other), you get a straight line.

The graph at the right is the function $y = 2^x$. Notice where the following points appear on the graph:

Domain	Range
0	1
1	2
2	4
3	8
4	16
5	32

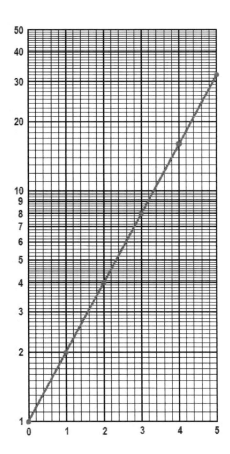

Notice also that you can use the graph to find intermediate values. For example, at x = 2.6, the graph shows that y = 6.06.

Use this space for summary and/or additional notes:

Summary: Math & Measurement

Unit: Math & Measurement

List the main ideas of this chapter in phrase form:

Write an introductory sentence that categorizes these main ideas.

Turn the main ideas into sentences, using your own words. You may combine multiple main ideas into one sentence.

Add transition words to make your writing clearer and rewrite your summary below.

Use this space for summary and/or additional notes:

Introduction: Matter

Unit: Matter

Topics covered in this chapter:

States of Matter 109
Properties of Matter 111
Phase Diagrams 115
Phases & Phase Changes 120
Conservation of Mass, Energy and Charge 123

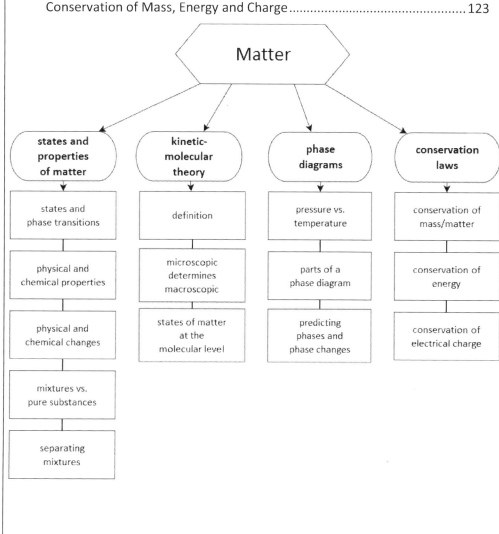

Use this space for summary and/or additional notes:

Big Ideas	Details	Unit: Matter

Introduction: Matter

Standards addressed in this chapter:

Massachusetts Curriculum Frameworks & Science Practices (2016):

HS-PS1-11(MA): Design strategies to identify and separate the components of a mixture based on relevant chemical and physical properties.

HS-PS2-8(MA): Use kinetic molecular theory to compare the strengths of electrostatic forces and the prevalence of interactions that occur between molecules in solids, liquids, and gases. Use the combined gas law to determine changes in pressure, volume, and temperature in gases.

Massachusetts Curriculum Frameworks (2006):

1.1 Identify and explain physical properties (e.g., density, melting point, boiling point, conductivity, malleability) and chemical properties (e.g., the ability to form new substances). Distinguish between chemical and physical changes.

1.2 Explain the difference between pure substances (elements and compounds) and mixtures. Differentiate between heterogeneous and homogeneous mixtures.

1.3 Describe the three normal states of matter (solid, liquid, gas) in terms of energy, particle motion, and phase transitions.

6.3 Using the kinetic molecular theory, describe and contrast the properties of gases, liquids, and solids. Explain, at the molecular level, the behavior of matter as it undergoes phase transitions.

6.4 Describe the law of conservation of energy. Explain the difference between an endothermic process and an exothermic process.

Use this space for summary and/or additional notes:

Big Ideas	Details
	# States of Matter

Unit: Matter

MA Curriculum Frameworks (2016): HS-PS1-11(MA)

MA Curriculum Frameworks (2006): 1.3, 6.3

Mastery Objective(s): (Students will be able to...)
- Define & describe states of matter and transitions between states.
- Classify matter according to its physical state.

Success Criteria:
- States of matter and transitions are described using correct vocabulary.
- States of matter are correctly identified based on observable properties.

Tier 2 Vocabulary: state, matter

Language Objectives:
- Explain the three common states of matter (solid, liquid and gas) and the properties of each.
- Explain the difference between a physical and chemical change. |
| | **Notes:**

<u>matter</u>: anything that has mass and takes up space (has volume).

    Examples of matter: anything you can touch or feel—solids, liquids, and gases.

    Examples of things that are *not* matter: forms of energy such as light, microwaves, radio waves, etc.

<u>state of matter</u>: the physical form the matter is in (solid, liquid, gas, or plasma)

<u>solid</u>: a state of matter in which the molecules are bonded (attached) to each other. Molecules in a solid move back and forth, but cannot break free from the other molecules. Solids have a definite shape and a definite volume.

<u>liquid</u>: a state of matter in which bonds between the molecules are continuously breaking and forming. Molecules in a liquid are free to move, but are attracted to nearby liquid molecules. Liquids have a definite volume, but not a definite shape. (Liquids take on the shape of their containers.)

<u>gas</u>: a state of matter in which the molecules are not bonded to one another. Molecules in a gas are free to move anywhere within the confines of their container. Gases have neither a definite shape nor a definite volume. (Gases expand to fill their containers.) |

Use this space for summary and/or additional notes:

States of Matter

plasma: a state in which the molecules have so much energy that they cannot hold onto all of their electrons. Charge is continuously flowing through the plasma and can often be seen as blue streaks (such as in a plasma globe).

Note that the distinctions between the phases can be subtle. For example, ketchup has a definite shape unless you wait for a long time, but it eventually takes on the shape of its container. As it turns out, ketchup is a liquid with a high viscosity (meaning that it resists flowing). Glass flows very slowly (windows that are centuries old are thicker at the bottom than at the top), but this is because of the movement of solid particles. (This is analogous to small pebbles settling to the bottom of a bucket of rocks.) Glass is therefore an amorphous (irregularly-shaped) solid, not a viscous liquid.

melting: the transition from a solid to a liquid.

freezing: the transition from a liquid to a solid.

boiling: the transition from a liquid to a gas.

condensation (or condensing): the transition from a gas to a liquid.

sublimation (or subliming): the transition from a solid directly to a gas.

deposition (or depositing): the transition from a gas directly to a solid.

Some properties of solids, liquids and gases and the processes of converting between them are summarized in the phase diagram below:

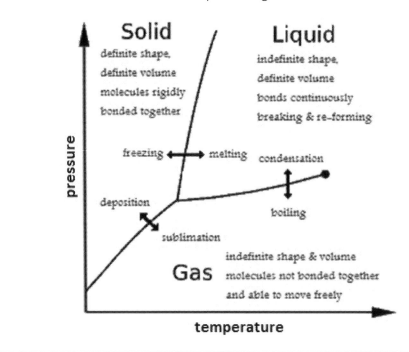

Use this space for summary and/or additional notes:

Big Ideas	Details
	# Properties of Matter

Unit: Matter
MA Curriculum Frameworks (2016): HS-PS1-11(MA)
MA Curriculum Frameworks (2006): 1.1, 1.2
Mastery Objective(s): (Students will be able to…)
- Describe the properties of mixtures and pure substances.
- Classify substances as heterogeneous or homogeneous mixtures, compounds, or elements according based on information about those substances.
- Identify methods of separating mixtures based on differences in properties.

Success Criteria:
- Substances are correctly identified as mixtures (heterogeneous or homogeneous), compounds or elements.
- Suitable methods for separating mixtures are chosen based on differences in chemical or physical properties.

Tier 2 Vocabulary: physical, chemical, property, mixture, compound, element
Language Objectives:
- Demonstrate understanding of the key terms "homogeneous mixture," "heterogeneous mixture," "compound," and "element." |
| | **Notes:**

physical properties: characteristics of the substance that can be measured or observed without changing the identity of the substance. *E.g.,* boiling point, freezing point, density, size, shape, color, *etc.*

chemical properties: characteristics of the substance having to do with how the atoms and molecules that make up substance can be combined with or changed into other substances. These properties can only be measured through changes to the identity of the substance. *E.g.,* chemical reactivity, flammability.

physical change: any change that alters the physical properties of the substance, such as freezing, boiling, tearing, crushing, *etc.*

chemical change: any change that alters the chemical properties (identity) of the substance, such as burning, cooking, rusting, decaying, *etc.* |

Use this space for summary and/or additional notes:

Properties of Matter

Note that the difference between a physical change and a chemical change can be subtle. For example, if you have a solution of sugar dissolved in water and you let the water evaporate, this would be a *physical* change because the sugar and water molecules are each unchanged by the process.

However, if you have a solution of salt dissolved in water and you evaporate the water, this would be a *chemical* change, because when salt dissolves in water, the ionic bonds between the sodium and chloride ions break and the ions remain separate while the salt is in solution. When you evaporate the water, ionic bonds form between the sodium and chloride ions, which creates the ionic compound sodium chloride.

There is no way to see this difference. This means you need to understand chemical bonds and the processes of forming and breaking them in order to be able to decide whether a change is physical or chemical!

Mixtures vs. Pure Substances

mixture: two or more different substances sharing the same space or volume. Mixtures can be separated based on differences in physical properties.

Mixtures can be:

homogeneous: every sample of the mixture is the same, no matter what part of the mixture it's taken from. (*homo* = same) *E.g.,* salt water. Gatorade

heterogeneous: samples taken from different parts of the mixture may be different. (*hetero* = different) *E.g.,* chocolate chip cookies, orange juice.

pure substance: a pure substance is a substance that cannot be separated or broken down by any physical change. A pure substances can be a:

compound: a substance made out of different kinds of atoms that are chemically bonded together. Compounds can be broken down through chemical changes. Anything that can be described by a chemical formula is a compound. *E.g.,* H_2O, $C_6H_{12}O_6$, $NaCl$ (table salt), C_3H_8O (rubbing alcohol).

element: a substance made out of only one kind of atom. Elements cannot be broken down through chemical changes. The periodic table lists all of the known elements according to their properties, which means any substance on the periodic table is an element. *E.g.,* iron, gold, oxygen, aluminum.

Use this space for summary and/or additional notes:

Properties of Matter

Separating Mixtures

Mixtures can be separated based on differences in the physical properties of the different substances that make up the mixture. Some processes used for separating mixtures include:

<u>filtration</u>: separating substances by size—larger ones are trapped on the filter and smaller ones can pass through.

<u>distillation</u>: separating substances that have different boiling points by heating to a temperature at which one boils and the other does not.

<u>evaporation</u>: evaporating or boiling off water (or another solvent) to leave behind a solid.

<u>crystallization</u>: separating substances that have different freezing points by letting one form a solid (freeze), but not the other.

<u>centrifugation</u>: separating substances according to their densities by spinning them at high speeds.

<u>chromatography</u>: separating substances by how quickly or slowly they move through another substance.

Homework Problems

For each of the following changes, state whether the change is chemical or physical and explain how you know.

1. water boiling

2. iron rusting

3. cooking an egg

4. breaking glass

5. tearing up a piece of paper

6. burning a piece of paper

Use this space for summary and/or additional notes:

Properties of Matter

Page: 114
Unit: Matter

Big Ideas	Details

7. making crushed ice in a blender

8. garbage turning into compost

9. leaves changing color

Classify the each of the following types of matter as a heterogeneous mixture, homogeneous mixture, compound, or element.

10. pure water (H_2O)

11. helium

12. chocolate-chip cookies

13. salt water

14. orange juice

15. 14-karat gold (note: pure gold is 24K)

16. carbonated soda

17. ice water (pure H_2O, but both liquid and solid)

18. aluminum

19. chicken noodle soup

20. chicken broth

21. glucose ($C_6H_{12}O_6$)

22. How would you separate a mixture of sugar, sand, and hollow plastic beads?

Use this space for summary and/or additional notes:

Chemistry 1

Mr. Bigler

Phase Diagrams

Unit: Matter

MA Curriculum Frameworks (2016): HS-PS1-11(MA)

MA Curriculum Frameworks (2006): 1.3, 6.3

Mastery Objective(s): (Students will be able to...)
- Identify the phase of a substance at any combination of temperature and pressure.
- Determine the melting and boiling points of a substance any pressure.

Success Criteria:
- Phases are correctly identified as solid, liquid, gas, supercritical fluid, *etc.,* in accordance with the temperature and pressure indicated on the phase diagram.
- Melting and boiling point temperatures are correctly identified for a substance from its phase diagram for a given pressure.
- The effects of a pressure or temperature change (*e.g.,* substance would melt, sublime, *etc.*) are correctly explained based on the phase diagram.

Tier 2 Vocabulary: phase, curve, fusion, solid, liquid, gas, vapor

Language Objectives:
- Explain the regions of a phase diagram and the relationship between each region and the temperature and pressure of the substance..

Notes:

The phase of a substance (solid, liquid, gas) depends on its temperature and pressure.

phase diagram: a graph showing the phase(s) present at different temperatures and pressures.

Use this space for summary and/or additional notes:

Phase Diagrams

Big Ideas | **Details** | Unit: Matter

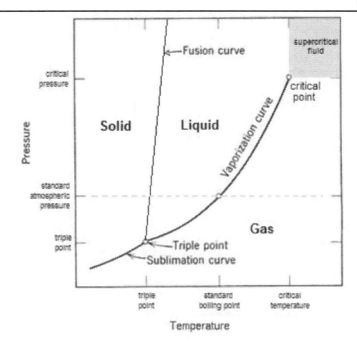

fusion curve: the set of temperatures and pressures at which a substance melts/freezes.

vaporization curve: the set of temperatures & pressures at which a substance vaporizes/condenses.

sublimation curve: the set of temperatures & pressures at which a substance sublimes/deposits.

triple point: the temperature and pressure at which a substance can exist simultaneously as a solid, liquid, and gas.

critical point: the highest temperature at which the substance can exist as a liquid. The critical point is the endpoint of the vaporization curve.

supercritical fluid: a substance whose temperature and pressure are above the critical point. The substance would be expected to be a liquid (due to the pressure), but the molecules have so much energy that the substance behaves more like a gas.

Use this space for summary and/or additional notes:

Phase Diagrams

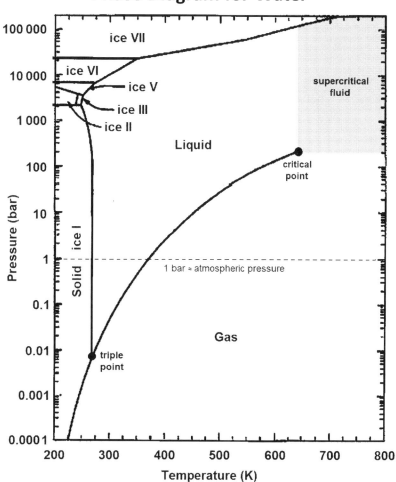

Note that pressure is on a logarithmic scale, and that standard atmospheric pressure is 1 bar ≈ 1 atm.

Note also that the temperature is in kelvin. To convert degrees Celsius to kelvin, add 273. (*e.g.*, 25 °C + 273 = 298 K.)

Notice that the slope of the fusion curve (melting/freezing line) is negative. This is because ice I is less dense than liquid water. At temperatures near the melting point and pressures less than about 2 000 bar, increasing the pressure will cause ice to melt. Water is one of the only known substances that exhibits this behavior.

Use this space for summary and/or additional notes:

Phase Diagrams

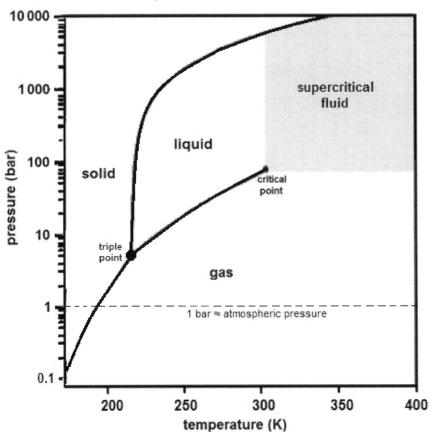

Notice that the pressure of the triple point for CO_2 is about 5 bar, which means CO_2 cannot be a liquid at atmospheric pressure. This is why dry ice (solid CO_2) sublimes directly from a solid to a gas.

Use this space for summary and/or additional notes:

Phase Diagrams

Homework Problems

Answer these questions based on the phase diagrams for water and carbon dioxide.

1. Approximately what pressure would be necessary to boil water at a temperature of 350 K?

2. What is the minimum pressure necessary for water to exist as a liquid at 350 K?

3. At approximately what temperature would water boil if the pressure is 10 bar?

4. What is the highest temperature at which carbon dioxide can exist as a liquid?

5. At 1.0 bar of pressure, what is the temperature at which carbon dioxide sublimes?

6. At room temperature (25 °C ≈ 300 K), what is the minimum pressure at which liquid carbon dioxide can exist?

7. Describe the phase transitions and temperatures for water going from 200 K to 400 K at a pressure of 0.1 bar.

8. Describe the phase transitions and temperatures for carbon dioxide going 200 K to 300 K at a pressure of 10 bar.

Use this space for summary and/or additional notes:

Phases & Phase Changes

Unit: Matter

MA Curriculum Frameworks (2016): HS-PS1-3, HS-PS2-8(MA)

MA Curriculum Frameworks (2006): 1.3, 6.3

Mastery Objective(s): (Students will be able to...)

- Compare observable states of matter and phase transitions with behavior at the molecular level.

Success Criteria:

- Descriptions include connectedness and motion of molecules.
- Descriptions include comparative descriptions of molecular speed.
- Descriptions relate molecular motion and speed to temperature.

Tier 2 Vocabulary: phase, solid, liquid, gas, vapor

Language Objectives:

- Explain phase changes in terms of changes in molecular behavior.

Notes:

macroscopic: objects or bulk properties of matter that we can observe directly.

microscopic: objects or properties of matter that are too small to observe directly.

In chemistry, the *macroscopic* properties of a substance are determined by *microscopic* interactions between the individual molecules.*

---

* In this text, the term "molecules" is frequently used to refer to the particles that make up a substance. A molecule is more properly a group of atoms that are covalently bonded together. A substance can be made of individual atoms, molecules, crystals, or other types of particles. This text uses the term "molecules" because the term gives most students a reasonably correct picture of entities that are firmly attached to each other and cannot be pulled apart by physical means.

Use this space for summary and/or additional notes:

Phases & Phase Changes

Page: 121
Unit: Matter

States of Matter

The following table shows interactions between the molecules and some observable properties for solids, liquids and gases. Note that the table includes heating curves, which will be discussed in more detail later in the course. For now, understand that a heating curve shows how the temperature changes as heat is added. Notice in particular that the temperature stays constant during melting and boiling.

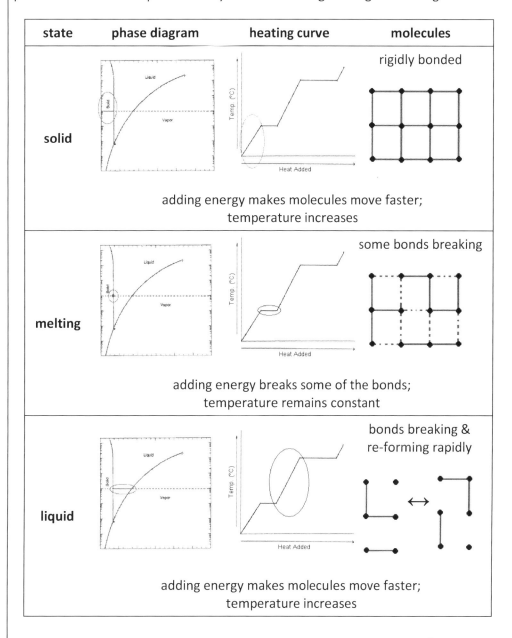

Use this space for summary and/or additional notes:

Phases & Phase Changes

Page: 122
Unit: Matter

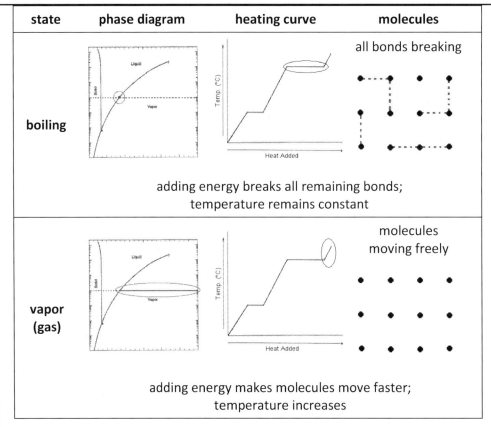

Note that because liquids are continually forming and breaking bonds, if a liquid molecule at the surface breaks its bonds with other liquids, it can "escape" from the attractive forces of the other liquid molecules and become a vapor molecule. This is how evaporation happens at temperatures well below the boiling point of the liquid.

Use this space for summary and/or additional notes:

Conservation of Mass, Energy and Charge

Unit: Matter

MA Curriculum Frameworks (2016): HS-PS1-3, HS-PS2-8(MA)

MA Curriculum Frameworks (2006): 6.4

Mastery Objective(s): (Students will be able to...)

- Explain conservation of mass, energy, and charge.
- Apply conservation of mass, energy, and charge to situations.

Success Criteria:

- Explanations account for mass, energy, and charge in a variety of changing situations.
- Students can set up and solve equations of the form *initial* + *change* = *final* in the context of problems involving conservation of mass, energy, and charge.

Tier 2 Vocabulary: conservation

Language Objectives:

- Explain what happens to mass, energy and charge in situations where these quantities are transferred from one object or system to another.

Notes:

conservation law: a statement that says that a quantity is the same before and after some change; a "before = after" statement.

Conservation of Mass/Matter: matter (mass) cannot be created or destroyed (except in a nuclear reaction), only changed in form. The *total* mass before any chemical or physical change equals the total mass after the change. Also the mass of *each element* (each kind of atom) before any chemical or physical change equals the mass of that same element afterwards.

Conservation of Energy: energy cannot be created or destroyed (except in nuclear reactions), only changed in form. The energy that was present before any change equals the energy that is present after the change.

Conservation of Electrical Charge: electrical charges cannot be created or destroyed, only transferred from one atom/molecule/etc. to another.

Use this space for summary and/or additional notes:

Conservation of Mass, Energy and Charge

Sample problem:

Q: 12.5 g of sodium hydrogen carbonate is added to a beaker containing 100. g of dilute hydrochloric acid. The reaction produces carbon dioxide gas, sodium hydroxide, water, and sodium chloride, according to the following chemical equation:

$$NaHCO_3 + HCl \rightarrow NaCl + H_2O + CO_2$$

If the contents of the beaker have a mass of 108.2 g after the reaction is complete, how much CO_2 gas escaped?

A: The total mass of the chemicals before the reaction must equal the total mass afterwards. The initial mass is 100 + 12.5 = 112.5 g. The mass afterwards must also be 112.5 g. If 108.2 g is still in the beaker, then the remaining mass, 112.5 − 108.2 = 4.3 g, must be the CO_2 that escaped.

Nuclear Reactions

In a nuclear reaction, mass can be converted to energy according to the formula:

$$E = mc^2$$

Where:

E = energy $\qquad$ m = mass $\qquad$ c = the speed of light

Note that the speed of light is a very large number: $3.00 \times 10^8 \, \frac{m}{s}$. Therefore, c^2 is even larger: $9.00 \times 10^{16} \, \frac{m^2}{s^2}$. This means that a very small amount of mass creates an extremely large amount of energy. This is where the energy in a nuclear explosion comes from—mass that was turned into energy.

This law is called the law of <u>Conservation of Mass and Energy</u>. This law (and its equation, $E = mc^2$) describes the only we know of that mass can be destroyed, and the only way we know of that energy can be created.

Use this space for summary and/or additional notes:

Conservation of Mass, Energy and Charge

Homework Problems

1. Suppose your breakfast contained 500 Calories of energy. Suppose you missed your bus (or your ride) and you had to walk 2 miles to school, which burned 200 Calories. How many Calories of energy from your breakfast are left for you to get through your morning classes?

    Answer: 300 Calories

2. In your car's gas tank, the following chemical reaction occurs:

    $$C_8H_{18} + O_2 \rightarrow CO_2 + H_2O$$

    The gasoline (C_8H_{18}) in a typical car's fuel tank weighs about 80 pounds. Burning that much gas uses about 300 pounds of oxygen from the Earth's atmosphere, and it produces 120 pounds of H_2O. How many pounds of CO_2 did that tank of gas produce?

    Answer: 260 lbs.

3. A nuclear reactor converts 4×10^{-9} kg of uranium to energy. How much energy is produced?

    Answer: 3.6×10^8 J or 360 MJ

Use this space for summary and/or additional notes:

Summary: Matter

Unit: Matter

List the main ideas of this chapter in phrase form:

Write an introductory sentence that categorizes these main ideas.

Turn the main ideas into sentences, using your own words. You may combine multiple main ideas into one sentence.

Add transition words to make your writing clearer and rewrite your summary below.

Use this space for summary and/or additional notes:

Introduction: Gases

Unit: Gases

Topics covered in this chapter:

Gases & Kinetic-Molecular Theory 129
Gas Conversion Factors 132
Gas Laws 134
Ideal Gas Law 144
Partial Pressures 148
Molecular Speed & Effusion 155

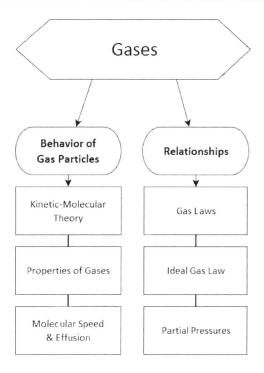

Use this space for summary and/or additional notes:

Introduction: Gases

Big Ideas	Details
	Standards addressed in this chapter:
	Massachusetts Curriculum Frameworks & Science Practices (2016):
	HS-PS2-8(MA) Use kinetic molecular theory to compare the strengths of electrostatic forces and the prevalence of interactions that occur between molecules in solids, liquids, and gases. Use the combined gas law to determine changes in pressure, volume, and temperature in gases.
	Massachusetts Curriculum Frameworks (2006):
	6.1 Using the kinetic molecular theory, explain the behavior of gases and the relationship between pressure and volume (Boyle's law), volume and temperature (Charles's law), pressure and temperature (Amonton's law), and the number of particles in a gas sample (Avogadro's hypothesis). Use the combined gas law to determine changes in pressure, volume, and temperature.
	6.2 Perform calculations using the ideal gas law. Understand the molar volume at 273 K and 1 atmosphere (STP).
	6.3 Using the kinetic molecular theory, describe and contrast the properties of gases, liquids, and solids. Explain, at the molecular level, the behavior of matter as it undergoes phase transitions.

Use this space for summary and/or additional notes:

Gases & Kinetic-Molecular Theory

Unit: Gases
MA Curriculum Frameworks (2016): HS-PS2-8(MA)
MA Curriculum Frameworks (2006): 6.3
Mastery Objective(s): (Students will be able to…)

- Explain how each aspect of Kinetic-Molecular Theory applies to gases.

Success Criteria:

- Descriptions account for behavior at the molecular level.
- Descriptions account for measurable properties, *e.g.,* temperature, pressure, volume, *etc.*

Tier 2 Vocabulary: kinetic, gas, ideal, real

Language Objectives:

- Explain how gas molecules behave and how their behavior relates to properties we can measure.

Notes:

Recall the following definitions:

- solid: molecules rigidly bonded (definite shape & volume)
- liquid: molecules bonded (definite volume), but loosely. Bonds continually breaking & re-forming (indefinite shape)
- gas: molecules not bonded (indefinite volume & shape)
- plasma: heat of surroundings > ionization energy, so electrons are loosely bonded & continually dissociate from and re-associate with ions. Electrical charge is fluid and in continual motion.

evaporation: conversion of liquid to gas.

boiling point: the temperature at which a liquid completely evaporates.

normal boiling point: the boiling point of a liquid when the pressure is 1 atm (average atmospheric pressure at sea level).

Use this space for summary and/or additional notes:

Kinetic-Molecular Theory

Kinetic-Molecular Theory (KMT) is a theory, developed by James C. Maxwell and Ludwig Boltzmann, that predicts the behavior of gases by modeling them as moving molecules. The theory states that:

- Gases are made of very large numbers of molecules
- Molecules are constantly moving (obeying Newton's laws of motion), and their speeds are constant
- Molecules are very far apart compared with their diameter
- Molecules collide with each other and walls of container in elastic collisions
- Molecules behaving according to KMT are not reacting[*] or exerting any other forces (attractive or repulsive) on each other.

temperature: a measure of the average kinetic energy of the molecules of a substance

ideal gas: a gas whose molecules behave according to KMT. Most gases are ideal under *some* conditions (but not all).

real gas: a gas that is not behaving according to KMT. This can occur with all gases, most commonly at temperatures and pressures that are close to the solid or liquid sections of the phase diagram for the substance.

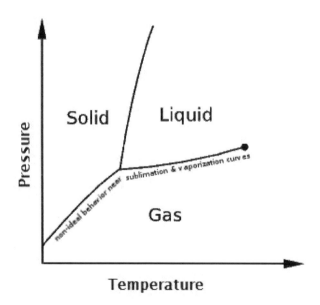

---

[*] Of course, reactions can occur, but chemical reactions are part of collision theory, which is separate from KMT.

Use this space for summary and/or additional notes:

Gases & Kinetic-Molecular Theory

Measurable Properties of Gases

All gases have the following properties that can be measured:

Property	Variable	Units	Description
amount	n	moles (mol)	amount of gas (1 mol = 6.02×10^{23} molecules)
volume	V	liters (L)	space that the gas takes up
temperature	T	kelvin (K)	ability to transfer heat (proportional to the average kinetic energy of the molecules)
pressure	P	bar, atm, kPa, mm Hg (torr), *etc.*	average force on the walls of the container due to collisions between the molecules and the walls

Notes about calculations:

- Temperature must be absolute, which means you *must* use Kelvin. A temperature of 0 in a gas laws calculation can only mean absolute zero.

- Pressures must be absolute. (For example, you can't use a tire gauge because it measures "gauge pressure," which is the difference between atmospheric pressure and the pressure inside the tire.) A pressure of 0 in a gas laws calculation can only mean that there are no molecules colliding with the walls.

Use this space for summary and/or additional notes:

Gas Conversion Factors

Unit: Gases

MA Curriculum Frameworks (2016): HS-PS2-8(MA)

MA Curriculum Frameworks (2006): 6.1, 6.2

Mastery Objective(s): (Students will be able to...)
- Choose conversion factors based on the units needed for a calculation or conversion.

Success Criteria:
- Conversion factor has the same units as other numbers in a chosen word problem or situation.

Tier 2 Vocabulary: conversion, absolute, standard, vacuum

Language Objectives:
- Explain and defend the choice of a conversion factor or constant for use in a problem involving gases.

Notes:

<u>absolute zero</u>: the temperature at which molecules are moving so slowly that they can't transfer energy to other molecules. Absolute zero is −273.15 °C = 0 K

<u>vacuum</u>: the absence of gas molecules. In a total vacuum, the Pressure = 0

"Standard Pressure" = 1 bar[*]

"Standard Temperature" = 0°C = 273.15 K

S.T.P. ("Standard Temperature and Pressure") = 0 °C and 1 bar.

"Room Temperature" = 25 °C = 298 K

1 mole of an ideal gas has a volume of 22.7 L at S.T.P.

---

[*] In 1982, the IUPAC defined standard pressure to be exactly 1 bar (= 100 kPa = 0.987 atm). However, many chemists and many standardized assessments still use 1 atm.

Use this space for summary and/or additional notes:

Gas Conversion Factors

Conversion Factors

Pressure:

1 atm ≡ 101.325 kPa ≡ 0.101 325 MPa ≡ 1.01325 bar
≡ 101 325 $\frac{N}{m^2}$ ≡ 101 325 Pa

1 atm ≡ 760 mm Hg ≡ 760 torr = 29.92 in. Hg

1 atm = 14.696 $\frac{lb.}{in.^2}$ = 14.696 psi ("psi" = "pounds per square inch")

Volume:

1 mL ≡ 1 cm^3

1 L ≡ 1000 m^3

Moles:

1 mol = 22.7 L at S.T.P.*

Use dimensional analysis to turn the molar mass of a compound (measured in $\frac{g}{mol}$) into a conversion factor between grams and moles.

Temperature:

K ≡ °C + 273.15   °F ≡ (1.8 × °C) + 32   °R ≡ °F + 459.67

The Gas Constant:

The gas constant R is a natural constant that appears in several relationships in chemistry, including the ideal gas law (which we will study in a subsequent class).

$$R = 0.0821 \frac{L \cdot atm}{mol \cdot K}$$

$$R = 8.31 \frac{L \cdot kPa}{mol \cdot K} = 8.31 \frac{J}{mol \cdot K} = 8.31 \times 10^{-3} \frac{kJ}{mol \cdot K}$$

$$R = 62.4 \frac{L \cdot torr}{mol \cdot K} \qquad R = 1.987 \frac{cal}{mol \cdot K} = 1.987 \frac{BTU}{lb\text{-}mol \cdot °R}$$

---

* Massachusetts assessments still use the outdated definition of S.T.P. The volume of one mole of an ideal gas at 1 atm and 0 °C is 22.4 L.

Use this space for summary and/or additional notes:

Gas Laws

Unit: Gases

MA Curriculum Frameworks (2016): HS-PS2-8(MA)

MA Curriculum Frameworks (2006): 6.1

Mastery Objective(s): (Students will be able to...)

- Qualitatively describe the relationship between any two of the quantities: *number of particles*, *temperature*, *pressure*, and *volume* in terms of Kinetic Molecular Theory (KMT).
- Quantitatively determine the *number of particles*, *temperature*, *pressure*, or *volume* in a before & after problem in which one or more of these quantities is changing.

Success Criteria:

- Descriptions relate behavior at the molecular level to behavior at the macroscopic level.
- Solutions have the correct quantities substituted for the correct variables.
- Chosen value of the gas constant has the same units as the other quantities in the problem.
- Algebra and rounding to appropriate number of significant figures is correct.

Tier 2 Vocabulary: ideal, law

Language Objectives:

- Identify each quantity based on its units and assign the correct variable to it.
- Understand and correctly use the terms "pressure," "volume," and "temperature," and "ideal gas."
- Explain the placement of each quantity in the ideal gas law.

Labs, Activities & Demonstrations:

- Vacuum pump (pressure & volume) with:
  - balloon (air vs. water)
  - shaving cream
- Absolute zero apparatus (pressure & temperature)
- Can crush (pressure, volume & temperature)

Use this space for summary and/or additional notes:

Gas Laws

Notes:

ideal gas: a gas that behaves as if each molecule acts independently, according to kinetic-molecular theory. Specifically, this means the molecules are far apart, and move freely in straight lines at constant speeds. When the molecules collide, the collisions are perfectly elastic, which means they bounce off each other with no energy or momentum lost.

Most gases behave ideally except at temperatures and pressures near the vaporization curve on a phase diagram. (*I.e.*, gases stop behaving ideally when conditions are close to those that would cause the gas to condense to a liquid or solid.)

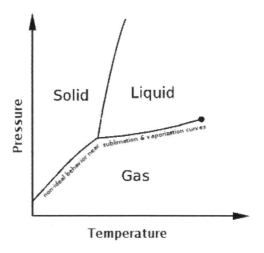

Use this space for summary and/or additional notes:

Gas Laws

Note about Subscripts and Variables

When variables appear more than once in an equation with different values each time, we use subscripts to group them. You have already seen this a few times in math, *e.g.*, in the formula for the slope of a line and the distance formula:

$$\text{slope: } m = \frac{y_2 - y_1}{x_2 - x_1} \qquad \text{distance: } d = \sqrt{(x_2 - x_1)^2 + (y_2 - y_1)^2}$$

In the above examples, the subscripts "1" and "2" are used to group the x and y values based on whether they refer to the first point (x_1, y_1), or the second one (x_2, y_2).

In chemistry, we use subscripts the same way. For example, if a gas is heated, that means the temperature is changing. We refer to the starting temperature (temperature #1) as T_1, and the ending temperature (temperature #2) as T_2. The same concept applies to other variables as well, such as moles (n), volume (V), and pressure (P).

Proportionalities

<u>directly proportional</u>: if two quantities are *directly* proportional, as one increases, the other increases proportionately.

If x and y are directly proportional, then $x \propto y$ which means $x = ky$ and $\frac{x}{y} = k$

where k is a constant. You should notice that x and y are either <u>numerator and denominator</u> in a fraction, or are on <u>opposite sides</u> of the equals sign.

<u>inversely proportional</u>: if two quantities are *inversely* proportional, as one increases, the other decreases proportionately.

If x and y are inversely proportional, then $x \propto \frac{1}{y}$ which means $xy = k$ where k is a constant. You should notice that x and y are on the <u>same side</u> of the equals sign.

Use this space for summary and/or additional notes:

Gas Laws

Avogadro's Principle

In 1811, Italian physicist Amedeo Avogadro (whose full name was Lorenzo Romano Amedeo Carlo Avogadro di Quaregna e di Cerreto) published the principle that equal volumes of an ideal gas at the same temperature and pressure must contain equal numbers of particles.

What did we do?	What happened?	What are the molecules doing?	Conclusion
put more (moles of) air into a balloon $n \uparrow$	the volume of the balloon got larger $V \uparrow$	crowding each other → pushing each other farther away	n and V are directly proportional. $\dfrac{V}{n} = \text{constant}$

If the pressure and temperature are constant, then for an ideal gas:

$$\frac{V_1}{N_1} = \frac{V_2}{N_2} \quad ^*$$

Because it is almost always more convenient to work with moles of gas (n) rather than particles (N), we can rewrite Avogadro's principle as:

$$\frac{V_1}{n_1} = \frac{V_2}{n_2}$$

---

* Avogadro's principle is usually stated $\dfrac{n_1}{V_1} = \dfrac{n_2}{V_2}$. I have inverted it in these notes so that the quantities in the numerator and denominator are the same as the quantities in the numerator and denominator of the combined gas law.

Use this space for summary and/or additional notes:

Gas Laws

Big Ideas | **Details** | Page: 138
Unit: Gases

Boyle's Law

In 1662, British physicist and chemist Robert Boyle published his findings that the pressure and volume of a gas were inversely proportional.

What did we do?	What happened?	What are the molecules doing?	Conclusion
decrease pressure by putting a balloon in a vacuum chamber $P\downarrow$	the volume of the air inside the balloon increased $V\uparrow$	colliding with less force → pushing each other less far away	P and V are inversely proportional. PV = constant

Therefore, if the temperature and the number of particles of gas are constant, then for an ideal gas:

$$P_1 V_1 = P_2 V_2$$

(Note that by convention, gas laws use subscripts "1" and "2" instead of "i" and "f".)

Charles' Law

In the 1780s, French physicist Jacques Charles discovered that the volume and temperature of a gas were directly proportional.

What did we do?	What happened?	What are the molecules doing?	Conclusion
cool gas by putting a soda can full of very hot air into cool water $T\downarrow$	the volume of the air got smaller and crushed the can $V\downarrow$	moving more slowly → pushing each other less far away	V and T are directly proportional. $\dfrac{V}{T}$ = constant

If pressure and the number of particles are constant, then for an ideal gas:

$$\frac{V_1}{T_1} = \frac{V_2}{T_2}$$

Use this space for summary and/or additional notes:

Chemistry 1 — Mr. Bigler

Gas Laws

Gay-Lussac's Law

In 1702, French physicist Guillaume Amontons discovered that there is a relationship between the pressure and temperature of a gas. However, precise thermometers were not invented until after Amontons' discovery, so it wasn't until 1808, over a century later, that French chemist Joseph Louis Gay-Lussac confirmed this law mathematically. The pressure law is most often attributed to Gay-Lussac, though some texts refer to it as Amontons' Law.

What did we do?	What happened?	What are the molecules doing?	Conclusion
increase temperature by heating a metal sphere full of air $T\uparrow$	the pressure of the air increased $P\uparrow$	moving faster → colliding with more force	P and T are directly proportional. $\dfrac{P}{T} = \text{constant}$

If volume and the number of particles are constant, then for an ideal gas:

$$\frac{P_1}{T_1} = \frac{P_2}{T_2}$$

The Combined Gas Law

We can combine each of the above principles. When we do this (keeping P and V in the numerator and n and T in the denominator for consistency), we get following relationship for an ideal gas:

$$\frac{P_1 V_1}{n_1 T_1} = \frac{P_2 V_2}{n_2 T_2} = \text{constant}$$

Note, however, that in most problems, the number of moles of gas remains constant. This means $n_1 = n_2$ and we can cancel it from the equation, which gives:

$$\frac{P_1 V_1}{T_1} = \frac{P_2 V_2}{T_2}$$

This equation is called the "combined gas law", which is used to solve most "before/after" problems involving ideal gases.

Use this space for summary and/or additional notes:

Gas Laws

When using the combined gas law, any quantity that is not changing may be cancelled out of the equation. (If a quantity is not mentioned in the problem, you can assume that it is constant and may be cancelled.)

This brings us to an important point about science problems: *If something is not mentioned in a problem, **always** assume that it doesn't affect the problem.* On a standardized test like MCAS or an AP® test, it's usually best to state those assumptions explicitly, because if your assumption is valid and you do the rest of the problem correctly, you will almost always receive some credit, *even if your assumption was different from what the person who wrote the problem intended.*

For example, suppose a problem doesn't mention anything about temperature. That means T is constant and you can cancel it. When you cancel T from both sides of the combined gas law, you get:

$$\frac{P_1 V_1}{\cancel{T_1}} = \frac{P_2 V_2}{\cancel{T_2}}$$ which simplifies to $P_1 V_1 = P_2 V_2$ (Boyle's Law)

Solving Problems Using the Combined Gas Law

You can use this method to solve any "before/after" gas law problem:

1. Determine which variables you have
2. Determine which values are *initial* (#1) *vs. final* (#2).
3. Start with the combined gas law and cancel any variables that are explicitly not changing or omitted (assumed not to be changing).
4. Substitute your numbers into the resulting equation and solve. (Make sure all initial and final quantities have the same units, and don't forget that temperatures *must* be in Kelvin!)

Use this space for summary and/or additional notes:

Gas Laws

Page: 141
Unit: Gases

Sample problem:

Q: A gas has a temperature of 25 °C and a pressure of 1.5 bar. If the gas is heated to 35 °C, what will the new pressure be?

A: 1. Find which variables we have.

   We have two temperatures (25 °C and 35 °C), and two pressures (1.5 bar and the new pressure that we're looking for).

   2. Find the action being done on the gas ("heated"). Anything that was true about the gas *before* the action is time "1", and anything that is true about the gas *after* the action is time "2".

   Time 1 ("before"):   **Time 2 ("after"):**

   $P_1 = 1.5$ bar $P_2 = P_2$

   $T_1 = 25\,°C + 273 = 298\,K$ $T_2 = 35\,°C + 273 = 308\,K$

   3. Set up the formula. We can cancel volume (*V*), because the problem doesn't mention it:

   $$\frac{P_1 \cancel{V_1}}{T_1} = \frac{P_2 \cancel{V_2}}{T_2} \quad \text{which gives us} \quad \frac{P_1}{T_1} = \frac{P_2}{T_2} \quad \text{(Gay-Lussac's Law)}$$

   4. Plug in our values and solve:

   $$\frac{1.5\,\text{bar}}{298\,K} = \frac{P_2}{308\,K} \quad \rightarrow \quad \boxed{P_2 = 1.55\,\text{bar}}$$

Use this space for summary and/or additional notes:

Chemistry 1 — Mr. Bigler

Gas Laws

Big Ideas | **Details** | **Unit: Gases**

Homework Problems

Solve these problems using one of the gas laws in this section. Remember to convert temperatures to Kelvin!

1. A sample of oxygen gas occupies a volume of 250. mL at a pressure of 740. torr. What volume will it occupy at 800. torr?

    Answer: 0.0272 m^3

2. A sample of O_2 is at a temperature of 40.0 °C and occupies a volume of 2.30 L. To what temperature should it be raised to occupy a volume of 6.50 L?

    Answer: 612 °C

3. H_2 gas was cooled from 150. °C to 50. °C. Its new pressure is 750 torr. What was its original pressure?

    Answer: 980 torr

4. A 2.00 L container of N_2 had a pressure of 3.20 atm. What volume would be necessary to decrease the pressure to 98.0 kPa?

    (*Hint: notice that the pressures are in different units. You will need to convert one of them so that both pressures are in either atm or kPa.*)

    Answer: 6.62 L

Use this space for summary and/or additional notes:

Gas Laws

Page: 143
Unit: Gases

Big Ideas | Details

5. A sample of air has a volume of 60.0 mL at S.T.P. What volume will the sample have at 55.0 °C and 745 torr?

   Answer: 73.5 mL

6. N_2 gas is enclosed in a tightly stoppered 500. mL flask at 20.0 °C and 1 atm. The flask, which is rated for a maximum pressure of 3.00 atm, is heated to 680. °C. Will the flask explode?

   Answer: P_2 = 3.25 atm. Yes, the flask will explode.

7. A scuba diver's 10. L air tank is filled to a pressure of 210 bar at a dockside temperature of 32.0 °C. When the diver is breathing the air underwater, the water temperature is 8.0 °C, and the pressure is 2.1 bar.

   a. What volume of air does the diver use?

   Answer: 921 L

   b. If the diver uses air at the rate of 8.0 L/min, how long will the diver's air last?

   Answer: 115 min

Use this space for summary and/or additional notes:

Chemistry 1

Mr. Bigler

Ideal Gas Law

Unit: Gases

MA Curriculum Frameworks (2016): HS-PS2-8(MA)

MA Curriculum Frameworks (2006): 6.2

Mastery Objective(s): (Students will be able to...)

- Describe the relationship between any two variables in the ideal gas law.
- Apply the ideal gas law to problems involving a sample of gas.

Success Criteria:

- Solutions have the correct quantities substituted for the correct variables.
- Chosen value of the gas constant has the same units as the other quantities in the problem.
- Algebra and rounding to appropriate number of significant figures is correct.

Tier 2 Vocabulary: ideal, law

Language Objectives:

- Identify each quantity based on its units and assign the correct variable to it.
- Explain the placement of each quantity in the ideal gas law.

Notes:

<u>ideal gas</u>: a gas that behaves according to Kinetic-Molecular Theory (KMT).

When we developed the combined gas law, before we cancelled the number of moles, we had the equation:

$$\frac{P_1 V_1}{n_1 T_1} = \frac{P_2 V_2}{n_2 T_2} = \text{constant}$$

Because P, V, n and T are all of the quantities needed to specify the conditions of an ideal gas, this expression must be true for *any ideal gas* under *any conditions*. If V is in liters, P is in kPa, n is in moles and T is in Kelvin, then the value of this constant is $8.31 \frac{\text{L·kPa}}{\text{mol·K}}$. This number is called "the gas constant", and is denoted by the variable R in equations.

Therefore, we can rewrite the above expression as:

$$\frac{P_1 V_1}{n_1 T_1} = \frac{P_2 V_2}{n_2 T_2} = R \quad \text{and therefore} \quad P_1 V_1 = n_1 R T_1 \quad \text{and} \quad P_2 V_2 = n_2 R T_2$$

Because this expression is true under any conditions, 1, 2, or whatever, we can drop the subscripts:

$$PV = nRT$$

Use this space for summary and/or additional notes:

Ideal Gas Law

Unit: Gases

Choosing a value of R

The purpose of the gas constant R is to convert the quantity nT from units of mol·K to units of pressure × volume. As we saw earlier in this chapter, the gas constant has different values, depending on the units it needs to cancel:

$$R = 0.0821 \frac{L \cdot atm}{mol \cdot K}$$

$$R = 8.31 \frac{L \cdot kPa}{mol \cdot K} = 8.31 \frac{J}{mol \cdot K} = 8.31 \times 10^{-3} \frac{kJ}{mol \cdot K}$$

$$R = 62.4 \frac{L \cdot torr}{mol \cdot K} \qquad R = 1.987 \frac{cal}{mol \cdot K} = 1.987 \frac{BTU}{lb\text{-}mol \cdot °R}$$

In order for the units in a problem to work out properly, *R needs to have exactly the same units as the values in the problem*. In the problems we will be solving, this usually means you need to look at the pressure units and *choose the value of R that has the same pressure unit as the pressure given in the problem*.

Amazing Fact:

In the metric system, the unit of pressure (the Pascal) is a combination of the S.I. units $\frac{kg}{m \cdot s^2}$, and volume has units of m³. This means that pressure times volume (*PV*) has S.I. units of $\frac{kg \cdot m^2}{s^2}$, which happens to equal Joules (the unit for energy).

Dimensional analysis tells us that if the units work, the resulting formula must be correct.

So, if *PV* is equivalent to energy, *nRT* *must also be* equivalent to energy. In fact, *PV* is the energy that the gas expends in doing work on its surroundings (such as by expanding a balloon or pushing on a piston), and *nRT* is the kinetic energy of the individual gas molecules.

In other words, the ideal gas law *PV = nRT* is simply the law of conservation of energy, applied to gases!

Use this space for summary and/or additional notes:

Ideal Gas Law

Page: 146
Unit: Gases

Solving Problems Using the Ideal Gas Law

If a gas behaves according to the ideal gas law, simply substitute the values for pressure, volume, number of moles (or particles), and temperature into the equation. Be sure your units are correct (especially that temperature is in Kelvin), and that you use the correct constant, depending on whether you know the number of particles or the number of moles of the gas.

Sample Problem:

A 3.50 mol sample of an ideal gas has a pressure of 1.20 atm and a temperature of 35 °C. What is its volume?

n above "3.50 mol", P above "1.20 atm", $T \rightarrow K$ below "35 °C", V circled below "volume"

Answer:

First, we need to declare our variables and choose a value for R based on the units:

$P = 1.20$ atm

$V = V$ (because we don't' know it yet)

$n = 3.50$ mol

$R = 0.0821 \frac{\text{L·atm}}{\text{mol·K}}$ (Choose this one because P is in atm.)

$T = 35\,°C + 273 = 308$ K

Then we substitute these numbers into the ideal gas law and solve:

$$PV = nRT$$

$$(1.20\,\text{atm})V = (3.50\,\text{mol})(0.0821\,\tfrac{\text{L·atm}}{\text{mol·K}})(308\,\text{K})$$

$$1.20\,V = 88.5$$

$$V = \frac{88.5}{1.20} = 73.75\,\text{L}$$

Use this space for summary and/or additional notes:

Ideal Gas Law

Big Ideas | **Details** | Page: 147
Unit: Gases

Homework Problems

Use the ideal gas law to solve the following problems. Be sure to choose the appropriate value for the gas constant and to convert temperatures to Kelvin.

1. A sample of 1.00 moles of oxygen at 50.0°C and 98.6 kPa occupies what volume?

    Answer: 27.2 L

2. If a steel cylinder with a volume of 1.50 ℓ contains 10.0 moles of oxygen, under what pressure is the oxygen if the temperature is 27.0°C?

    Answer: 164 atm = 125 000 torr = 16 600 kPa

3. In a gas thermometer, the pressure of 0.500 L of helium is 113.30 kPa at a temperature of −137 °C. How many moles of gas are in the sample?

    Answer: 0.050 mol

4. A sample of 4.25 mol of hydrogen at 20.0 °C occupies a volume of 25.0 L. Under what pressure is this sample?

    Answer: 4.09 atm = 3 108 torr = 414 kPa

Use this space for summary and/or additional notes:

Partial Pressures

Unit: Gases

MA Curriculum Frameworks (2016): HS-PS1-7

MA Curriculum Frameworks (2006): N/A

Mastery Objective(s): (Students will be able to...)

- Calculate partial pressures based on conservation of matter.

Success Criteria:

- Solutions have the correct quantities substituted for the correct variables.
- Mole fractions are paired correctly with their partial pressures.
- If the problem requires the ideal gas law, chosen value of the gas constant has the same units as the other quantities in the problem.
- Algebra and rounding to appropriate number of significant figures is correct.

Tier 2 Vocabulary: mole

Language Objectives:

- Describe the pairing of each gas with its mole fraction and pressure.

Notes:

<u>Partial Pressure:</u> the partial pressure of a gas is the amount of pressure that would result from *only* the molecules of that gas. The partial pressure of a substance is denoted by the variable P (for pressure) and the chemical formula of the substance as a subscript. For example, the partial pressure of carbon dioxide in a sample would be denoted by P_{CO_2}.

<u>Dalton's Law of Partial Pressures:</u> the sum of all of the partial pressures in a sealed container equals the total pressure.

$$P = P_T = P_1 + P_2 + P_3 + \ldots$$

(To make things more clear, we will use P_T to mean the total pressure.)

<u>mole fraction</u> (χ): the fraction of the total moles (or molecules) that are the compound of interest. For example, if we have 20 moles of gas, and 9 moles are N_2, the mole fraction of N_2 is:

$$\chi_{N_2} = \frac{9 \, mol \, N_2}{20 \, mol \, total} = 0.45$$

Use this space for summary and/or additional notes:

Partial Pressures

Suppose we had the following tank, with a total pressure of 1.00 atm:

[Diagram of a tank containing a mixture of N_2 and O_2 molecules]

If we ignore all of the molecules except for nitrogen, the tank would look like this:

[Diagram of the same tank showing only N_2 molecules]

If 45 % of the molecules are nitrogen (χ_{N_2} = 0.45), then the pressure just from these nitrogen molecules (the partial pressure of nitrogen) must be 0.45 times the total pressure of 1 atm. This means:

$$P_{N_2} = \chi_{N_2} P_T$$

$$P_{N_2} = (0.45)(1 \text{ atm}) = 0.45 \text{ atm}$$

Similarly, because 55 % of the molecules are oxygen, this means:

$$\chi_{O_2} = 0.55$$

$$P_{O_2} = \chi_{O_2} P_T \quad P_{O_2} = (0.55)(1 \text{ atm}) = 0.55 \text{ atm}$$

Note that the two partial pressures add up to the total pressure:

$$P_T = P_{N_2} + P_{O_2} = 0.45 \text{ atm} + 0.55 \text{ atm} = 1 \text{ atm}$$

Use this space for summary and/or additional notes:

Partial Pressures

Using Dalton's Law with the Ideal Gas Law

Recall the two tanks from our example. Assuming N_2 and O_2 are behaving like ideal gases, the ideal gas law must be true in both tanks.

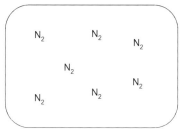

$P = P_T \quad n = n_T$

$P_T V = n_T RT$

$P = P_{N_2} \quad n = n_{N_2}$

$P_{N_2} V = n_{N_2} RT$

In other words, the ideal gas law can be used either with the total moles and total pressure, or with the moles of one specific gas and the partial pressure of that gas.

Use this space for summary and/or additional notes:

| Big Ideas | Details | Partial Pressures | Page: 151 Unit: Gases |

Vapor Pressure

<u>vapor pressure</u> (P_{vap}) the partial pressure of a substance due to evaporation.

Because liquids are continually forming and breaking bonds, when a liquid molecule at the surface breaks its bonds with other liquids, it can escape the attractive forces of the other liquid molecules and become a vapor molecule. The tendency for molecules to do this, when expressed as a partial pressure, is called the vapor pressure.

Vapor pressure is a function of the kinetic energy of the molecules, which means vapor pressure increases with temperature. At the boiling point, all of the molecules have enough energy to enter the gas phase. This means that at the boiling point, the vapor pressure must be equal to the ambient (atmospheric) pressure.

The following table shows the vapor pressure of water at different temperatures.

Vapor Pressure of Water

Temp (°C)	P_{vap} (kPa)	Temp (°C)	P_{vap} (kPa)	Temp (°C)	P_{vap} (kPa)
0.01	0.61173	30	4.2455	70	31.176
1	0.65716	35	5.6267	75	38.563
4	0.81359	40	7.3814	80	47.373
5	0.87260	45	9.5898	85	57.815
10	1.2281	50	12.344	90	70.117
15	1.7056	55	15.752	95	84.529
20	2.3388	60	19.932	100	101.32
25	3.1691	65	25.022	105	120.79

Relative humidity is the actual partial pressure of water in air as a percentage of its vapor pressure.

For example, suppose air at 30 °C (86 °F) has a partial pressure of 2.8 kPa. The vapor pressure of air at 30 °C is 5.6 kPa. 2.8 kPa is half of 5.6 kPa, so the relative humidity would be 50 %.

Use this space for summary and/or additional notes:

Big Ideas	Details	
	Partial Pressures	Page: 152
		Unit: Gases

Sample problem:

A 12.0 L tank of gas has a temperature of 30.0 °C and a total pressure of 1.75 atm. If the partial pressure of oxygen in the tank is 0.350 atm, how many moles of oxygen are in the tank? How many total moles of gas are in the tank?

Solution:

For oxygen:

P_{O_2} = 0.350 atm $\qquad$ n = n

V = 12.0 L $\qquad$ R = 0.0821 $\frac{L \cdot atm}{mol \cdot K}$

$\qquad\qquad\qquad$ T = 30.0 °C + 273 = 303.0 K

$$P_{O_2} V = n_{O_2} RT$$

$$(0.350)(12.0) = n_{O_2} (0.0821)(303.0)$$

$$n_{O_2} = 0.169 \text{ mol}$$

You could figure out the total moles two ways. One is to use the ideal gas law on the total moles:

P = 1.75 atm $\qquad$ n = n

V = 12.0 L $\qquad$ R = 0.0821 $\frac{L \cdot atm}{mol \cdot K}$

$\qquad\qquad\qquad$ T = 30.0 °C + 273 = 303.0 K

$$PV = nRT$$

$$(1.75)(12.0) = n (0.0821)(303.0)$$

$$n = 0.844 \text{ mol}$$

Use this space for summary and/or additional notes:

Partial Pressures

The other way to find the total moles is to use the mole fraction and the partial pressure:

$$P_{O_2} = \chi_{O_2} P_T$$

We know that

$$P_{O_2} = 0.350 \text{ atm}$$

$$P_T = 1.75 \text{ atm}$$

$$0.350 \text{ atm} = \chi_{O_2}(1.75 \text{ atm})$$

$$\chi_{O_2} = \frac{0.350 \text{ atm}}{1.75 \text{ atm}} = 0.200$$

Now that we know the mole fraction of O_2, we can figure out the total moles:

$$\chi_{O_2} = \frac{n_{O_2}}{n_T}$$

$$0.200 = \frac{0.169 \text{ mol } O_2}{n_T}$$

$$n_T = \frac{0.169}{0.200} = 0.845 \text{ mol}$$

Homework Problems

1. A 5 L container contains 0.125 mol of O_2 and 1.000 mol of He at 65 °C. What is the partial pressure of each gas? What is the total pressure?

   Answer: 6.24 atm

Use this space for summary and/or additional notes:

Partial Pressures

Page: 154
Unit: Gases

2. A 50 L gas cylinder contains 186 mol of N_2 and 140 mol of O_2. If the temperature is 24 °C, what is the total pressure in the cylinder?

   Answer: 159 atm

3. A sample of O_2 gas is collected by water displacement at 25 °C. If the atmospheric pressure in the laboratory is 100.7 kPa and the vapor pressure of water is 3.17 kPa at 25 °C, what is the partial pressure of the O_2 gas in the sample?

   Answer: 97.5 kPa

4. Two flasks are connected with a stopcock. The first flask has a volume of 5 liters and contains nitrogen gas at a pressure of 0.75 atm. The second flask has a volume of 8 ℓ and contains oxygen gas at a pressure of 1.25 atm. When the stopcock between the flasks is opened and the gases are free to mix, what will the (total) pressure be in the resulting mixture?

   Answer: 1.058 atm

Use this space for summary and/or additional notes:

Big Ideas	Details
	# Molecular Speed & Effusion

Unit: Gases
MA Curriculum Frameworks (2016): N/A (optional topic)
MA Curriculum Frameworks (2006): N/A
Mastery Objective(s): (Students will be able to...)
- Determine the velocities of gas molecules from their molar masses.

Success Criteria:
- Solutions use the equation appropriate for the information given.
- Solutions have the correct quantities substituted for the correct variables.
- Algebra and rounding to appropriate number of significant figures is correct.

Tier 2 Vocabulary: root, mean, square

Language Objectives:
- Describe the pairing of each gas with its mole fraction and pressure. |
| | **Notes:**

effusion: the escape of a gas through a small opening.

Graham's Law of Effusion: lighter molecules move faster and heavier molecules move more slowly. If you have a small opening, such as a leak in a tank, the lighter molecules will escape faster.

temperature: a measure of the average kinetic energy of the molecules in a substance

root mean square speed: the average speed of the gas molecules. "Root mean square" means this average is calculated by determining the average of the squares of the speeds, then taking the square root of the result. |

Use this space for summary and/or additional notes:

Molecular Speed & Effusion

Unit: Gases

Root Mean Square Speed of Gas Molecules

The root mean square velocity (v_{rms}) of gas molecules is given by the formula[*]:

$$v_{rms} = \sqrt{\frac{3000\, RT}{M}}$$

where M is the molar mass of the gas in $\frac{g}{mol}$, R is the gas constant ($8.31 \frac{J}{mol \cdot K}$), and T is the temperature in Kelvin. The quantity v_{rms} comes out in units of $\frac{m}{s}$.

Sample Problem:

A tank is filled with a mixture of helium and oxygen at a temperature of 25 °C. Calculate the root mean square speed (v_{rms}) of the oxygen molecules. (Note: oxygen has a molar mass of $32 \frac{g}{mol}$.)

$$T = 25°C + 273 = 298\,K \quad \text{and} \quad M_{O_2} = 32 \tfrac{g}{mol}$$

$$v_{rms} = \sqrt{\frac{3000\, RT}{M}}$$

$$v_{rms} = \sqrt{\frac{3000\,(8.31 \tfrac{J}{mol \cdot K})(298\,K)}{32 \tfrac{g}{mol}}}$$

$$v_{rms} = \sqrt{232{,}161} = 482 \tfrac{m}{s}$$

---

[*] This formula is usually given as $v_{rms} = \sqrt{\frac{3RT}{M}}$, but this requires that M be expressed in $\frac{kg}{mol}$. Because we are not deriving the formula in this class, it makes sense to change the factor from 3 to 3000 in order to keep molar mass expressed in the more familiar units of $\frac{g}{mol}$.

Use this space for summary and/or additional notes:

Molecular Speed & Effusion

Graham's Law

From physics, the formula for kinetic energy is:

$$E_{kinetic} = \tfrac{1}{2}mv^2$$

where m is the mass and v is the velocity (speed).

If two molecules have the same temperature, then they have the same kinetic energy. If molecule #1 has mass m_1 and velocity v_1 and molecule #2 has mass m_2 and velocity v_2, then the kinetic energies are:

$$E_{kinetic} = \tfrac{1}{2}m_1 v_1^2 = \tfrac{1}{2}m_2 v_2^2$$

If we cancel the ½ from both sides, and rearrange so the masses are on one side and the speeds are on the other, we get:

$$\frac{v_2^2}{v_1^2} = \frac{m_1}{m_2}$$

Taking the square root of both sides:

$$\frac{v_2}{v_1} = \sqrt{\frac{m_1}{m_2}}$$

This formula is called Graham's Law, named after Thomas Graham who first proposed it.

For the purpose of these calculations, the mass of a molecule is its molar mass, which is simply the same number of grams as the molecule's atomic mass.

Because the rate of effusion (r) is the velocity divided by time, the units of time cancel, giving $\dfrac{r_2}{r_1} = \dfrac{v_2}{v_1}$

Use this space for summary and/or additional notes:

Molecular Speed & Effusion

Unit: Gases

Sample Problem:

If the speed of oxygen molecules in a tank is $482 \frac{m}{s}$, calculate the speed of helium molecules in the same tank. (O_2 has a molar mass of $32.0 \frac{g}{mol}$ and helium has a molar mass of $4.01 \frac{g}{mol}$.)

$$M_{O_2} = 32.0 \tfrac{g}{mol} \text{ and } M_{He} = 4.01 \tfrac{g}{mol}$$

$$\frac{v_{He}}{v_{O_2}} = \sqrt{\frac{m_{O_2}}{m_{He}}}$$

$$\frac{v_{He}}{482} = \sqrt{\frac{32.0}{4.01}}$$

$$\frac{v_{He}}{482} = \sqrt{8.00} = 2.82$$

$$v_{He} = (2.82)(482) = 1360 \tfrac{m}{s}$$

Use this space for summary and/or additional notes:

Molecular Speed & Effusion

Homework Problems

1. The temperature of the air in a room is 25.0 °C. Determine the root mean square speed of the nitrogen molecules (molar mass $28.0 \frac{g}{mol}$) in the air in that room.

    Answer: $515 \frac{m}{s}$

2. If the N₂ molecules (which have a molar mass of $28.0 \frac{g}{mol}$) in a sample of air have a root mean square speed of $450 \frac{m}{s}$, what is the speed of the O₂ molecules (which have a molar mass of $32.0 \frac{g}{mol}$)?

    Answer: $421 \frac{m}{s}$

3. A balloon is filled with 1.0 mole each of helium (molar mass $4.0 \frac{g}{mol}$) and oxygen (molar mass $32 \frac{g}{mol}$). If half of the gas in the balloon is allowed to escape, how many moles each of helium and oxygen are left in the balloon?

    (*Hint: Find the ratio of the speeds of the two molecules. This will be the same as the ratio of the number of moles of each gas that escapes. The reciprocal will be the ratio of the number of moles of each gas left in the balloon.*)

    Answer: 0.26 mol He, 0.74 mol O₂

Use this space for summary and/or additional notes:

Summary: Gases

Unit: Gases

Big Ideas	Details
	Summary: Gases
	Unit: Gases
	List the main ideas of this chapter in phrase form:
	Write an introductory sentence that categorizes these main ideas.
	Turn the main ideas into sentences, using your own words. You may combine multiple main ideas into one sentence.
	Add transition words to make your writing clearer and rewrite your summary below.
	Use this space for summary and/or additional notes:

Introduction: Atomic Structure

Unit: Atomic Structure

Topics covered in this chapter:

Atomic Structure 163
History of Atomic Theory 166
Conservation of Mass; Definite & Multiple Proportions 173
Fundamental Forces 175
The Standard Model 177
Average Atomic Mass 182

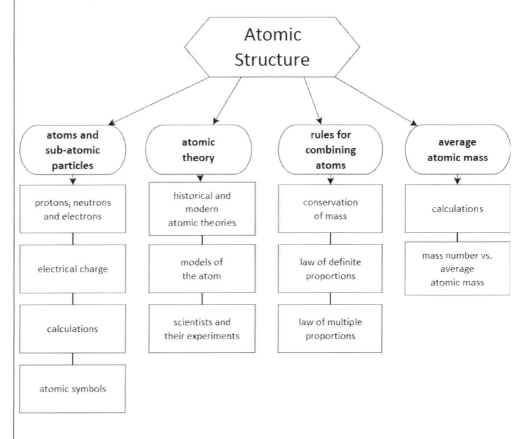

Use this space for summary and/or additional notes:

Introduction: Atomic Structure

Unit: Atomic Structure

Standards addressed in this chapter:

Massachusetts Curriculum Frameworks & Science Practices (2016):

HS-PS1-1: Use the periodic table as a model to predict the relative properties of main group elements, including ionization energy and relative sizes of atoms and ions, based on the patterns of electrons in the outermost energy level of each element. Use the patterns of valence electron configurations, core charge, and Coulomb's law to explain and predict general trends in ionization energies, relative sizes of atoms and ions, and reactivity of pure elements.

Massachusetts Curriculum Frameworks (2006):

2.1 Recognize discoveries from Dalton (atomic theory), Thomson (the electron), Rutherford (the nucleus), and Bohr (planetary model of atom), and understand how each discovery leads to modern theory.

2.2 Describe Rutherford's "gold foil"• experiment that led to the discovery of the nuclear atom. Identify the major components (protons, neutrons, and electrons) of the nuclear atom and explain how they interact.

2.3 Interpret and apply the laws of conservation of mass, constant composition (definite proportions), and multiple proportions.

Use this space for summary and/or additional notes:

Big Ideas	Details	Atomic Structure	Unit: Atomic Structure

Atomic Structure

Unit: Atomic Structure

MA Curriculum Frameworks (2016): HS-PS1-1

MA Curriculum Frameworks (2006): 2.2

Mastery Objective(s): (Students will be able to...)

- Identify subatomic particles and their locations within the atom.

Success Criteria:

- Protons and neutrons are correctly located in the nucleus.
- Electrons are correctly located outside the nucleus.
- Relative masses of protons, neutrons and electrons are correct.
- Chemical symbols are written correctly with correct value and placement of atomic symbol, atomic number, atomic mass and charge.

Tier 2 Vocabulary: nucleus, charge

Language Objectives:

- Correctly describe the parts of the atom and their locations within the atom.

Notes:

atom: the smallest piece of an element that retains the properties of that element.

nucleus: a dense region in the center of an atom. The nucleus is made of protons and neutrons, and contains almost all of an atom's mass.

proton: a subatomic particle found in the nucleus of an atom. It has a charge of +1, and a mass of 1 atomic mass unit (amu).

neutron: a subatomic particle found in the nucleus of an atom. It has no charge (is neutral), and has a mass of 1 amu.

electron: a subatomic particle found *outside* the nucleus of an atom. It has charge of −1 and a mass of 0 amu (really about $1/_{2000}$ amu). Atoms can gain, lose, or share electrons in chemical reactions.

charge: electrical charges can be positive or negative. Opposite cancel each other out, so the charge of an atom is the difference between how many positive charges (protons) it has, and how many negative charges (electrons) it has. For example, a chlorine atom with 17 protons (+17) and 18 electrons (−18) would have a charge of −1. (The difference is 1, and it's negative because it has more negatives than positives.)

Use this space for summary and/or additional notes:

Atomic Structure

Big Ideas	Details

neutral atom: an atom with a charge of zero (positives = negatives).

ion: an atom or molecule that has a charge (either positive or negative), because it has either gained or lost electrons. This means it has either more negatives (electrons) than positives (protons), or more positives (protons) than negatives (electrons).

atomic number (Z): the identity of an atom is based on the amount of (positive) charge in its nucleus. (This works because particles from the nucleus cannot be given to or shared with another atom.) The atomic number is the number of protons in the nucleus. Each element has a unique atomic number.

mass number (A): the total number of protons + neutrons in the nucleus of an atom. (Protons and neutrons each have a mass of almost exactly 1 amu, and the electrons are so small that their mass is negligible.) Generally equal to the whole number that is closest to the atomic mass.

atomic mass: the actual mass of an atom; the sum of the masses of its protons, neutrons, and electrons (minus a small amount of mass that is converted to energy to hold the atom together). Always close in value to the mass number.

isotopes: atoms of the same element (same atomic number = same # of protons), but that have different numbers of neutrons (and therefore different mass numbers) from each other.

Isotopes are described by their mass numbers. For example, carbon-12 (^{12}C) has 6 protons and 6 neutrons, which gives it a mass number of 12. Carbon-14 (^{14}C) has 6 protons and 8 neutrons, which gives it a mass number of 14.

element symbol: a one- or two-letter abbreviation for an element. (New elements are given temporary three-letter symbols.) The first letter in an element symbol is always capitalized. Other letters in an element symbol are always lower case. *This is important to remember.* For example, Co is the element cobalt, but CO is the compound carbon monoxide, which contains the elements carbon and oxygen.

Use this space for summary and/or additional notes:

Atomic Structure

chemical symbol: a shorthand notation that shows information about an element, including its element symbol, atomic number, mass number, and charge. For example, the symbol for a magnesium-25 ion with a +2 charge would be:

$$^{25}_{12}Mg^{2+}$$

This notation shows the element symbol for magnesium (Mg) in the center, the atomic number (12, because it has 12 protons) on the bottom left, the mass number (25, because it has 12 protons + 13 neutrons = 25 amu) on the top left, and the charge (2+, which means it has somehow lost two of its electrons) on the top right. By convention, ions are labeled with the charge number before the charge sign, so we write 2+ instead of +2.

Homework Problems

Each row in this table is like a Sudoku puzzle. For each row, use the numbers given, the relationships between the columns, and the periodic table of the elements to fill in the rest of the row. Use the first row as an example.

symbol	atomic #	mass #	protons	neutrons	electrons	charge
$^{41}_{20}Ca^{2+}$	20	41	20	21	18	+2
B				6	5	
		56	24			0
Ca^{2+}				20		
	60			84	57	
		207			80	+2
				0	0	+1
Kr		84				0
			35	39	36	

Use this space for summary and/or additional notes:

History of Atomic Theory

Unit: Atomic Structure
MA Curriculum Frameworks (2016): HS-PS1-1
MA Curriculum Frameworks (2006): 2.1
Mastery Objective(s): (Students will be able to...)
- Give a timeline for the development of atomic theory.
- Explain how each discovery changed our model of the atom.

Success Criteria:
- Discoveries are in the correct chronological order.
- Descriptions explain how each new discovery added to or changed the model of the atom.

Tier 2 Vocabulary: model,

Language Objectives:
- Correctly describe the parts of the atom and their locations within the atom.

Notes:

<u>atomic theory</u>: a theory that explains behavior of chemical elements based on the atoms that they are made of, and the composition of those atoms.

Modern Atomic Theory

The current model (theory) of the atom is the <u>quantum mechanical model</u>. It states that:

- The atom contains a nucleus at the center. The nucleus contains most of the mass of the atom. The nucleus consists of:
    - protons (positively charged)
    - neutrons (neutral)
- The atom contains electrons (negatively charged) outside the nucleus.
- Electrons can be added to or removed from atoms. An atom that has gained or lost electrons is called an ion.
- Each electron in an atom is confined to one of several specific regions around the nucleus (called orbitals), but each electron may move freely within its orbital. These orbitals are regions, but they do not have solid boundaries; each electron remains within its orbital because of a balance of forces, which are determined by how much energy the electron has.

Use this space for summary and/or additional notes:

Big Ideas	Details

History of Atomic Theory

Historical Development of Atomic Theory

Democritus: ancient Greek philosopher. Credited with the first theory of atoms (~400 B.C.E.). The theory of Democritus held that everything is composed of "atoms", which are physically, but not geometrically, indivisible; that between atoms, there lies empty space; that atoms are indestructible, and have always been and always will be in motion; that there is an infinite number of atoms and of kinds of atoms, which differ in shape, size and mass.

John Dalton: English chemist, physicist and meteorologist. Credited with the first theory that described what atoms are and how they behave:

Dalton's Atomic Theory (1807-08):

- everything is made of atoms
- atoms of the same element are identical; atoms of different elements are different
- atoms are not created or destroyed in chemical reactions. Chemical reactions are simple rearrangements of the atoms into different compounds.
- every sample (molecule) of a compound contains the same atoms in the same proportions ("Law of Constant Composition")
- atoms in compounds occur in simple, whole-number ratios ("Law of Multiple Proportions")

J.J. Thomson: English physicist. Discovered the electron (1897). His experiment was to apply an electric current to a gas. This created cathode rays—rays of negatively-charged electric particles, which appeared to come from the cathode (positive electric terminal). Thomson determined that these particles (which he named "corpuscles") came from the atoms that the cathode was made of. This discovery was important because it was the first evidence that atoms were divisible. Thomson received the Nobel Prize for Physics in 1906 for this discovery.

"plum pudding" model: (1904) J.J. Thomson compared the atom with a bowl of plum pudding with raisins. (Plum pudding, which is a lot like oatmeal, is popular in England.) The "pudding" was the positively charged substance that most of the atom was made of, and the "raisins" were the negatively-charged electrons (which he called "corpuscles"). Thomson published this theory in 1907 in a book called *The Corpuscular Theory of Matter*.

Use this space for summary and/or additional notes:

History of Atomic Theory

Unit: Atomic Structure

planetary model: (early 1900s) The atom was compared with a miniature solar system. In the 1906 physics textbook, *A First Course in Physics,* authors Robert Millikan and Henry Gale credited Thomson with this model (which at the time was called *electron theory*):

> "But since the atoms are probably electrically neutral, it is necessary to assume that they contain equal amounts of positive and negative electricity. Since, however, no evidence has yet appeared to show that positively charged electrons ever become detached from molecules, Thomson brings forward the hypothesis that perhaps the positive charges constitute the nucleus of the atom about the center of which the negative electrons are rapidly rotating.

> "According to this hypothesis, then, an atom is a sort of infinitesimal solar system whose members, the electrons, are no bigger with respect to the diameter of the atom than is the earth with respect to the diameter of the earth's orbit. Furthermore, according to this hypothesis, it is the vibrations of these electrons which give rise to light and heat waves; it is the streaming through conductors of electrons which have become detached from atoms which constitutes an electric current in a metal; it is an excess of electrons upon a body which constitutes a static negative charge, and a deficiency of electrons which constitutes a positive charge."

Robert Millikan: American physicist. Measured the electrical charge on an electron based on the rate that oil drops fall through an electric field (1909). The common factor in all of the measurements must be the basic particle of electric charge—the electron. Millikan received the Nobel Prize for Physics in 1923 for this discovery.

Use this space for summary and/or additional notes:

History of Atomic Theory

Unit: Atomic Structure

Ernest Rutherford: New Zealand-born British physicist. Supervised an experiment that corroborated electron theory and showed the existence of a dense, positively-charged nucleus (1909). In Rutherford's experiment, positively-charged alpha particles were allowed to pass through a thin sheet of gold foil.

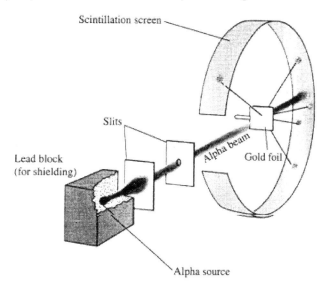

Most of the particles passed through, as Rutherford had expected. However, a few were deflected sharply, as if they had interacted with a dense object with a strong positive charge. Rutherford described the result as follows:

> "It was quite the most incredible event that has ever happened to me in my life. It was almost as incredible as if you fired a 15-inch shell at a piece of tissue paper and it came back and hit you. On consideration, I realized that this scattering backward must be the result of a single collision, and when I made calculations I saw that it was impossible to get anything of that order of magnitude unless you took a system in which the greater part of the mass of the atom was concentrated in a minute nucleus. It was then that I had the idea of an atom with a minute massive centre, carrying a charge."

Conclusions from Rutherford's Experiment:

- Most of the atom is empty space.
- The atom has a dense, positively-charged nucleus in the center.
- Nearly all of the mass of the atom is in the nucleus.

Many people believe that the idea of a positively-charged nucleus originated with Rutherford. Electron theory actually predated Rutherford's gold foil experiment by several years (as stated above), but the experiment was the confirmation that enabled the theory to gain acceptance.

Use this space for summary and/or additional notes:

History of Atomic Theory

Bohr model: (1913) Danish physicist Niels Bohr hypothesized that electrons moved around the nucleus as in the planetary model, and the distance of each electron from the nucleus was determined by the amount of energy it had. The energy was quantized, so only specific orbits were allowed. These quantum values of energy could be described by a quantum number (n).

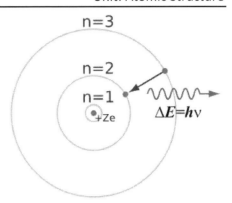

Bohr's model gained wide acceptance, because it combined three prominent theories of the time: electron theory, spectroscopy and quantum theory.

Even though the Bohr model of the atom has been superseded by the quantum mechanical model, the Bohr model is frequently taught today in elementary and middle school science classes because it is easier to visualize and because it relates the atom to the solar system, which is already familiar.

The Bohr model is described in more detail in the section "The Bohr Model of the Hydrogen Atom," which begins on page 216.

quantum mechanics: in 1900, German physicist Max Planck postulated that absorption and emission of energy that produces light occurs in discrete packets called "quanta".

photoelectric effect: In 1905, German physicist Albert Einstein discovered that energy from light could cause electrons to be emitted from a metal, that the energy from this light agreed with Planck's equation, that there was a certain minimum amount of energy specific to each metal that was required to drive off the electrons, and that this energy was quantized—the energy needed to release the electrons was all-or-nothing. Einstein received the Nobel Prize in Physics in 1921 for this discovery.

Use this space for summary and/or additional notes:

History of Atomic Theory

Big Ideas | Details | Unit: Atomic Structure

Louis de Broglie: In 1924, French physicist Louis de Broglie suggested that matter can act as both a particle and a wave. He theorized that the reason that only integer values for quantum numbers were possible was because as the electron orbits the nucleus, its path must be an integer multiple of the wavelength:

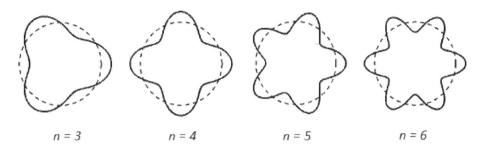

$n = 3$ $n = 4$ $n = 5$ $n = 6$

Erwin Schrödinger: Austrian physicist. Expressed de Broglie's hypothesis in mathematical form (the Schrödinger wave equation) and used it to predict the quantum energies of atoms (1926).

The solutions to Schrödinger's equation defined additional integer quantum numbers (ℓ and m) that specified the arrangements of electrons within the atom. These solutions supported the idea that an electron is either able to be detected (present), or unable to be detected (absent), as would be the case for a wave that is detectable at an antinode, but not at a node. Schrödinger's equations resulted in maps of regions around the nucleus of an atom (later named "orbitals," based on the probabilities of finding an electron in the different regions as a function of the energy of the electron.

Schrödinger received the Nobel Prize for Physics in 1933 for his work.

Sir James Chadwick: British physicist. Discovered the neutron (in 1932), which accounted for previously unexplained mass within an atom. His experiment was to collide alpha particles into beryllium, which caused neutral particles with the same mass as a proton to be ejected. Because these particles were neutrally charged, Chadwick named them neutrons. Chadwick received the Nobel Prize for Physics in 1935 for this discovery.

Use this space for summary and/or additional notes:

History of Atomic Theory

Unit: Atomic Structure

Homework

Make a timeline of how the theory of the atom developed, including the models of Democritus, Dalton, the "plum pudding" model, the planetary model, Bohr, de Broglie and Schrödinger. For each entry, your timeline should include:

- a sketch of what the atom might have looked like according to the model
- the year the model was proposed
- the name(s) of the scientist(s) credited with the model
- a 1–2 sentence description of the model

Here is an example of what the timeline entry for Democritus might look like:

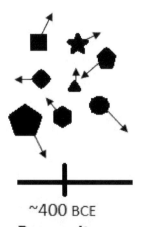

~400 BCE

Democritus

Everything is made of indestructible atoms of different sizes, shapes and masses. These atoms are in constant motion.

Use this space for summary and/or additional notes:

Conservation of Mass; Definite & Multiple Proportions

Unit: Atomic Structure

MA Curriculum Frameworks (2016): HS-PS1-1

MA Curriculum Frameworks (2006): 2.3

Mastery Objective(s): (Students will be able to...)

- Explain the laws of conservation of mass, definite proportions, and multiple proportions.
- Solve problems relating to the conservation of mass.

Success Criteria:

- Explanations account for observations about the way atoms combine.
- Solutions account for all mass before and after some change.

Tier 2 Vocabulary: conservation

Language Objectives:

- Explain the laws of conservation of mass, definite proportions, and multiple proportions.

Notes:

<u>conservation of mass</u>: matter (mass) can neither be created nor destroyed, only changed in form. All of the mass that was present before a chemical or physical change took place is present after the change.

This law holds for the total mass, and also individually for the mass of each type of atom (element).

For example, in the chemical equation:

$$HCl + NaOH \rightarrow NaCl + H_2O$$

1. The combined mass of HCl and NaOH before the reaction is equal to the combined mass of NaCl and H_2O produced by the reaction.

2. The mass of each element before the reaction is equal to the mass of that same element after. For example, the number of grams of chlorine in the HCl that reacts is equal to the grams of chlorine in the NaCl produced.

Use this space for summary and/or additional notes:

Conservation of Mass; Definite & Multiple Proportions

Unit: Atomic Structure

Law of Constant Composition (Law of Definite Proportions): the same compound always contains atoms of the same elements in the same proportions by mass. *E.g.*, water (H_2O) always contains 11 % hydrogen and 89 % oxygen by mass.

The Law of Constant Composition was part of Dalton's theory of atoms, first published in 1803.

Note also that the reverse is not necessarily true—very different compounds can have the same atoms in the same proportions, and even the exact same chemical formulas. For example, the compounds ethyl acetate and butyric acid both have the same chemical formula ($C_4H_8O_2$). However, ethyl acetate smells like nail polish, whereas butyric acid smells like a combination of rancid butter and vomit.

Law of Multiple Proportions: elements always combine in simple, whole-number ratios. (This works whether you're comparing atoms or masses.) For example, copper and chlorine can combine to form $CuCl$ or $CuCl_2$, but they won't combine to form ratios like $Cu_{1.7}Cl_{4.83}$.

There is a joke whose punch line depends on the law of multiple proportions:

> A chemist and her friend walk into a bar. The chemist tells the bartender, "I'd like a glass of H_2O, please." Her friend says, "I'd like H_2O too." Both drink, and the friend dies.

The basis of the punch line is that "H_2O too" sounds like "H_2O_2," which is hydrogen peroxide.

The Law of Multiple Proportions was also first proposed by John Dalton in 1803 as part of his theory of atoms.

While the chemistry that we will study this year depends on the laws of constant composition and multiple proportions, there are a few unusual compounds whose elemental composition can vary from sample to sample. One example is the iron oxide wüstite, which can contain between 0.83 and 0.95 iron atoms for every oxygen atom, and thus contains anywhere between 23 % and 25 % oxygen.

Use this space for summary and/or additional notes:

Fundamental Forces

Unit: Atomic Structure

MA Curriculum Frameworks (2016): N/A

MA Curriculum Frameworks (2006): N/A

Mastery Objective(s): (Students will be able to…)

- Name, describe, and give relative magnitudes of the four fundamental forces of nature.

Success Criteria:

- Descriptions & explanations are accurate and account for observed behavior.

Tier 2 Vocabulary: strong, weak

Language Objectives:

- Describe the fundamental forces, how strong they are and what they act on.

Notes:

All forces in nature ultimately come from one of the following four forces:

Use this space for summary and/or additional notes:

Fundamental Forces

Big Ideas	Details	Mastery Objective(s): (Students will be able to...)

strong force (or "strong nuclear force" or "strong interaction"): an attractive force between quarks. The strong force holds the nuclei of atoms together. The energy comes from converting mass to energy.
Effective range: about the size of the nucleus of an average-size atom.

weak force (or "weak nuclear force" or "weak interaction"): the force that causes protons and/or neutrons in the nucleus to become unstable and leads to beta nuclear decay. This happens because the weak force causes an up or down quark to change its flavor. (This process is described in more detail in the section on **Error! Reference source not found.** of particle chemistry, starting **Error! Bookmark not defined.**.) **Strength**: 10^{-6} to 10^{-7} times the strength of the strong force.
Effective range: about $\frac{1}{3}$ the diameter of an average nucleus.

electromagnetic force: the force between electrical charges. If the charges are the same ("like charges")—both positive or both negative—the particles repel each other. If the charges are different ("opposite charges")—one positive and one negative—the particles attract each other.
Strength: about $\frac{1}{137}$ as strong as the strong force.
Effective range: ∞, but gets smaller as $(distance)^2$.

gravitational force: the force that causes masses to attract each other. Usually only observable if one of the masses is very large (like a planet).
Strength: only 10^{-39} times as strong as the strong force.
Effective range: ∞, but gets smaller as $(distance)^2$.

Use this space for summary and/or additional notes:

Chemistry 1 — Mr. Bigler

The Standard Model

Unit: Atomic Structure
MA Curriculum Frameworks (2016): N/A
MA Curriculum Frameworks (2006): N/A
Mastery Objective(s): (Students will be able to…)

- Describe the fundamental particles that make up atoms and other matter and non-matter.

Success Criteria:

- Explanations correctly describe aspects of the standard model.

Tier 2 Vocabulary: standard, color
Language Objectives:

- Describe particles of the standard model and their properties.

Notes:

The Standard Model is a theory of particle physics that:

- identifies the particles that matter is ultimately comprised of
- describes properties of these particles, including their mass, charge, and spin
- describes interactions between these particles

The Standard Model dates to the mid-1970s, when the existence of quarks was first experimentally confirmed. Physicists are still discovering new particles and relationships between particles, so the model and the ways it is represented are evolving, much like atomic theory and the Periodic Table of the Elements was evolving at the turn of the twentieth century. The table and the model described in these notes represent our understanding, as of 2018. By the middle of this century, the Standard Model may evolve to a form that is substantially different from the way we represent it today.

The Standard Model in its present form does not incorporate dark matter, dark energy, or gravitational attraction.

Use this space for summary and/or additional notes:

The Standard Model

The Standard Model is often presented in a table, with rows, columns, and color-coded sections used to group subsets of particles according to their properties.

As of 2018, the standard model is represented by a table similar to this one:

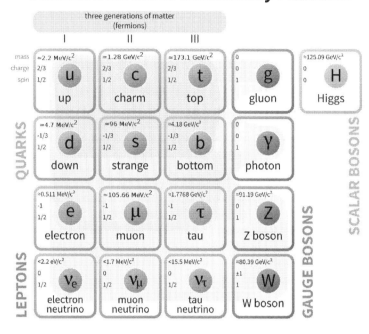

Quarks

Quarks are particles that participate in strong interactions (sometimes called the "strong force") through the action of "color charge" (which will be described later). Because protons and neutrons (which make up most of the mass of an atom) are made of three quarks each, quarks are the subatomic particles that make up most of the ordinary matter in the universe. (Dark matter, which accounts for 84.5% of the total matter in the universe, is made from other types of subatomic particles.)

- quarks have color charge (*i.e.*, they interact via the strong force)
- quarks have spin of $+\frac{1}{2}$
- "up-type" quarks have a charge of $+\frac{1}{2}$; "down-type" quarks have a charge of $+\frac{1}{2}$.

There are six flavors[*] of quarks: up and down, charm and strange, and top and bottom. Originally, top and bottom quarks were called truth and beauty.

---

[*] Yes, "flavors" really is the correct term. Blame the 1960s.

Use this space for summary and/or additional notes:

The Standard Model

Leptons

Leptons are the smaller particles that make up most matter. The most familiar lepton is the electron. Leptons participate in "electroweak" interactions, meaning combinations of the electromagnetic and weak forces.

- leptons do not have color charge (*i.e.*, they do not interact via the strong force)
- leptons have spins of $+\frac{1}{2}$
- electron-type leptons have a charge of −1; neutrinos do not have a charge.
- neutrinos oscillate, which makes their mass indefinite.

Gauge Bosons

Gauge bosons are the particles that carry force—their interactions are responsible for the fundamental forces of nature: the strong force, the weak force, the electromagnetic force and the gravitational force. The hypothetical particle responsible for the gravitational force is the graviton, which has not yet been detected (as of 2018).

- photons are responsible for the electromagnetic force.
- gluons are responsible for the strong interaction (strong force)
- W and Z bosons are responsible for the weak interaction (weak force)

Scalar Bosons

At present, the only scalar boson we know of is the Higgs boson, which is responsible for mass.

Particle Interactions

The interactions of these particles can be confusing. The diagram to the right, which shows which particles interact with which, may be helpful.

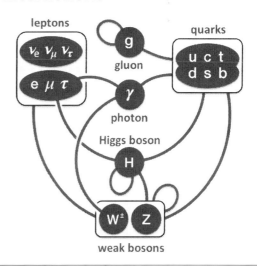

Use this space for summary and/or additional notes:

The Standard Model

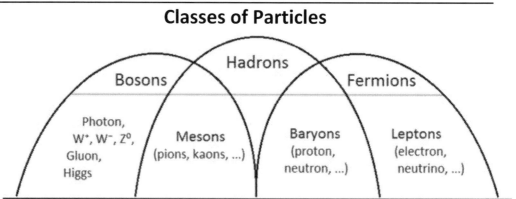

Fermions

Quarks and leptons are fermions. Fermions are described by Fermi-Dirac statistics and obey the Pauli exclusion principle (which states that no two particles in an atom may have the same exact set of quantum numbers—which are numbers that describe the energy states of the particle).

Fermions are the building blocks of matter. They have a spin of ½, and each fermion has its own antiparticle. (The antiparticle of a fermion is identical to its corresponding particle, but has a charge of the opposite sign. Antiparticles have the same name as the corresponding particle with the prefix "anti-"; for example, the antiparticle of a tau neutrino is a tau antineutrino. Note, however, that for historical reasons an antielectron is usually called a positron.)

Bosons

Bosons are described by Bose-Einstein statistics, have integer spins and do not obey the Pauli Exclusion Principle. Interactions between boson are responsible for forces and mass.

Each of the fundamental bosons is its own antiparticle, except for the W^- boson (whose antiparticle is the W^+ boson).

Use this space for summary and/or additional notes:

The Standard Model

Unit: Atomic Structure

Hadrons

Hadrons are a special class of strongly-interacting composite particles (meaning that they are comprised of multiple individual particles). Hadrons can be bosons or fermions. Hadrons composed of strongly-interacting fermions are called baryons; hadrons composed of strongly-interacting bosons are called mesons.

Baryons

The most well-known baryons are protons and neutrons, which each comprised of three quarks. Protons are made of two up quarks and one down quark ("uud"), and carry a charge of +1. Neutrons are made of one up quark and two down quarks ("udd"), and carry a charge of zero.

Some of the better-known baryons include:

- nucleons (protons & neutrons).
- hyperons, *e.g.,* the Λ, Σ, Ξ, and Ω particles. These contain one or more strange quarks, and are much heavier than nucleons.
- various charmed and bottom baryons.
- pentaquarks, which contain four quarks and an antiquark.

Mesons

Ordinary mesons are comprised of a quark plus an antiquark. Examples include the pion, kaon, and the J/Ψ. Mesons mediate the residual strong force between nucleons.

Some of the exotic mesons include:

- tetraquarks, which contain two quarks and two antiquarks.
- glueball, a bound set of gluons with no quarks.
- hybrid mesons, which contain one or more quark/antiquark pairs and one or more gluons.

Use this space for summary and/or additional notes:

Color Charge

Color charge is the property that is responsible for the strong nuclear interaction. All electrons and fermions (particles that have half-integer spin quantum numbers) must obey the Pauli Exclusion Principle, which states that no two particles within the same larger particle (such as a hadron or atom) can have identical sets of quantum numbers. For electrons, (as you learned in chemistry), if two electrons share the same orbital, they need to have opposite spins. In the case of quarks, all quarks have a spin of $+\frac{1}{2}$, so in order to satisfy the Pauli Exclusion Principle, if a proton or neutron contains three quarks, there has to be some other quantum property that has different values for each of those quarks. This property is called "color charge" (or sometimes just "color[*]").

The "color" property has three values, which are called "red," "green," and "blue" (named after the primary colors of light). When there are three quarks in a subatomic particle, the colors have to be different, and have to add up to "colorless". (Recall that combining each of the primary colors of light produces white light, which is colorless.)

Quarks can exchange color charge by emitting a gluon that contains one color and one anticolor. Another quark absorbs the gluon, and both quarks undergo color change. For example, suppose a blue quark emits a blue antigreen gluon:

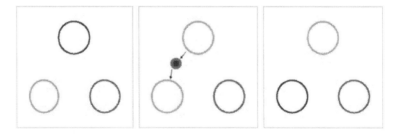

You can imagine that the quark sent away its own blue color (the "blue" in the "blue antigreen" gluon). Because it also sent out antigreen, it was left with green so it became a green quark. Meanwhile, the antigreen part of the gluon finds the green quark and cancels its color. The blue from the blue antigreen gluon causes the receiving quark to become blue. After the interaction, the particle once again has one red, one green, and one blue quark, which means color charge is conserved.

---

[*] Just like "spin" is the name of a property of energy that has nothing to do with actual spinning, "color" is a property that has nothing to do with actual color. In fact, quarks couldn't possibly have actual color—the wavelengths of visible light are thousands of times larger than quarks!

Use this space for summary and/or additional notes:

Average Atomic Mass

Unit: Atomic Structure

MA Curriculum Frameworks (2016): N/A

MA Curriculum Frameworks (2006): N/A

Mastery Objective(s): (Students will be able to...)

- Calculate the average atomic mass of an atom from percent abundance data.

Success Criteria:

- Solutions correctly turn masses into percentages.
- Algebra and rounding to appropriate number of significant figures is correct.

Tier 2 Vocabulary: abundance

Language Objectives:

- Explain the laws of conservation of mass, definite proportions, and multiple proportions.

Notes:

<u>mass number</u>: the mass of *one individual atom* (protons + neutrons). Always a whole number.

<u>abundance</u>: the percentage of atoms of an element that are one specific isotope.

<u>average atomic mass</u>: the estimated weighted *average* of the mass numbers *of all of the atoms* of a particular element on Earth.

Analogy: average atomic mass works the same way as class average on a test.

1. Multiply each score times the number of students who got it.
2. Add up the number for each score to get the total points.
3. Divide the total by the number of students to get class average.

Use this space for summary and/or additional notes:

Average Atomic Mass

Unit: Atomic Structure

Problem:

The atomic mass and abundance of the two stable isotopes of carbon are:

Isotope	Atomic Mass (amu)	Relative Abundance
carbon-12	12.000 000	98.93 %
carbon-13	13.003 355	1.07 %

What is the average atomic mass of carbon?

How to solve:

1. Convert percent abundances to fractions (divide by 100).
2. Multiply the fractional abundance times the atomic mass for each isotope.
3. Add up the sub-total from each isotope to get the total atomic mass.
4. Check that your average atomic mass is in between the smallest and largest.

Answer:

1. Convert abundances to fractions

   98.93 % ÷ 100 = 0.9893   1.07 % ÷ 100 = 0.0107

2. Multiply abundance x mass # for each isotope

   0.9893 x 12.000 000 = 11.8716

   0.0107 x 13.003 355 = 0.1391

3. Add up the number from each isotope to get the total

   11.8716 + 0.1391 = 12.0107

4. Check that your answer is in between the mass number of the smallest isotope and the mass number of the largest one.

   Yes, 12.0107 is between 12 and 13.

Use this space for summary and/or additional notes:

	Average Atomic Mass	
Big Ideas	Details	Unit: Atomic Structure

Homework Problems

Calculate the average atomic mass of each of the following elements, based on the percent abundance of their isotopes. For each element, your answers should agree with the atomic mass listed on the periodic table.

Because you can look up the answers, you must show how to set up the calculations in order to receive credit.

1. bromine

isotope	atomic mass (amu)	relative abundance
$^{79}_{35}Br$	78.9184	50.69 %
$^{81}_{35}Br$	80.9163	49.31 %

2. boron

isotope	atomic mass (amu)	relative abundance
$^{10}_{5}B$	10.0129	19.9 %
$^{11}_{5}B$	11.0093	80.1 %

Use this space for summary and/or additional notes:

Average Atomic Mass

3. chlorine

isotope	atomic mass (amu)	relative abundance
$^{35}_{17}\text{Cl}$	34.9689	75.78 %
$^{37}_{17}\text{Cl}$	36.9659	24.22 %

4. magnesium

isotope	atomic mass (amu)	relative abundance
$^{24}_{12}\text{Mg}$	23.9850	78.99 %
$^{25}_{12}\text{Mg}$	24.9858	10.00 %
$^{26}_{12}\text{Mg}$	25.9826	11.01 %

Use this space for summary and/or additional notes:

Big Ideas	Details
	Summary: Atomic Structure **Unit:** Atomic Structure List the main ideas of this chapter in phrase form: Write an introductory sentence that categorizes these main ideas. Turn the main ideas into sentences, using your own words. You may combine multiple main ideas into one sentence. Add transition words to make your writing clearer and rewrite your summary below.

Use this space for summary and/or additional notes:

Introduction: Nuclear Chemistry

Unit: Nuclear Chemistry

Topics covered in this chapter:

Radioactive Decay 191
Nuclear Equations... 196
Mass Defect & Binding Energy 199
Half-Life... 201
Nuclear Fission & Fusion.. 204
Practical Uses for Nuclear Radiation 207

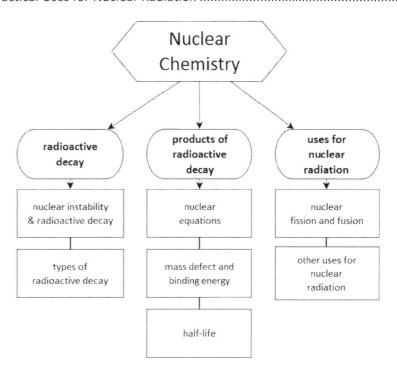

Use this space for summary and/or additional notes:

Introduction: Nuclear Chemistry

Standards addressed in this chapter:

Massachusetts Curriculum Frameworks & Science Practices (2016):

The Nuclear Chemistry unit is not included in the NGSS standards for high school chemistry. Massachusetts moved the topic to the high school physics course, which means this unit may be omitted from a chemistry course that is intended to follow the 2016 Massachusetts Curriculum Frameworks.

Massachusetts Curriculum Frameworks (2006):

2.5 Identify the three main types of radioactive decay (alpha, beta, and gamma) and compare their properties (composition, mass, charge, and penetrating power).

2.6 Describe the process of radioactive decay by using nuclear equations, and explain the concept of half-life for an isotope (for example, C-14 is a powerful tool in determining the age of objects).

2.7 Compare and contrast nuclear fission and nuclear fusion.

Use this space for summary and/or additional notes:

Big Ideas	Details	Radioactive Decay

Radioactive Decay

Unit: Nuclear Chemistry

MA Curriculum Frameworks (2016): N/A (HS-PS1-8 in physics frameworks)

MA Curriculum Frameworks (2006): 2.5

Mastery Objective(s): (Students will be able to...)

- Determine the products of alpha, beta-minus, and beta-plus radioactive decay and electron capture.
- Predict the most likely form of radioactive decay for an isotope based on its position relative to the band of stability on a proton-neutron graph.

Success Criteria:

- Form of radioactive decay correctly identified.
- Products of decay correctly identified.
- Correct nuclear equation.

Tier 2 Vocabulary: decay

Language Objectives:

- Explain the processes of alpha, beta-plus and beta-minus radioactive decay and electron capture.
- Understand and correctly use the terms "radioactive decay," "nuclear instability," "alpha decay," "beta decay," "gamma rays," and "penetrating power."

Notes:

<u>nuclear instability</u>: When something is unstable, it is likely to change. If the nucleus of an atom is unstable, changes can occur that affect the number of protons and neutrons in the atom.

Note that when this happens, the nucleus ends up with a different number of protons. This causes the atom to literally turn into an atom of a different element. When this happens, the physical and chemical properties instantaneously change into the properties of the new element!

<u>radioactive decay</u>: the process by which the nucleus of an atom changes, transforming the element into a different element or isotope.

<u>nuclear equation</u>: an equation describing (through chemical symbols) what happens to an atom as it undergoes radioactive decay.

Use this space for summary and/or additional notes:

Causes of Nuclear Instability

Two of the causes of nuclear instability are:

Size

because the strong force acts over a limited distance, when nuclei get too large (more than 82 protons), it is no longer possible for the strong force to keep the nucleus together indefinitely. The form of decay that results from an atom exceeding its stable size is called alpha (α) decay.

The Weak Nuclear Force

The weak force is caused by the exchange (absorption and/or emission) of W and Z bosons. This causes a down quark to change to an up quark or vice-versa. The change of quark flavor has the effect of changing a proton to a neutron, or a neutron to a proton. (Note that the action of the weak force is the only known way of changing the flavor of a quark.) The form of decay that results from the action of the weak force is called beta (β) decay.

band of stability: isotopes with a ratio of protons to neutrons that results in a stable nucleus (one that does not spontaneously undergo radioactive decay). This observation suggests that the ratio of up to down quarks within the nucleus is somehow involved in preventing the weak force from causing quarks to change flavor.

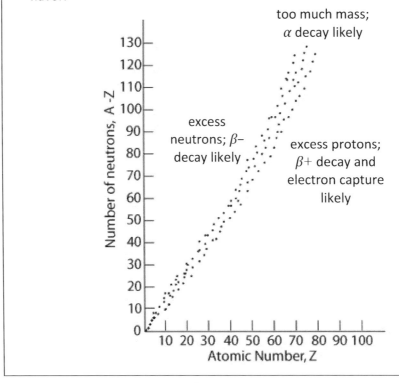

Use this space for summary and/or additional notes:

Radioactive Decay

Big Ideas | **Details** | Unit: Nuclear Chemistry

<u>alpha (α) decay</u>: a type of radioactive decay in which the nucleus loses two protons and two neutrons (an alpha particle). An alpha particle is a $^{4}_{2}He^{2+}$ ion (the nucleus of a helium-4 atom), with two protons, a mass of 4 amu, and a charge of +2. For example:

$$^{238}_{92}U \rightarrow {}^{234}_{90}Th + {}^{4}_{2}He$$

Atoms are most likely to undergo alpha decay if they have an otherwise stable proton/neutron ratio but a large atomic number.

Alpha decay has never been observed in atoms with an atomic number less than 52 (tellurium), and is rare in elements with an atomic number less than 73 (tantalum).

<u>Net effects of α decay:</u>

- Atom loses 2 protons and 2 neutrons (atomic number goes down by 2 and mass number goes down by 4)
- An α particle (a $^{4}_{2}He^{+2}$ ion) is ejected from the nucleus at high speed.

<u>beta minus (β−) decay</u>: a type of radioactive decay in which a neutron is converted to a proton and the nucleus ejects a high speed electron ($^{0}_{-1}e$).

Note that a neutron consists of one up quark and two down quarks (udd), and a proton consists of two up quarks and one down quark (uud). When β− decay occurs, the weak force causes one of the quarks changes its flavor from down to up, which causes the neutron (uud) to change into a proton (udd). Because a proton was gained, the atomic number increases by one. However, because the proton used to be a neutron, the mass number does not change. For example:

$$^{32}_{15}P \rightarrow {}^{32}_{16}S + {}^{0}_{-1}e$$

Atoms are likely to undergo β− decay if they have too many neutrons and not enough protons to achieve a stable neutron/proton ratio. Almost all isotopes that are heavier than isotopes of the same element within the band of stability (because of the "extra" neutrons) undergo β− decay.

<u>Net effects of β− decay:</u>

- Atom loses 1 neutron and gains 1 proton (atomic number goes up by 1; mass number does not change)
- A β− particle (an electron) is ejected from the nucleus at high speed.

Note that a β− particle is assigned an atomic number of −1. *This does not mean an electron is some sort of "anti-proton".* The −1 is just used to make the equation for the number of protons work out in the nuclear equation.

Use this space for summary and/or additional notes:

Chemistry 1 | Mr. Bigler

Radioactive Decay

Unit: Nuclear Chemistry

beta plus (β+) decay: a type of radioactive decay in which a proton is converted to a neutron and the nucleus ejects a high speed antielectron (positron, $^{0}_{+1}e$).

With respect to the quarks, β+ decay is the opposite of β– decay When β+ decay occurs, one of the quarks changes its flavor from up to down, which changes the proton (uud) into a neutron (udd). Because a proton was lost, the atomic number decreases by one. However, because the neutron used to be a proton, the mass number does not change. For example:

$$^{23}_{12}Mg \rightarrow {}^{23}_{11}Na + {}^{0}_{+1}e$$

Atoms are likely to undergo β+ decay if they have too many protons and not enough neutrons to achieve a stable neutron/proton ratio. Almost all isotopes that are lighter than the isotopes of the same element that fall within the band of stability ("not enough neutrons") undergo β+ decay.

Net effects of β+ decay:

- Atom loses 1 proton and gains 1 neutron (atomic number goes down by 1; mass number does not change)
- A β+ particle (an antielectron or positron) is ejected from the nucleus at high speed.

electron capture (sometimes called "K-capture"): when the nucleus of the atom "captures" an electron from the innermost shell (the K-shell) and incorporates it into the nucleus. This process is exactly the reverse of β– decay; during electron capture, a quark changes flavor from up to down, which changes a proton (uud) into a neutron (udd):

$$^{23}_{12}Mg + {}^{0}_{-1}e \rightarrow {}^{23}_{11}Na$$

Note that β+ decay and electron capture produce the same products. Electron capture can sometimes (but not often) occur without β+ decay. However, β+ decay is *always* accompanied by electron capture.

Atoms are likely to undergo electron capture (and usually also β+ decay) if they have too many protons and not enough neutrons to achieve a stable neutron/proton ratio. Almost all isotopes that are lighter than the isotopes of the same element that fall within the band of stability undergo electron capture, and usually also β+ decay.

Net effects of electron capture:

- An electron is absorbed by the nucleus.
- Atom loses 1 proton and gains 1 neutron (atomic number goes down by 1; mass number does not change)

Use this space for summary and/or additional notes:

Radioactive Decay

Unit: Nuclear Chemistry

gamma (γ) rays: most radioactive decay produces energy. Some of that energy is emitted in the form of gamma rays, which are very high energy photons of electromagnetic radiation. (Radio waves, visible light, and X-rays are other types of electromagnetic radiation.) Because gamma rays are waves (which have no mass), they can penetrate far into substances and can do a lot of damage. Because gamma rays are not particles, emission of gamma rays does not change the composition of the nucleus.

All of the types of radioactive decay mentioned in these notes also produce γ rays. This means to be complete, we would add gamma radiation to each of the radioactive decay equations described above:

$$^{238}_{92}U \rightarrow ^{234}_{90}Th + ^{4}_{2}He + ^{0}_{0}\gamma \qquad ^{32}_{15}P \rightarrow ^{32}_{16}S + ^{0}_{-1}e + ^{0}_{0}\gamma$$

$$^{23}_{12}Mg \rightarrow ^{23}_{11}Na + ^{0}_{+1}e + ^{0}_{0}\gamma \qquad ^{23}_{12}Mg + ^{0}_{-1}e \rightarrow ^{23}_{11}Na + ^{0}_{0}\gamma$$

penetrating power: the distance that radioactive particles can penetrate into/through another substance is directly related to the velocity of the emission (faster = more penetrating) and inversely related to the mass of the emission (heaver = less penetrating):

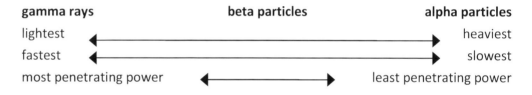

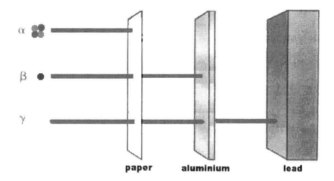

Note also that denser substances (such as lead) do a better job of blocking and absorbing radioactive emissions. This is why lead is commonly used as shielding for experiments involving radioactive substances.

Use this space for summary and/or additional notes:

Nuclear Equations

Unit: Nuclear Chemistry
MA Curriculum Frameworks (2016): N/A (HS-PS1-8 in physics frameworks)
MA Curriculum Frameworks (2006): 2.6
Mastery Objective(s): (Students will be able to...)
- Write & solve nuclear equations.

Success Criteria:
- Equations include the correct product of decay (α, $\beta-$ or $\beta+$ particle)
- Equations include the correct starting material(s) and/or product(s).

Tier 2 Vocabulary: decay

Language Objectives:
- Explain the equations for radioactive decay and how to calculate the products.

Notes:

nuclear equation: a chemical equation describing the process of an isotope undergoing radioactive decay. For example:

$$^{238}_{92}U \rightarrow {}^{234}_{90}Th + {}^{4}_{2}He$$

In a nuclear equation, the number of protons (atomic number) and the total mass (mass number) are conserved on both sides of the arrow. If you look at the bottom (atomic) numbers, and replace the arrow with an = sign, you would have the following:

$$92 = 90 + 2$$

Similarly, if you look at the top (mass) numbers, and replace the arrow with an = sign, you would have:

$$238 = 234 + 4$$

Use this space for summary and/or additional notes:

Nuclear Equations

Unit: Nuclear Chemistry

Sample problems:

Q: What are the products of beta-minus ($\beta-$) decay of ^{131}I?

A: A $\beta-$ particle is an electron, which we write as $_{-1}^{0}e$ in a nuclear equation. This means ^{131}I decays into some unknown particle plus $_{-1}^{0}e$. The equation is:

$$^{131}_{53}I \rightarrow\, ^{m}_{p}X + {^{0}_{-1}e}$$

We can write the following equations for the atomic and mass numbers:

Atomic #s: $53 = p + -1 \rightarrow p = 54$; therefore X is Xe

Mass #s: $131 = m + 0 \rightarrow m = 131$

Therefore, particle X is $^{131}_{54}Xe$ So our final answer is:

The two products of decay in this reaction are $^{131}_{54}Xe$ and $_{-1}^{0}e$.

Q: Which particle was produced in the following radioactive decay reaction:

$$^{212}_{86}Rn \rightarrow\, ^{208}_{84}Po + {^{m}_{p}X}$$

A: The two equations are:

Atomic #s: $86 = 84 + p \rightarrow p = 2$; therefore X is He

Mass #s: $212 = 208 + m \rightarrow m = 4$

Therefore, particle X is $^{4}_{2}He$, which means it is an α particle.

Use this space for summary and/or additional notes:

Nuclear Equations

Unit: Nuclear Chemistry

Homework Problems

For these problems, you will need to use a periodic table and radioactive decay information from "Table U. Selected Radioisotopes" on page 554 of your Chemistry Reference Tables.

Give the nuclear equation(s) for radioactive decay of the following:

1. ^{222}Rn

2. ^{85}Kr

3. ^{220}Fr

4. ^{37}K

5. ^{3}H

Give the starting material for the following materials produced by radioactive decay:

6. Alpha (α) decay resulting in $^{267}_{108}$Hs

7. Beta-minus ($\beta-$) decay resulting in $^{185}_{75}$Re

Use this space for summary and/or additional notes:

Big Ideas	Details
	# Mass Defect & Binding Energy

Unit: Nuclear Chemistry

MA Curriculum Frameworks (2016): N/A (HS-PS1-8 in physics frameworks)

MA Curriculum Frameworks (2006): N/A

Mastery Objective(s): (Students will be able to…)

- Explain where the energy that powers the strong force comes from.
- Calculate the "missing" mass and convert it to energy.

Success Criteria:

- Theoretical mass accounts for mass of all protons, neutrons and electrons.
- Mass defect is calculated correctly.
- Einstein's equation ($E = mc^2$) is properly applied to the mass defect to calculate energy.

Tier 2 Vocabulary: defect

Language Objectives:

- Explain the concept of "missing" mass and how it converts to energy.

Notes:

<u>binding energy</u>: the energy that holds the nucleus of an atom together through the strong nuclear force

The binding energy comes from the small amount of mass (the mass defect) that was turned into a large amount of energy, given by the equation:

$$E = mc^2$$

where E is the binding energy, m is the mass defect, and c is the speed of light ($3 \times 10^8 \, \frac{m}{s}$, which means c^2 is a very large number).

<u>mass defect</u>: the difference between the actual mass of an atom, and the sum of the masses of the protons, neutrons, and electrons that it contains.

- A proton has a mass of 1.6726×10^{-27} kg = 1.0073 amu
- A neutron has a mass of 1.6749×10^{-27} kg = 1.0087 amu
- An electron has a mass of 9.1094×10^{-31} kg = 0.0005486 amu

To calculate the mass defect, total up the masses of each of the protons, neutrons, and electrons in an atom. The actual (observed) atomic mass of the atom is always *less* than this number. The "missing mass" is called the mass defect.

Use this space for summary and/or additional notes:

Mass Defect & Binding Energy

Unit: Nuclear Chemistry

Sample problem:

Q: Find the mass defect of 1 mole of uranium-238.

A: $^{238}_{92}U$ has 92 protons, 146 neutrons, and 92 electrons. This means the total mass of one atom of $^{238}_{92}U$ should theoretically be:

92 protons × 1.0073 amu = 92.6704 amu
146 neutrons × 1.0087 amu = 147.2661 amu
92 electrons × 0.000 5486 amu = 0.0505 amu
92.6704 + 147.2661 + 0.0505 = 239.9870 amu

The actual observed mass of one atom of $^{238}_{92}U$ is 238.0003 amu.

The mass defect is therefore 239.9870 − 238.0003 = 1.9867 amu.

One mole of $^{238}_{92}U$ would have a mass of 238.0003 g, and therefore a total mass defect of 1.9867 g (which is 0.001 9867 kg).

Because $E = mc^2$, that means the binding energy of one mole of $^{238}_{92}U$ is:

$$0.001\,9867\,\text{kg} \times (3.00 \times 10^8)^2 = 1.79 \times 10^{14}\ \text{J}$$

In case you don't realize just how large that number is, the binding energy of just 238 g (1 mole) of $^{238}_{92}U$ would be enough energy to heat every house on Earth for an entire winter!

Use this space for summary and/or additional notes:

Half-Life

Unit: Nuclear Chemistry

MA Curriculum Frameworks (2016): N/A (HS-PS1-8 in physics frameworks)

MA Curriculum Frameworks (2006): 2.6

Mastery Objective(s): (Students will be able to...)

- Determine the amount of radioactive material remaining after an integer number of half-lives.
- Determine the amount of time that has elapsed based on the fraction of radioactive material remaining (*e.g.*, carbon dating).

Success Criteria:

- Solutions use the appropriate equation for the information given.
- Solutions have the correct quantities substituted for the correct variables.
- Algebra and rounding to appropriate number of significant figures is correct.

Tier 2 Vocabulary: half-life, decay

Language Objectives:

- Explain how exponential decay works.

Notes:

The atoms of radioactive elements are unstable, and they spontaneously decay (change) into atoms of other elements.

For any given atom, there is a certain probability, P, that it will undergo radioactive decay in a given amount of time. The half-life, τ, is how much time it would take to have a 50 % probability of the atom decaying. If you start with n atoms, after one half-life, half of them ($0.5n$) will have decayed.

Amount of Material Remaining

If we start with 32 g of ^{53}Fe, which has a half-life (τ) of 8.5 minutes, we would observe the following:

# minutes	0	8.5	17	25.5	34
# half lives	0	1	2	3	4
amount left	32 g	16 g	8 g	4 g	2 g

Use this space for summary and/or additional notes:

Half-Life

Big Ideas | **Details** | Page: 202
Unit: Nuclear Chemistry

Finding the Time that has Passed (integer number of half-lives)

If the amount you started with divided by the amount left is an exact power of two, you have an integer number of half-lives and you can make a table.

Sample problem:

Q: If you started with 64 g of ^{131}I, how long would it take until there was only 4 g remaining? The half-life (τ) of ^{131}I is 8.07 days.

A: $\dfrac{64}{4} = 16$ which is a power of 2, so we can simply make a table:

# half lives	0	1	2	3	4
amount remaining	64 g	32 g	16 g	8 g	4 g

From the table, after 4 half-lives, we have 4 g remaining.

The half-life (τ) of ^{131}I is 8.07 days.

$$8.07 \times 4 = 32.3 \text{ days}$$

Finding the amount remaining and time that has passed for a non-integer number of half-lives requires logarithms, and is beyond the scope of this course.

Homework Problems

For these problems, you will need to use half-life information from "Table U. Selected Radioisotopes" on page 554 of your Chemistry Reference Tables.

1. If a lab had 128 g of ^{3}H waste 49 years ago, how much of it would be left today? (*Note: you may round off to a whole number of half-lives.*)

    Answer: 8 g

Use this space for summary and/or additional notes:

Half-Life

Unit: Nuclear Chemistry

2. Suppose a student stole a 20. g sample of ^{42}K at 8:30am on Friday. When the student was called down to the vice principal's office on Monday at the convenient time of 10:54am, how much of the ^{42}K was left?

   Answer: 0.31 g

3. If a school wants to dispose of small amounts of radioactive waste, they can store the materials for ten half-lives, and then dispose of the materials as regular trash.

   a. If we had a sample of ^{32}P, how long would we need to store it before disposing of it?

      Answer: 143 days

   b. If we had started with 64 g of ^{32}P, how much ^{32}P would be left after ten half-lives? Approximately what fraction of the original amount would be left?

      Answer: 0.063 g; approximately $\frac{1}{1000}$ of the original amount.

4. If the carbon in a sample of human bone contained only one-fourth of the expected amount of ^{14}C, how old is the sample?

   (Hint: pretend you started with 1 g of ^{14}C and you have 0.25 g remaining.)

   Answer: 11 460 years

Use this space for summary and/or additional notes:

Big Ideas	Details	Nuclear Fission & Fusion	Page: 204
			Unit: Nuclear Chemistry

Nuclear Fission & Fusion

Unit: Nuclear Chemistry

MA Curriculum Frameworks (2016): N/A (HS-PS1-8 in physics frameworks)

MA Curriculum Frameworks (2006): 2.7

Mastery Objective(s): (Students will be able to...)
- Identify nuclear processes as "fission" or "fusion".
- Describe the basic construction and operation of fission-based and fusion-based nuclear reactors.

Success Criteria:
- Descriptions account for how the energy is produced and how the radiation is contained.

Tier 2 Vocabulary: fusion, nuclear

Language Objectives:
- Explain how fission-based and fusion-based nuclear reactors work.

Notes:

Fission

<u>fission</u>: splitting of the nucleus of an atom, usually by bombarding it with a high-speed neutron.

When atoms are split by bombardment with neutrons, they can divide in hundreds of ways. For example, when ^{235}U is hit by a neutron, it can split more than 200 ways. Three examples that have been observed are:

$$^{1}_{0}n + ^{235}_{92}U \rightarrow ^{90}_{37}Rb + ^{144}_{55}Cs + 2\,^{1}_{0}n$$

$$^{1}_{0}n + ^{235}_{92}U \rightarrow ^{87}_{35}Br + ^{146}_{57}La + 3\,^{1}_{0}n$$

$$^{1}_{0}n + ^{235}_{92}U \rightarrow ^{72}_{30}Br + ^{160}_{62}Sm + 4\,^{1}_{0}n$$

Note that each of these bombardments produces more neutrons. A reaction that produces more fuel (in this case, neutrons) than it consumes will accelerate. This self-propagation is called a <u>chain reaction</u>.

Note also that the neutron/proton ratio of ^{235}U is about 1.5. The stable neutron/proton ratio of each of the products would be approximately 1.2. This means that almost all of the products of fission reactions have too many neutrons to be stable, which means they will themselves undergo β− decay.

Use this space for summary and/or additional notes:

Nuclear Fission & Fusion

Nuclear Fission Reactors

In a nuclear reactor, the heat from a fission reaction is used to heat water. The radioactive hot water from the reactor (under pressure, so it can be heated well above 100 °C without boiling) is used to boil clean (non-radioactive) water. The clean steam is used to turn a turbine, which generates electricity.

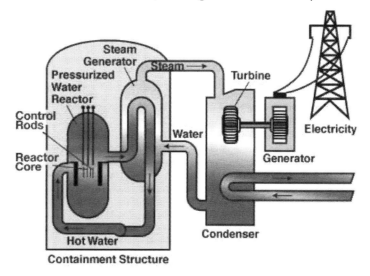

The inside of the reactor looks like this:

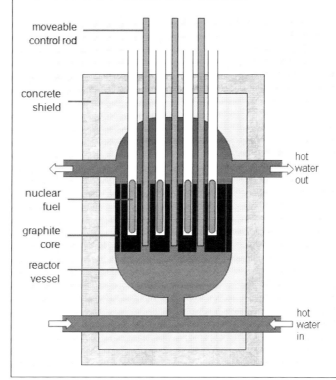

The fuel is the radioactive material (such as ^{235}U) that is undergoing fission. The graphite in the core of the reactor is used to absorb some of the neutrons. The moveable control rods are adjusted so they can absorb some or all of the remaining neutrons as desired. If the control rods are all the way down, all of the neutrons are absorbed and no heating occurs. When the reactor is in operation, the control rods are raised just enough to make the reaction proceed at the desired rate.

Use this space for summary and/or additional notes:

Fusion

<u>fusion</u>: the joining together of the nuclei of two atoms, accomplished by colliding them at high speeds.

Nuclear fusion reactions occur naturally on stars (such as the sun), and are the source of the heat and energy that stars produce.

On the sun, fusion occurs between atoms of deuterium (^{2}H) to produce helium:

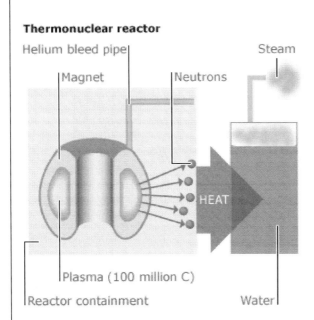

$$^2_1H + {^2_1}H \rightarrow {^4_2}He$$

The major challenge in building nuclear fusion reactors is the high temperatures produced—on the order of 10^6–10^9 °C. In a tokamak fusion reactor, the starting materials are heated until they become plasma—a sea of highly charged ions and electrons. The highly charged plasma is kept away from the sides by powerful electromagnets.

At the left is a schematic of the ITER tokamak reactor currently under construction in southern France.

MIT has a smaller tokamak reactor at its Plasma Science & Fusion Center. The MIT reactor is able to conduct fusion reactions lasting for only a few seconds; if the reaction continued beyond this point, the current in the electromagnets that is necessary to generate the high magnetic fields required to confine the reaction would become hot enough to melt the copper wire and fuse the coils of the electromagnet together.

After each "burst" (short fusion reaction), the electromagnets in the MIT reactor need to be cooled in a liquid nitrogen bath (−196 °C) for fifteen minutes before the reactor is ready for the next burst.

Use this space for summary and/or additional notes:

Practical Uses for Nuclear Radiation

Unit: Nuclear Chemistry

MA Curriculum Frameworks (2016): N/A (HS-PS1-8 in physics frameworks)

MA Curriculum Frameworks (2006): 2.6

Mastery Objective(s): (Students will be able to...)

- Identify & describe practical (peaceful) uses for nuclear radiation.

Success Criteria:

- Descriptions give examples and explain how radiation is essential to the particular use.

Tier 2 Vocabulary: radiation

Language Objectives:

- Explain how radiation makes certain scientific procedures possible.

Notes:

While most people think of the dangers and destructive power of nuclear radiation, there are a lot of other uses of radioactive materials:

Power Plants: nuclear reactors can generate electricity in a manner that does not produce CO_2 and other greenhouse gases.

Cancer Therapy: nuclear radiation can be focused in order to kill cancer cells in patients with certain forms of cancer. Radioprotective drugs are now available that can help shield non-cancerous cells from the high-energy gamma rays.

Radioactive Tracers: chemicals made with radioactive isotopes can be easily detected in complex mixtures or even in humans. This enables doctors to give a patient a chemical with a small amount of radioactive material and track the progress of the material through the body and determine where it ends up. It also enables biologists to grow bacteria with radioactive isotopes and follow where those isotopes end up in subsequent experiments.

Use this space for summary and/or additional notes:

Practical Uses for Nuclear Radiation

Irradiation of Food: food can be exposed to high-energy gamma rays in order to kill germs. These gamma rays kill all of the bacteria in the food, but do not make the food itself radioactive. (Gamma rays cannot build up inside a substance.) This provides a way to create food that will not spoil for months on a shelf in a store. There is a lot of irrational fear of irradiated food in the United States, but irradiation is commonly used in Europe. For example, irradiated milk will keep for months on a shelf at room temperature without spoiling.

Carbon Dating: Because ^{14}C is a long-lived isotope (with a half-life of 5 700 years), the amount of ^{14}C in archeological samples can give an accurate estimate of their age. One famous use of carbon dating was its use to prove that the Shroud of Turin (the supposed burial shroud of Jesus Christ) was actually made between 1260 C.E. and 1390 C.E.

Smoke Detectors: In a smoke detector, ^{241}Am emits positively-charged alpha particles, which are directed towards a metal plate. This steady flow of positive charges completes an electrical circuit. If there is a fire, smoke particles neutralize positive charges. This makes the flow of charges through the electrical circuit stop, which is used to trigger the alarm.

Use this space for summary and/or additional notes:

Summary: Nuclear Chemistry

Unit: Nuclear Chemistry

List the main ideas of this chapter in phrase form:

Write an introductory sentence that categorizes these main ideas.

Turn the main ideas into sentences, using your own words. You may combine multiple main ideas into one sentence.

Add transition words to make your writing clearer and rewrite your summary below.

Use this space for summary and/or additional notes:

Introduction: Electronic Structure

Unit: Electronic Structure

Topics covered in this chapter:

The Electron	214
The Bohr Model of the Hydrogen Atom	216
The Quantum-Mechanical Model of the Atom	220
Waves	223
Electron Energy Transitions	226
Orbitals	228
Electron Configurations	233
Exceptions to the Aufbau Principle	242
Valence Electrons	244

Use this space for summary and/or additional notes:

Introduction: Electronic Structure

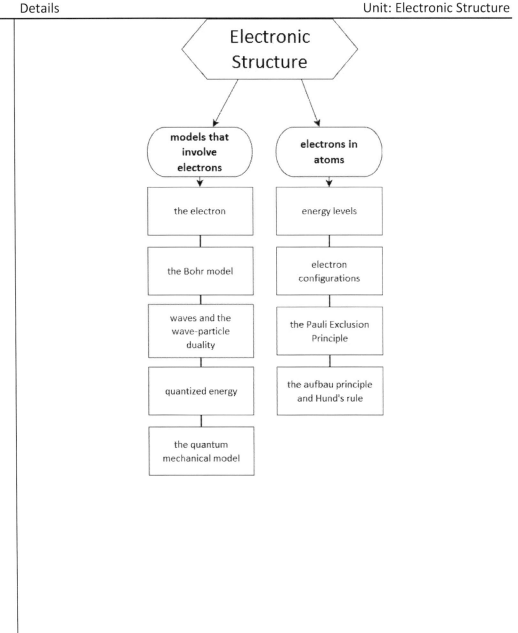

Use this space for summary and/or additional notes:

Introduction: Electronic Structure

Unit: Electronic Structure

Standards addressed in this chapter:

Massachusetts Curriculum Frameworks & Science Practices (2016):

HS-PS1-1 Use the periodic table as a model to predict the relative properties of main group elements, including ionization energy and relative sizes of atoms and ions, based on the patterns of electrons in the outermost energy level of each element. Use the patterns of valence electron configurations, core charge, and Coulomb's law to explain and predict general trends in ionization energies, relative sizes of atoms and ions, and reactivity of pure elements.

Massachusetts Curriculum Frameworks (2006):

2.4 Write the electron configurations for the first twenty elements of the periodic table.

3.3 Relate the position of an element on the periodic table to its electron configuration and compare its reactivity to the reactivity of other elements in the table.

Use this space for summary and/or additional notes:

The Electron

Unit: Electronic Structure
MA Curriculum Frameworks (2016): HS-PS1-1
MA Curriculum Frameworks (2006): 2.2
Mastery Objective(s): (Students will be able to...)

- Describe & explain the particle *vs.* wave nature of electrons.

Success Criteria:

- Descriptions successfully communicate accurate information about electrons and their behavior.

Tier 2 Vocabulary: charge
Language Objectives:

- Explain scientific information about electrons.

Notes:

electron: a small subatomic particle found outside the nucleus of an atom.

mass $= 9.11 \times 10^{-31}$ kg $= 9.11 \times 10^{-28}$ g
       $= {}^{1}/_{1836}$ of the mass of a proton

charge $= -1.6022 \times 10^{-19}$ coulomb
       $= -1$ elementary charge

radius $= 2.8179 \times 10^{-15}$ m

electric current (electricity): electrons moving from one place to another.

Protons and neutrons remain in the nucleus of their atom (except for nuclear decay), but electrons can be removed from one atom and added to another.

ion: an atom (or group of atoms that functions like a single atom) that has an electric charge because it has either gained or lost electrons.

Use this space for summary and/or additional notes:

The Electron

Unit: Electronic Structure

Because an electron has mass (though it's very small—about $1/1836$ of the mass of a proton or neutron), this means electrons are particles, and all of the equations that apply to motion of solid particles also apply to electrons.

However, an electromagnetic wave is a wave of electricity, and electricity is made of electrons that are moving. This means that moving electrons are also waves—they move through empty space, carrying energy with them. Therefore, all of the equations that apply to waves also apply to electrons.

> This means that an electron must be *both* a wave *and* a particle at the same time.

Use this space for summary and/or additional notes:

The Bohr Model of the Hydrogen Atom

Unit: Electronic Structure

MA Curriculum Frameworks (2016): HS-PS1-1

MA Curriculum Frameworks (2006): 2.2

Mastery Objective(s): (Students will be able to...)

- Describe developments that led to the Bohr model of the atom.
- Describe & explain the Bohr model of the atom.
- Explain how the quantum mechanical model of the atom grew out of the Bohr model.

Success Criteria:

- Descriptions successfully communicate developments prior to the Bohr model that were incorporated into the model.
- Descriptions successfully communicate accurate information about the Bohr model and how it describes the behavior of atoms.

Tier 2 Vocabulary: model

Language Objectives:

- Explain scientific information about the Bohr mechanical model of the atom.

Notes:

Significant Developments Prior to 1913

Atomic Theory

Significant developments in atomic theory are described in the "History of Atomic Theory" section, which begins on page 166. The most significant advances were the discovery of the electron and the planetary model of the atom.

Early Quantum Theory

<u>"Old" Quantum Theory (ca. 1900)</u>: sub-atomic particles obey the laws of classical mechanics, but that only certain "allowed" states are possible.

Use this space for summary and/or additional notes:

The Bohr Model of the Hydrogen Atom

Unit: Electronic Structure

Spectroscopy

<u>Balmer Formula</u> (1885): Swiss mathematician and physicist Johann Balmer devised an empirical equation to relate the emission lines in the visible spectrum for the hydrogen atom.

<u>Rydberg Formula</u> (1888): Swedish physicist Johannes Rydberg developed a generalized formula that could describe the wave numbers of all of the spectral lines in hydrogen (and similar elements).

There are several series of spectral lines for hydrogen, each of which converge at different wavelengths. Rydberg described the Balmer series in terms of a pair of integers (n_1 and n_2, where $n_1 < n_2$), and devised a single formula with a single constant (now called the Rydberg constant) that relates them.

$$\frac{1}{\lambda_{vac}} = R_H \left(\frac{1}{n_1^2} - \frac{1}{n_2^2} \right)$$

The value of Rydberg's constant is $\dfrac{m_e e^4}{8\varepsilon_o^2 h^3 c} = 10\,973\,731.6 \text{ m}^{-1} \approx 1.1 \times 10^7 \text{m}^{-1}$

where m_e is the rest mass of the electron, e is the elementary charge, ε_o is the electrical permittivity of free space, h is Planck's constant, and c is the speed of light in a vacuum.

Use this space for summary and/or additional notes:

The Bohr Model of the Hydrogen Atom

Rydberg's equation was later found to be consistent with other series discovered later, including the Lyman series (in the ultraviolet region; first discovered in 1906) and the Paschen series (in the infrared region; first discovered in 1908).

Those series and their converging wavelengths are:

Series	Wavelength	n_1	n_2
Lyman	91 nm	1	$2 \rightarrow \infty$
Balmer	365 nm	2	$3 \rightarrow \infty$
Pasch7en	820 nm	3	$4 \rightarrow \infty$

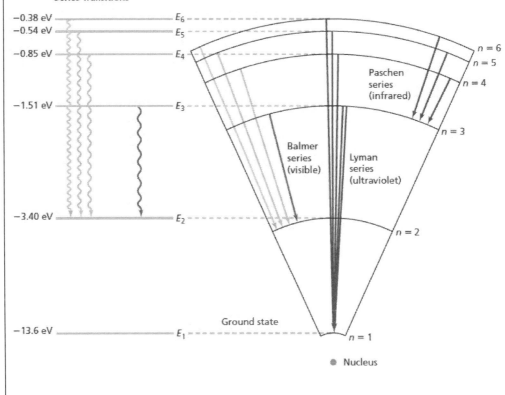

Use this space for summary and/or additional notes:

The Bohr Model of the Hydrogen Atom

Bohr's Model of the Atom (1913)

In 1913, Danish physicist Niels Bohr combined atomic, spectroscopy, and quantum theories into a single theory. Bohr hypothesized that electrons moved around the nucleus as in Rutherford's model, but that these electrons had only certain allowed quantum values of energy, which could be described by a quantum number (n). The value of that quantum number was the same n as in Rydberg's equation, and that using quantum numbers in Rydberg's equation could predict the wavelengths of light emitted when the electrons gained or lost energy by moved from one quantum level to another.

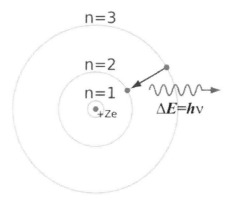

Bohr's model gained wide acceptance, because it related several prominent theories of the time. The theory worked well for hydrogen, giving a theoretical basis for Rydberg's equation. Bohr defined the energy released when an electron descended to an energy level using an integer quantum number (n) and Rydberg's constant:

$$E_n = -\frac{R_H}{n^2}$$

Bohr received the Nobel Prize in physics in 1922 for his contributions to quantum and atomic theory.

Although the Bohr model worked well for hydrogen, the equations could not be solved exactly for atoms with more than one electron, because of the additional effects that electrons exert on each other (e.g., via the Coulomb force, $F_c = \frac{kq_1q_2}{r^2}$).

Use this space for summary and/or additional notes:

The Quantum-Mechanical Model of the Atom

Unit: Electronic Structure
MA Curriculum Frameworks (2016): HS-PS1-1
MA Curriculum Frameworks (2006): 2.2
Mastery Objective(s): (Students will be able to...)
- Describe & explain the quantum mechanical model of the atom.

Success Criteria:
- Descriptions successfully communicate accurate information about the quantum mechanical model of the atom and how it describes the behavior of atoms.

Tier 2 Vocabulary: mechanical

Language Objectives:
- Explain scientific information about the quantum mechanical model of the atom.

Notes:

quantum: a discrete increment (plural: *quanta*) If a quantity is *quantized*, it means that only certain values for that quantity are possible.

Because an electron behaves like a wave, it can only absorb energy in quanta that correspond to exact multiples of its wavelength.

Neils Bohr was the first to realize that because atomic spectral emissions are quantized, electron energy levels must also be quantized. (See the "Historical Development of Atomic Theory" section on page 167 for more details.)

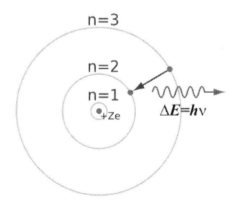

Recall that in the Bohr model of the atom, an electron's (quantum) energy level determined its distance from the nucleus.

What actually happens is not that simple. If an electron is in a particular energy level, the Bohr model may predict its *average* distance from the nucleus, but the electron is also a particle, so it has some freedom to move closer or farther.

Use this space for summary and/or additional notes:

The Quantum-Mechanical Model of the Atom

Unit: Electronic Structure

In 1925, Austrian physicist Erwin Schrödinger found that by treating each electron as a unique wave function, the energies of the electrons could be predicted by the mathematical solutions to a wave equation. The use of Schrödinger's wave equation to construct a probability map for where the electrons can be found in an atom is the basis for the modern quantum-mechanical model of the atom.

To understand the probability map, it is important to realize that because the electron acts as a wave, it is detectable when the amplitude of the wave is nonzero, but not detectable when the amplitude is zero. This makes it appear as if the electron is teleporting from place to place around the atom. If you were somehow able to take a time-lapse picture of an electron as it moves around the nucleus, the picture might look something like the diagram to the right, where each dot is the location of the electron at an instant in time.

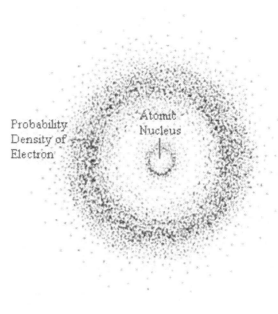

Notice that there is a region close to the nucleus where the electron is unlikely to be found, and a ring a little farther out where there is a much higher probability of finding the electron. As you get farther and farther from the nucleus, Schrödinger's equation predicts different shapes for these probability distributions. These regions of high probability are called "orbitals," because of their relation to the orbits originally suggested by the planetary model.

The quantum mechanical model of the atom is based on this combination of quantized energy levels and probabilities.

Use this space for summary and/or additional notes:

The Quantum-Mechanical Model of the Atom

Unit: Electronic Structure

The claims of the quantum mechanical model of the atom are:

- The electrons orbiting an atom behave like waves as well as particles, and their behavior can be described by Schrödinger's wave equation, which has integer solutions called quantum numbers.

- The energies and therefore locations of electrons within an atom cannot be determined exactly, but there are regions with a high probability of finding an electron (called "orbitals"), and other regions with a low probability of finding an electron.

- The energy of each electron (and therefore its probable location) can be described by a unique set of quantum numbers. No two electrons can have the same energies, which are described by the electron's unique set of quantum numbers, which means no two electrons can have the exact same set of quantum numbers. This is called the *Pauli exclusion principle*, named after the Swiss-American physicist Wolfgang Pauli.

- Electrons move within their orbitals at speeds near the speed of light.

- An electron is constrained to stay within its orbital because of its energy. If the electron absorbs energy, it can move to a higher-energy orbital. If an unoccupied lower-energy orbital is available, the electron can release energy (in the form of a photon, which can be observed as light) and move to the lower-energy orbital.

- The shape of any given orbital (the region where there is high probability of finding an electron) depends on all of the forces that affect that electron. Some of these forces relate to the energy characteristics of the specific orbital, but other forces can include electrostatic repulsion of other electrons within the atom, electrostatic repulsion of the electrons in ionically-bonded atoms, or the sharing of electrons in a covalently-bonded atom. In real atoms, the shapes of orbitals are continuously changing as all of the electrons within the atom repel one another as they move at near-light speed.

Use this space for summary and/or additional notes:

Big Ideas	Details	Waves	Page: 223 Unit: Electronic Structure

Waves

Unit: Electronic Structure

MA Curriculum Frameworks (2016): HS-PS1-1

MA Curriculum Frameworks (2006): N/A

Mastery Objective(s): (Students will be able to…)

- Explain what waves are and how they propagate.
- Describe the relative energies of different waves based on their frequencies and positions within the electromagnetic spectrum.
- Calculate the frequency and wavelength of electromagnetic waves.

Success Criteria:

- Descriptions are accurate and backed up by evidence.
- Calculations are correct.
- Algebra and rounding to appropriate number of significant figures is correct.

Tier 2 Vocabulary: wave, spectrum

Language Objectives:

- Explain scientific information about waves and the electromagnetic spectrum.

Notes:

wave: an energy disturbance that travels from one place to another.

medium: the substance that a wave travels through. Electromagnetic waves (including light) can travel without a medium.

- The wave travels through the medium.
- All (or nearly all) of the energy passes through the medium—the medium doesn't absorb it.

Some examples of waves:

Type of Wave	Medium
sound	air (or water, solids)
ocean	water
electromagnetic (*e.g.*, light, radio)	none

Use this space for summary and/or additional notes:

Waves

Unit: Electronic Structure

Big Ideas	Details

wavelength (λ): the length of the wave, measured from a specific point in the wave to the same point in the next wave. unit = distance (m, cm, nm, *etc.*)

frequency: (f or v) the number of waves that travel past a point in a given time. Symbol = f; unit = $1/\text{time}$ = Hz

$$\text{speed} = \lambda f$$

Electromagnetic waves (such as light, radio waves, etc.) travel at a constant speed—the speed of light. The speed of light is a constant, and is denoted by the letter "c" in equations.

$$c = 3.00 \times 10^8 \,\text{m}/\text{s} = 186{,}000 \text{ miles per second}$$

The energy (E) that a wave carries equals a constant times the frequency. (Think of it as the number of bursts of energy that travel through the wave every second.) For electromagnetic waves (including light), the constant is Planck's constant (named after the physicist Max Planck), which is denoted by a script h in equations. So the equation is:

$$E = hf$$

where $h = 6.63 \times 10^{-34}$ J·s = Planck's constant

Louis de Broglie: French physicist. Showed that any object with momentum (*i.e.,* has mass and is moving) creates a wave as it moves.

Large objects with a lot of momentum (such as people) create waves with wavelengths that are far too small to detect.

Small objects (such as electrons) create waves with wavelengths in the visible part of the spectrum. This is why we can see the light produced by electrons as they move.

Use this space for summary and/or additional notes:

spectrum: the set of all possible wavelengths. Visible light is one set of wavelengths that are part of the full electromagnetic spectrum.

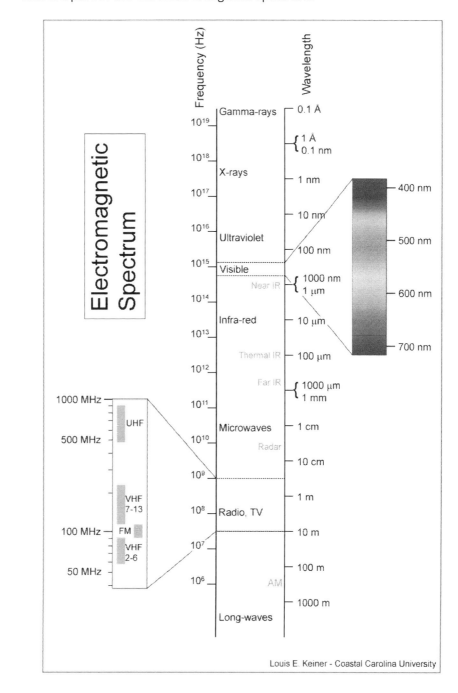

Use this space for summary and/or additional notes:

Electron Energy Transitions

Unit: Electronic Structure

MA Curriculum Frameworks (2016): HS-PS1-1

MA Curriculum Frameworks (2006): N/A

Mastery Objective(s): (Students will be able to...)

- Explain the meaning of the lines in emission spectra.

Success Criteria:

- Descriptions include colors, approximate wavelengths, and relationship between number of lines and number of electrons.

Tier 2 Vocabulary: emission, spectrum, ground, excited

Language Objectives:

- Explain scientific information about energy transmissions and line spectra.

Notes:

quantum: a discrete quantity of energy that cannot be divided.

ground state: the lowest available energy level for an electron.

excited state: a higher energy level than the ground state.

emission spectrum: the wavelengths (colors) of light emitted by an element when its electrons are "excited" (raised to a higher energy state) and then allowed to return to the ground state.

spectroscope: a device that separates colors of light based on their wavelengths.

Use this space for summary and/or additional notes:

Electron Energy Transitions

If you were to look at a glass tube filled with hydrogen gas that was energized with electricity, the gas in the tube would appear to be blue, because the electrons are energized, and the energy of the light they emit as they return to the $n = 2$ quantum energy level corresponds with a blue color.

If you were to split the light emitted by hydrogen into its component colors using a spectroscope, you would see the following:

Atoms with more electrons have a larger number of possible transitions, each with different energies. This results in more lines in their emission spectrum, as with iron:

photon: a single "piece" (particle/wave) of light.

luminescence: light that is not generated by high temperatures alone. (In fact, it usually occurs at low temperatures. Causes include electrical energy and chemical reactions.)

fluorescence: a type of luminescence that occurs when electrons of an element are excited and return immediately to the ground state, giving off a photon. The wavelength of the photon given off is usually different from the wavelength of the photon used to raise the electrons to the excited state.

phosphorescence: "glow-in-the-dark" luminescence—a type of luminescence that occurs when electrons are excited, but cannot return directly to the ground state. The indirect path is slower, which cause the material to "glow" for a longer period of time (in some cases, hours).

triboluminescence: a form of luminescence in which light is generated by breaking asymmetrical bonds in a crystal. In the case of wintergreen Life Savers, when the sugar crystals are crushed, the positive and negative charges get separated. The voltage between them causes a spark. Normally, these sparks are in the ultraviolet part of the spectrum and cannot be seen; however, wintergreen oil (methyl salicylate) is fluorescent. It absorbs the photons of ultraviolet light and emits photons of blue light, which we can see.

Use this space for summary and/or additional notes:

Orbitals

Unit: Electronic Structure

MA Curriculum Frameworks (2016): HS-PS1-1

MA Curriculum Frameworks (2006): 2.4, 3.3

Mastery Objective(s): (Students will be able to...)
- Explain the energy hierarchy of quantum levels, sub-levels, and orbitals.
- Explain how the hierarchy of quantum levels, sub-levels and orbitals corresponds with positions on the periodic table of the elements.

Success Criteria:
- Descriptions relate principal quantum number to period and sub-level to region of the periodic table.

Tier 2 Vocabulary: level, sub-level

Language Objectives:
- Explain the energy hierarchy of quantum levels, sub-levels and orbitals.

Notes:

orbital: a region in an atom (outside the nucleus) with a high probability of finding an electron.

These regions are called *orbitals* because these regions are what replaced the spherical orbits in the much simpler planetary model.

Note that orbitals are not physical objects with boundaries. They are simply the space that an electron moves around in because of its energy and the external forces on it. When an electron approaches the "boundary" of its orbital, the forces pulling it back are strong enough to overcome the energy that the electron has, and it cannot get farther away.

The locations and geometric shapes of these orbitals are the solutions to the wave equation, a complex mathematical formula that would require mathematics far too advanced for a high school class. Instead, we will categorize orbitals using a hierarchy that is divided according to energy levels.

Use this space for summary and/or additional notes:

Orbitals

Unit: Electronic Structure

Big Ideas	Details

Energy Level Hierarchy

Electrons have energy levels, which roughly correspond to the values of *n* in the Rutherford-Bohr model. Each of those levels has one or more kinds of sublevels. Each sublevel has one or more orbitals, and each of those orbitals can hold up to two electrons. It's easiest to think of the hierarchy as an outline:

1. energy level (1, 2, 3, …)
    a. sub-level (s, p, d, f, …)
        i. orbital
            a. individual electron

Energy Level

The main or principal level is a measure of total distance from the nucleus. The levels are numbered 1-7. The periodic table of the elements is arranged so that the Period (row number) that an element is in equals the energy level of the electron that has the most energy.

For example, helium is in period (row) #1. That means its highest energy level is 1, which means both of its electrons have to be in level 1.

Sulfur is in period (row) #3, which means it has electrons in levels 1, 2, and 3.

Use this space for summary and/or additional notes:

Orbitals

Page: 230
Unit: Electronic Structure

Sub-Levels & Orbitals

There are four types of sub-levels. Each type has a specific number of orbitals with specific shapes. Each of those orbitals can hold up to 2 electrons.

There are four kinds of sub-levels: **s, p, d,** and **f**[*]. The shapes of their orbitals are:

type of sub-level	shape(s) of orbital(s)	total # of orbitals	total # of electrons
s		1	2
p		3	6
d		5	10
f		7	14

Notice that each sub-level has an odd number of orbitals.

---

[*] The letters come from words that described the characteristics of the atomic spectra. **s** stood for "sharp," **p** for "principal," **d** for "diffuse," and **f** for "fundamental."

Use this space for summary and/or additional notes:

Orbitals

Unit: Electronic Structure

Arrangement of Levels and Sublevels

Each energy level in an atom has at least one of these kinds of sub-levels—each level contains an *s* sub-level, each level starting with level 2 contains *s* and *p* sub-levels, each level starting with level 3 contains s, p, and d sub-levels, *etc.*

Level	1	2	3	4	5	6	7
# of sub-levels	1	2	3	4	5	6	7
types of sub-levels	s	s, p	s, p, d	s, p, d, f	s, p, d, f, (g)	s, p, d, f, (g), (h)	s, p, d, f, (g), (h), (i)

Note that *g*, *h*, and *i* are in parentheses because no atom is large enough that it actually has any electrons in *g*, *h*, or *i* sublevels, but mathematically we know that those sub-levels will exist if we "discover" (create) a large enough atom.

Note also that these sub-levels overlap. For example, the levels and sublevels in a sodium (Na) atom might look like the following:

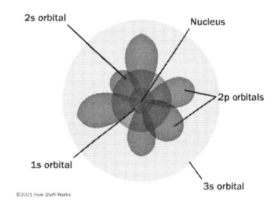

Use this space for summary and/or additional notes:

Orbitals

Page: 232
Unit: Electronic Structure

Sub-levels and the Periodic Table

Note that the sub-level of the electron that has the highest energy corresponds with the location of the element on the periodic table:

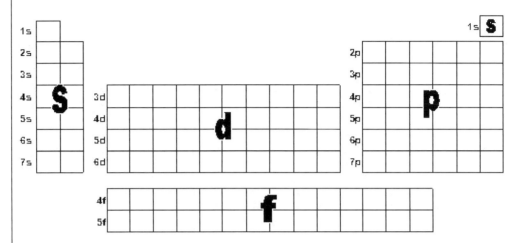

Notice, for example, that the "**s** section" of the periodic table is two columns wide. This is because each **s** sub-level has one orbital that can hold two electrons.

type of sub-level	# orbitals	× 2 =	# electrons	# columns on P.T.
s	1		2	2
p	3		6	6
d	5		10	10
f	7		14	14

Use this space for summary and/or additional notes:

Electron Configurations

Unit: Electronic Structure

MA Curriculum Frameworks (2016): HS-PS1-1

MA Curriculum Frameworks (2006): 2.4, 3.3

Mastery Objective(s): (Students will be able to…)

- Write the ground-state electron configuration for any element on the periodic table.

Success Criteria:

- Levels and sublevels are in the correct order.
- Each sublevel has the correct number of electrons.

Tier 2 Vocabulary: level, spin

Language Objectives:

- Explain the parts of an electron configuration.

Notes:

The electron configuration for an element is a list of all of the energy sub-levels that have electrons in them (in the ground state) in order from lowest to highest energy, and the number of electrons they contain.

For example, consider a neutral nitrogen atom with its seven electrons.

- The first two electrons occupy the 1s sublevel—the one with the lowest energy. We denote these two electrons as $1s^2$.
- The next two electrons occupy the 2s sublevel. We denote these two electrons as $2s^2$.
- The last three electrons are in the 2p sublevel. We denote these three electrons as $2p^3$.
- The complete electron configuration for nitrogen is therefore $1s^2\ 2s^2\ 2p^3$.

If this already makes sense, great! The next few pages explain where these numbers come from in more detail.

Use this space for summary and/or additional notes:

Electron Configurations

Big Ideas | **Details** | Page: 234
Unit: Electronic Structure

The following chart shows each orbital from 1s through 4p. Orbitals with the lowest energy are shown at the bottom, and those with the highest energy are at the top.

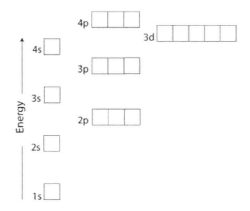

This can be a little confusing, because the levels overlap. For example, notice that the 3d sub-level is higher in energy than the 4s sub-level.

This means that the electrons will fill sub-levels in the following order:

1s, 2s, 2p, 3s, 3p, 4s, 3d, 4p, …

Use this space for summary and/or additional notes:

Electron Configurations

You can use the periodic table as the "road map":

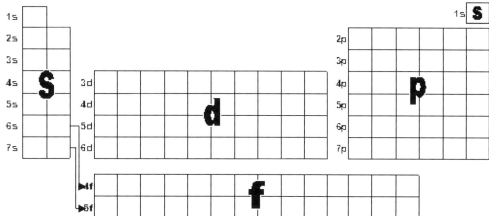

To determine the electron configuration, imagine that you are placing one electron in each element's box until you "use up" all of the electrons. The last box is the element that you are writing the electron configuration for.

The two columns on the left* correspond with the "s" sublevels. The six columns on the right correspond with the "p" sublevels. The ten columns of the transition metals correspond with the "d" sublevels. The fourteen columns below the rest of the table correspond with the "f" sublevels.

As you move through the positions in order, you are moving through the sub-levels from lowest to highest energy.

Remember that the "s" sub-levels start with 1s, the "p" sub-levels start with 2p, the "d" sub-levels start with 3d, and the "f" sub-levels start with 4f. The "gotchas" are:

- The 3d sub-level is in row 4, right after 4s.
- The 4f sub-level is in row 6, right after 6s.

---

* For the purpose of electron configurations, helium should be in the "s" block, next to hydrogen.

Use this space for summary and/or additional notes:

Electron Configurations

Unit: Electronic Structure

Writing Electron Configurations

An element has electrons that correspond with <u>each</u> of the available slots, from the beginning of the periodic table (where hydrogen is located) up to where that element is located.

If we were to represent an electron as an arrow, we could represent two electrons in a 1s sub-level like this: $\frac{\uparrow\downarrow}{1s}$. The 1s sub-level has one orbital, which is represented by the one blank. The two electrons are represented as arrows. Because two electrons sharing an orbital have opposite spins, we represent them with one arrow pointing up and the other arrow pointing down.

We could represent five electrons in a 2p orbital like this: $\frac{\uparrow\downarrow\;\uparrow\downarrow\;\uparrow}{2p}$. The 2p sub-level has 3 orbitals, represented by the 3 blanks. Two of those orbitals have two electrons in them, and the third one has only one electron.

We could represent all 13 of the electrons in aluminum like this:

$$\frac{\uparrow\downarrow}{1s} \quad \frac{\uparrow\downarrow}{2s} \quad \frac{\uparrow\downarrow\;\uparrow\downarrow\;\uparrow\downarrow}{2p} \quad \frac{\uparrow\downarrow}{3s} \quad \frac{\uparrow\;\_\;\_}{3p}$$

This diagram shows the <u>electron configuration</u> of aluminum.

<u>electron configuration</u>: a description of which levels and sub-levels the electrons in an element are occupying.

Notice that we have to show all three of the orbitals (blanks) in the 3p sub-level, even if some of those orbitals don't have any electrons in them.

<u>ground state</u>: when all of the electrons in an atom are in the lowest-energy sublevel that has an available "slot".

<u>Pauli Exclusion Principle</u>: every electron in an atom has a different quantum state from every other electron. In plain English, this means that something has to be different about each electron, whether it's the level, sub-level, which orbital it's in, or its spin.

<u>aufbau principle</u>: in the ground state, each electron in an atom will occupy the lowest available energy state. This means that you start with the lowest sub-level (1s) and work your way up until you've placed all the electrons.

Use this space for summary and/or additional notes:

Electron Configurations

Unit: Electronic Structure

Hund's Rule: electrons don't pair up in orbital until they have to. (This is kind of like siblings not wanting to share a room if there's an empty room available.) For example, the electron configuration for nitrogen would be:

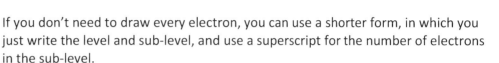

Wrong: ↑↓ ↑↓ ↑↓ ↑ __ X
         1s 2s   2p

Right: ↑↓ ↑↓ ↑ ↑ ↑ ☺
        1s 2s   2p

If you don't need to draw every electron, you can use a shorter form, in which you just write the level and sub-level, and use a superscript for the number of electrons in the sub-level.

For example, $\frac{↑↓}{1s}$ would become $1s^2$, and $\frac{↑↓\ ↑↓\ ↑↓}{2p}$ would become $2p^6$.

The electron configuration for aluminum would go from the orbital notation version:

↑↓ ↑↓ ↑↓ ↑↓ ↑↓ ↑↓ ↑ __ __
1s 2s   2p   3s   3p

to the "standard" version:

$$1s^2\ 2s^2\ 2p^6\ 3s^2\ 3p^1$$

Use this space for summary and/or additional notes:

Electron Configurations

Unit: Electronic Structure

The shorter version can still get tediously long for elements with a lot of electrons. For example, the electron configuration for gold (Au) is:

$$1s^2\ 2s^2\ 2p^6\ 3s^2\ 3p^6\ 4s^2\ 3d^{10}\ 4p^6\ 5s^2\ 4d^{10}\ 5p^6\ 6s^2\ 4f^{14}\ 5d^9$$

To shorten this even more, you're allowed to use the element in the *last* column of a row as an abbreviation for all of the electrons through the end of that row.

In our example, gold (Au) is in the 6th row of the periodic table:

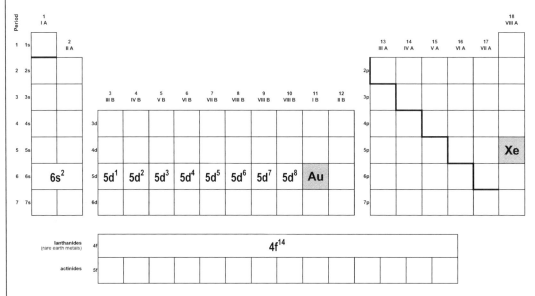

This means we're allowed to start from xenon (Xe) at the end of the previous (5th) row, and add on only the parts that come after Xe. This gives us:

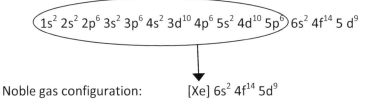

Noble gas configuration: $[Xe]\ 6s^2\ 4f^{14}\ 5d^9$

This notation is called the <u>noble gas configuration</u>, because the elements in the last column (the ones you start from) are called the noble gases.

Use this space for summary and/or additional notes:

Electron Configurations

Homework Problems

Give the electron configuration (orbital notation—with the arrows) for each of the following elements:

1. carbon

2. potassium

3. silicon

4. silver

For each of the following electron configurations, name the element.

5. $\underline{\uparrow\downarrow}\ \underline{\uparrow\downarrow}\ \underline{\uparrow\downarrow}\,\underline{\uparrow\downarrow}\,\underline{\uparrow}$
   1s   2s   2p

6. $\underline{\uparrow\downarrow}\ \underline{\uparrow\downarrow}\ \underline{\uparrow\downarrow}\,\underline{\uparrow\downarrow}\,\underline{\uparrow\downarrow}\ \underline{\uparrow\downarrow}\ \underline{\uparrow\downarrow}\,\underline{\uparrow\downarrow}\,\underline{\uparrow\downarrow}\ \underline{\uparrow\downarrow}\ \underline{\uparrow\downarrow}\,\underline{\uparrow\downarrow}\,\underline{\uparrow\downarrow}\,\underline{\uparrow}\,\underline{\uparrow}$
   1s   2s   2p   3s   3p   4s   3d

Each of the following electron configurations has something wrong with it. For each one:

- State what the mistake is.
- Re-write the electron configuration correctly, keeping the total number of electrons the same.

7. $\underline{\uparrow\downarrow}\ \underline{\uparrow\downarrow}\,\underline{\uparrow}\,\underline{\uparrow}$
   1s   2p

Electron Configurations

8. $\underline{\uparrow\downarrow}$ $\underline{\uparrow\downarrow}$ $\underline{\uparrow\downarrow}\,\underline{\uparrow\downarrow}\,\underline{}$
      1s   2s   2p

9. $\underline{\uparrow\downarrow}$ $\underline{\uparrow\downarrow}$ $\underline{\uparrow\downarrow}\,\underline{\uparrow\downarrow}\,\underline{\uparrow\downarrow}$ $\underline{\uparrow\downarrow}$ $\underline{\uparrow\downarrow}\,\underline{\uparrow\downarrow}\,\underline{}$
      1s   2s   2p   3s   3p

For each of the following elements, give the "standard" electron configuration (*e.g.*, $1s^2\ 2s^2\ 2p^6\ 3s^1$).

10. boron (B)

11. phosphorus (P)

12. vanadium (V)

13. strontium (Sr)

For each of the following electron configurations, give the element.

14. $1s^2\ 2s^2\ 2p^6\ 3s^2$

15. $1s^2\ 2s^2\ 2p^6\ 3s^2\ 3p^6\ 4s^2\ 3d^{10}\ 4p^6\ 5s^2\ 4d^6$

Use this space for summary and/or additional notes:

Electron Configurations

For each of the following elements, give the "noble gas" electron configuration (*e.g.,* [Ar] $4s^2\ 3d^5$).

16. zirconium (Zr)

17. platinum (Pt)

18. dysprosium (Dy)

19. gallium (Ga)

Use this space for summary and/or additional notes:

Big Ideas	Details	Unit: Electronic Structure

Exceptions to the Aufbau Principle

Unit: Electronic Structure

MA Curriculum Frameworks (2016): HS-PS1-1

MA Curriculum Frameworks (2006): 2.4, 3.3

Mastery Objective(s): (Students will be able to...)

- Predict which elements are likely to be exceptions to the aufbau principle.
- Explain why exceptions to the aufbau principle occur as you approach the middle and end of the d and f sub-levels.

Success Criteria:

- Predictions match observed electron configurations.

Language Objectives:

- Explain exceptions to the aufbau principle.

Notes:

Remember from Hund's Rule that electrons like to spread out.

Atoms are the most stable when their electrons are the most evenly distributed within the atom's energy levels and sub-levels. This means that elements with completely filled principal (numbered) energy levels are the most stable.

- The "noble gases" (the last column of the periodic table) already have all of their principal energy levels completely filled with electrons. This makes them very stable, because they do not need to react with other atoms to get their electrons into a more stable configuration. This is why noble gases almost never react with anything.
- Other elements gain, lose, or share electrons (in chemical reactions) in order to end up with electron configurations that are like the nearest noble gas on the periodic table.

Atoms with p, d, and f sub-levels that are exactly half full are more stable than atoms with slightly more or fewer electrons in their p, d, and f sub-levels. This makes those atoms slightly more stable (and therefore less reactive) than other atoms. For example:

- Nitrogen ([He] $2s^2\ 2p^3$), which has an exactly half-filled 2p sub-level, is chemically less reactive than oxygen ([He] $2s^2\ 2p^4$).
- Manganese ([Ar] $4s^2\ 3d^5$), which has an exactly half-full 3d sub-level, is chemically less reactive than iron ([Ar] $4s^2\ 3d^6$).

Use this space for summary and/or additional notes:

Exceptions to the Aufbau Principle

Big Ideas | **Details** | Unit: Electronic Structure

In fact, elements with a d or f sub-level that is one electron away from being half full will usually "borrow" one electron from the nearest s sub-level, because the half-filled d or f sub-level is more stable than the full s sub-level.

- Chromium "borrows" one of its 4s electrons to make its 3d sub-level exactly half full. This means that instead of having predicted electron configuration of [Ar] $4s^2\ 3d^4$, it is observed to have the electron configuration [Ar] $4s^1\ 3d^5$. This happens because a half-filled 4s sub-level plus a half-filled 3d sub-level is more stable than a completely filled 4s sub-level plus a 3d sub-level with 4 electrons in it.

- Copper "borrows" one of its 4s electrons to make its 3d sub-level completely full. This means that instead of having predicted electron configuration of [Ar] $4s^2\ 3d^9$, it is observed to have the electron configuration [Ar] $4s^1\ 3d^{10}$. Again, this happens because a half-filled 4s sub-level plus a completely filled 3d sub-level is more stable than a completely filled 4s sub-level plus a 3d sub-level with 9 electrons in it.

There are a significant number of other exceptions to the aufbau principle. Clearly, atoms do not care about the periodic table!

Use this space for summary and/or additional notes:

Valence Electrons

Unit: Electronic Structure

MA Curriculum Frameworks (2016): HS-PS1-1

MA Curriculum Frameworks (2006): 3.3

Mastery Objective(s): (Students will be able to…)

- Determine the number of valence electrons for representative elements.
- Draw Lewis dot diagrams for representative elements.

Success Criteria:

- Elements are drawn with the correct number of valence electrons.
- Dots representing electrons are spread around the element symbol in an appropriate fashion.

Language Objectives:

- Explain what valence electrons are and how to determine how many an element has.

Notes:

<u>valence electrons</u>: the outer electrons of an atom that are available to participate in chemical reactions.

In most atoms, these are the electrons in the s and p sub-levels of the highest (numbered) energy level.

For example, phosphorus (P) has the electron configuration: $1s^2\ 2s^2\ 2p^6\ 3s^2\ 3p^3$, or [Ne] $3s^2\ 3p^3$. The highest energy level is level 3.

The $3s^2\ 3p^3$ at the end of its electron configuration tells us that phosphorus has 2 electrons in the 3s sub-level plus 3 in the 3p sub-level, for a total of 5 electrons in level 3. This means that phosphorus has 5 valence electrons.

Note that only electrons in s and p sub-levels can be valence electrons. For example, arsenic (As) has the electron configuration [Ar] $4s^2\ 3d^{10}\ 4p^3$. The highest energy level is 4, so only the electrons in level 4 count. Arsenic has 2 electrons in the 4s sub-level, and 3 electrons in the 4p sub-level, for a total of 5 valence electrons. The 10 electrons in the 3d sub-level are *not* in the highest level, so they don't count.

Use this space for summary and/or additional notes:

Valence Electrons

Big Ideas | **Details**

Recall that full sub-levels give an atom extra stability. This means noble gases (the elements in the last column of the periodic table) are the most stable elements because all of their sub-levels are filled. This is why noble gases almost never react with other elements.

Because noble gases have all sub-levels filled, this means they have "full" valence shells. Helium has 2 valence electrons (because it has only a 1s sub-level), and all other noble gases have 8 valence electrons (because their highest-numbered s sub-level is full with 2 electrons, and their highest-numbered p sub-level is full with 6 electrons, for a total of 8.)

For other elements, <u>the atoms can become much more stable if they can form ions with filled valence shells</u>, which would give the ion the same electron configuration as a noble gas.

For example, phosphorus ([Ne] $3s^2\ 3p^3$) has 5 valence electrons. It could have a full valent shell by gaining 3 more electrons to fill its 3p sub-level (which would give it the same electron configuration as argon), or by losing 5 electrons (which would give it the same electron configuration as neon). Because it is easier to gain 3 electrons than to lose 5, phosphorus is most likely to gain 3 electrons, which means it's most likely to form an ion with a −3 charge.

Potassium ([Ar] $4s^1$) has only one valence electron. Potassium could either lose 1 electron (which would give it the same electron configuration as argon), or gain 7 electrons (which would give it the same electron configuration as krypton). Because it is easier to lose 1 electron than to gain 7, potassium is most likely to lose 1 electron, which means it's most likely to form an ion with a +1 charge.

Transition Metals

Because the energy of an s sub-level is so close to the energy of the d sub-level of the next lower energy level, transition metals can easily shift electrons between these s and d sub-levels. This means they can have different numbers of valence electrons, depending on the situation. For example, copper can have the electron configuration [Ar] $4s^2\ 3d^9$, or [Ar] $4s^1\ 3d^{10}$, meaning that copper can have either one or two valence electrons. This explains why copper is observed to sometimes form a +1 ion, and other times a +2 ion.

Use this space for summary and/or additional notes:

Valence Electrons

Unit: Electronic Structure

Group Numbers

You can read the number of valence electrons that an element has directly from the periodic table, using the group numbers. For the "representative elements" (s and p block elements), the number of valence electrons is the last digit of the group number. Transition metals generally have two valence electrons, though there are exceptions. (See the section on "Exceptions to the Aufbau Principle" starting on page 242 for an explanation.)

Lewis Dot Diagrams

A Lewis dot diagram is a representation of an element surrounded by its valence electrons. The diagram consists of the element symbol (from the periodic table), with dots on the top, bottom, and sides representing the *s* and *p* sub-levels of its valence shell.

For example, aluminum has 3 valence electrons. The orbital-notation electron configuration for aluminum is:

$$\underline{\uparrow\downarrow}\ \underline{\uparrow\downarrow}\ \underline{\uparrow\downarrow}\,\underline{\uparrow\downarrow}\,\underline{\uparrow\downarrow}\ \underline{\uparrow\downarrow}\ \underline{\uparrow}\ \underline{\ }\ \underline{\ }$$
$$1s\quad 2s\quad\ \ 2p\quad\ \ \ 3s\quad\ \ 3p$$

Its Lewis dot diagram is ·Al:

Notice that it shows three dots representing the 3 valence electrons.

The dots are placed in singles or pairs on the top, bottom, left, and right of the element symbol. The convention is to place the first two valence electrons (the ones in the s sub-level) to the right of the element symbol, and the remaining valence electrons (the ones in the p sub-levels) on the top, left, and bottom. Start with one dot on the top, left, and bottom, and then pair them up one at a time. (This corresponds with Hund's Law, which says that electrons in the p sub-level do not pair up until they have to.)

In our example, the Lewis dot diagram for aluminum has two dots on the right representing the two electrons in the 3s sub-level, and one dot on the left for the one electron in the 3p sub-level.

Use this space for summary and/or additional notes:

Valence Electrons

| Big Ideas | Details | Unit: Electronic Structure |

Nitrogen has 5 valence electrons. Its orbital-notation electron configuration is:

$$\underline{\uparrow\downarrow} \quad \underline{\uparrow\downarrow} \quad \underline{\uparrow}\;\underline{\uparrow}\;\underline{\uparrow}$$
$$\;1s \quad\;\; 2s \quad\quad\; 2p$$

Its Lewis dot diagram would be ·N:
(with dots on top and bottom)

Again, notice that there are 2 dots on the right for the 2s sub-level, and one dot each on the top, bottom, and left sides for the one electron in each of the orbitals of the 2p sub-level.

Neon has 8 valence electrons. Its orbital-notation electron configuration is:

$$\underline{\uparrow\downarrow} \quad \underline{\uparrow\downarrow} \quad \underline{\uparrow\downarrow}\;\underline{\uparrow\downarrow}\;\underline{\uparrow\downarrow}$$
$$\;1s \quad\;\; 2s \quad\quad\; 2p$$

Its Lewis dot diagram is :Ne:
(with dots on top and bottom)

Use this space for summary and/or additional notes:

Valence Electrons

Homework Problems

Fill in the chart below. Use the first row as an example.

Element	Electron Configuration	Group #	Valence Electrons	Lewis Dot	Nearest Noble Gas	Charge of Ion
N	[He] $2s^2 2p^3$	15	5	·N:	Ne	−3
O						
Na						
P						
Ar						
Al						
Br						
B						
Ca						
C						
Cl						

Use this space for summary and/or additional notes:

Big Ideas	Details
	Summary: Electronic Structure **Unit:** Electronic Structure List the main ideas of this chapter in phrase form: Write an introductory sentence that categorizes these main ideas. Turn the main ideas into sentences, using your own words. You may combine multiple main ideas into one sentence. Add transition words to make your writing clearer and rewrite your summary below.

Use this space for summary and/or additional notes:

Introduction: Periodicity

Unit: Periodicity

Topics covered in this chapter:

Development of the Periodic Table 253
Regions of the Periodic Table 258
Ionization Energy 263
Electronegativity 267
Atomic & Ionic Radius 270

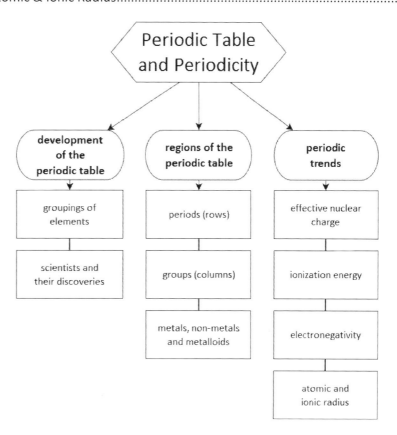

Use this space for summary and/or additional notes:

Introduction: Periodicity

Standards addressed in this chapter:

Massachusetts Curriculum Frameworks & Science Practices (2016):

HS-PS1-1 Use the periodic table as a model to predict the relative properties of main group elements, including ionization energy and relative sizes of atoms and ions, based on the patterns of electrons in the outermost energy level of each element. Use the patterns of valence electron configurations, core charge, and Coulomb's law to explain and predict general trends in ionization energies, relative sizes of atoms and ions, and reactivity of pure elements.

Massachusetts Curriculum Frameworks (2006):

3.1 Explain the relationship of an element's position on the periodic table to its atomic number. Identify families (groups) and periods on the periodic table.

3.2 Use the periodic table to identify the three classes of elements: metals, nonmetals, and metalloids.

3.3 Relate the position of an element on the periodic table to its electron configuration and compare its reactivity to the reactivity of other elements in the table.

3.4 Identify trends on the periodic table (ionization energy, electronegativity, and relative sizes of atoms and ions).

Use this space for summary and/or additional notes:

Development of the Periodic Table

Unit: Periodicity

MA Curriculum Frameworks (2016): HS-PS1-1

MA Curriculum Frameworks (2006): 3.1

Mastery Objective(s): (Students will be able to…)

- Describe how our understanding of groups of elements with similar properties developed into the modern periodic table.
- List scientists who contributed to our understanding of periodicity and their contributions.

Success Criteria:

- Descriptions include specific developments and prior developments that they supplanted.

Language Objectives:

- Describe how the modern periodic table developed over time.

Notes:

The Fifth Element

In ancient times, the world was believed to be made of four elements: earth, air, fire, and water. Every substance on Earth was thought to be made of one of these four elements.

When sulfur was discovered, it could be detected in the presence of each of the other four elements. For this reason, philosophers eventually decided that sulfur must be a fifth element.

Use this space for summary and/or additional notes:

Development of the Periodic Table

Chemical Symbols

By the early 1800s, more than twenty separate elements were known. Swedish chemist Jöns Jacob Berzelius developed a system of notation in which each element was given a one- or two-letter symbol based on its Latin name. This is the system in use today, though newer symbols are based on the elements' names in English rather than Latin. Some of Berzelius's names and symbols included:

Symbol	Latin Name	English Name	Symbol	Latin Name	English Name
Ag	argentum	silver	Na	natrium	sodium
Au	aurum	gold	Pb	plumbum	lead
Cu	cuprum	copper	Sb	stibium	antimony
Fe	ferum	iron	Sn	stannum	tin
Hg	hydrargyrum	mercury	W	wolfram	tungsten
K	kalium	potassium			

Berzelius also developed a system of notation for chemical compounds with the number of atoms of an element denoted as a superscript (*e.g.*, H^2O). Later in the 19th century, the superscript was changed to a subscript, in order to avoid confusing the number of atoms in a molecule with exponents, electrical charges, *etc.*, so H^2O became H_2O.

Early Classifications of the Elements

<u>Johnann Wolfgang Döbereiner</u>: German chemist; classified elements into triads (groups of three) with similar properties (1817).

<u>John Newlands</u>: British chemist; arranged the elements in order of increasing atomic mass. Proposed the *law of octaves* (1829).

<u>law of octaves</u>: when the elements are arranged by increasing atomic mass, every 8th element has similar properties. This worked for the first 20 elements.

Use this space for summary and/or additional notes:

Development of the Periodic Table

Big Ideas | Details | Unit: Periodicity

Valence Numbers

<u>valence number</u>: the "combining power" of an element, meaning the number of atoms that would need to combine with that element in order to "satisfy" it. Initially, the valence number was the largest number of atoms of oxygen that could form a compound with the element. Note that valence numbers were in use in the early 1800s, almost a century before electrons were discovered.

Valence Number	Metals	Valence Number	Metals	Valence Number	Metals
I.	K, Na, Li, Ca, Ba, Sr, Mg	III.	Cu, Pb, Sb, Bi, U, Ti, Ce, Te	V.	Hg, Ag, Au, Pt, Pd, Rh, Os, Ir
II.	Mn, Fe, Zn, Sn, Cd, Co, Ni	IV.	As, Mo, Cr, V, W, Ta	VI.	Be, Zr, Y, Th, Al, Si

We now define the valence number to be the number of electrons that an atom has in its outer (valent) shell.

The Periodic Table

<u>Julius Lothar Meyer</u>: German chemist. Published the first periodic table (1864), with 28 elements arranged in order of increasing atomic mass and grouped according to valence numbers.

<u>Dmitri Mendeleev</u>: Russian chemist, considered the author of the modern periodic table. Mendeleev's table was published in 1869, with elements arranged by increasing atomic mass, and grouped in columns by similar chemical & physical properties (including valence number).

Use this space for summary and/or additional notes:

Development of the Periodic Table

Mendeleev's Periodic Table

(Highlights within vertical columns indicate Döbereiner's triads.)

Period	Group I	II	III	IV	V	VI	VII	VIII
1	H=1							
2	Li=7	Be=9.4	B=11	C=12	N=14	O=16	F=19	
3	Na=23	Mg=24	Al=27.3	Si=28	P=31	S=32	Cl=35.5	
4	K=39	Ca=40	?=44	Ti=48	V=51	Cr=52	Mn=55	Fe=56; Co=59; Ni=59
5	Cu=63	Zn=65	?=68	?=72	As=75	Se=78	Br=80	
6	Rb=85	Sr=87	?Yt=88	Zr=90	Nb=94	Mo=96	?=100	Ru=104; Rh=104; Pd=106
7	Ag=108	Cd=112	In=113	Sn=118	Sb=122	Te=125	J=127	
8	Cs=133	Ba=137	?Di=138	?Ce=140				
9								
10			?Er=178	?La=180	Ta=182	W=184		Os=195; Ir=197; Pt=198
11	Au=199	Hg=200	Tl=204	Pb=207	Bi=208			
12				Th=231		U=240		

Mendeleev correctly predicted the existence and chemical and physical properties of undiscovered elements gallium ("eka-aluminum", labeled "?=68" above), and germanium ("eka-silicon", labeled "?=72" above). Mendeleev's periodic table gained significant credibility when gallium and germanium were discovered (during Mendeleev's lifetime) and were found to have the properties that he predicted.

Mendeleev's group numbers were chosen first based on the elements' chemical and physical properties, and second by increasing atomic mass. Tellurium (Te) was later found to have an average atomic mass of 128 (heavier than iodine, which was labeled "J" on Mendeleev's table), but Mendeleev kept the elements where they were, because tellurium is more like the other elements in group 6, and iodine is more like the other elements in group 7.

periodic law: when the elements are arranged in order of increasing atomic number, their properties repeat in regular intervals (periods). (This is the modern version of the law of octaves.) The periodic table is arranged so that each row represents one of these periods.

Use this space for summary and/or additional notes:

Development of the Periodic Table

Big Ideas | **Details** | **Unit: Periodicity**

Henry Moseley: British chemist; rearranged the elements by increasing nuclear charge (atomic number) instead of atomic mass (1913). This arrangement resulted in elements with similar properties falling into groups without the exceptions (such as the positions of iodine *vs.* tellurium) that occurred in Mendeleev's table. The modern periodic table is Moseley's table, extended to include elements that have been discovered since his death.

Use this space for summary and/or additional notes:

Regions of the Periodic Table

Unit: Periodicity

MA Curriculum Frameworks (2016): HS-PS1-1

MA Curriculum Frameworks (2006): 3.1, 3.2

Mastery Objective(s): (Students will be able to...)

- Identify regions of the periodic table by name.
- Describe the properties of different groups (families) of elements.

Success Criteria:

- Regions of the periodic table are identified correctly.
- Descriptions of properties are correct.

Tier 2 Vocabulary: period, group, family

Language Objectives:

- Name each of the regions of the periodic table.

Notes:

period: a row of the periodic table. Properties of the elements are *periodic*, meaning that they repeat after a specific interval. Elements in the same period have their highest energy electrons in the same principal energy level.

group (family): a column of the periodic table. Elements in the same group have the same number of valence electrons, and therefore have similar chemical and physical properties.

diatomic elements: elements whose natural state is a molecule that has two atoms of the element. There are seven diatomic elements:
H_2, N_2, O_2, F_2, Cl_2, Br_2, and I_2.

Use this space for summary and/or additional notes:

Regions of the Periodic Table

Metals, Non-Metals & Metalloids

<u>metals</u>: elements to the left of and below the "stairstep line."

<u>non-metals</u>: elements to the right of and above the "stairstep line."

<u>metalloids</u>: elements that exhibit both metallic and non-metallic character. These are most of the elements that touch the "stairstep line". (All except for Al and Po).

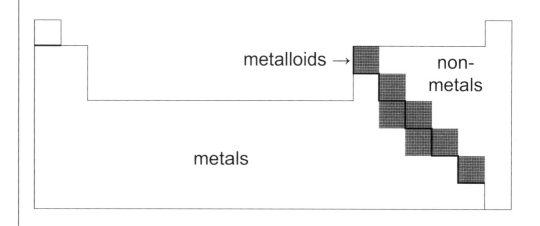

Use this space for summary and/or additional notes:

Regions of the Periodic Table

Big Ideas | **Details**

properties of metals:

- shiny
- high density
- good conductors of heat & electricity
- malleable & ductile (can be reshaped by hammering, bending and stretching)
- high melting & boiling points
- most have 3 or fewer valence electrons
- tend to form positive ions

properties of non-metals:

- dull
- low density
- poor conductors of heat & electricity
- brittle
- low melting & boiling points
- most have 4 or more valence electrons
- tend to form negative ions

properties of metalloids:

Metalloids can have properties "in between," or can have some properties like metals and others like non-metals.

Use this space for summary and/or additional notes:

Regions of the Periodic Table

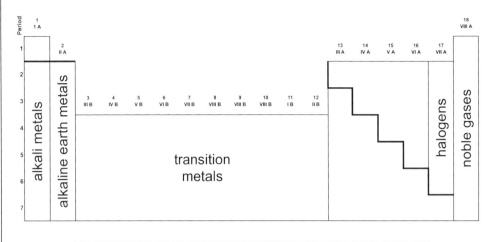

Groups ("Families") of Elements

alkali metals: elements in group 1 (IA) of the periodic table.

- 1 valence electron (form +1 ions)
- very reactive
- soft
- very high melting & boiling points
- ions are soluble in water

alkaline earth metals: elements in group 2 (IIA) of the periodic table.

- 2 valence electrons (form +2 ions)
- reactive, though not as much as group I metals
- very high melting & boiling points
- ions are not soluble in water

Use this space for summary and/or additional notes:

Regions of the Periodic Table

Unit: Periodicity

<u>transition metals</u>: elements in the center section (groups 3–12) of the periodic table.

- have a partially-filled *d* sub-level
- form colored ions when dissolved in water
- "officially" have 2 valence electrons, but can shift electrons into and out of *s* and *d* sub-levels. Often form more than one kind of ion.
- transition metals with several unpaired electrons in their *d* or *f* sub-levels are paramagnetic (are attracted to a magnet).
- most are shiny, hard metals with high melting & boiling points

<u>inner transition metals</u>: elements in the "f block" of the periodic table. (The "extra" section below the rest of the table.)

- are part of the transition metals
- have a partially-filled *f* sub-level
- officially have 2 valence electrons, but can shift electrons between *s*, *d*, and *f* sub-levels. Usually form ions with +3 charges.
- are rare

<u>noble gases</u>: elements in group 18 (VIIIA) of the periodic table.

- 8 valence electrons (except for He which has 2)—full valence shells
- do not form ions
- do not react with other compounds
- gases
- extremely low melting & boiling points. (In fact, helium cannot be made into a solid even at absolute zero, except at extremely high pressures.)

<u>halogens</u>: elements in group 17 (VIIA) of the periodic table.

- 7 valence electrons (form −1 ions)
- reactive
- diatomic (atoms in pairs) in their natural state: F_2, Cl_2, Br_2, I_2
- low melting & boiling points. (F & Cl are gases at room temp; Br is a liquid, and I is a solid, but will melt in your hand.)
- form salts that are soluble in water (except for fluorine—fluoride salts are not soluble in water.)

Homework

Color and label the regions of the periodic table on an actual periodic table (with elements and data).

Use this space for summary and/or additional notes:

Ionization Energy

Unit: Periodicity
MA Curriculum Frameworks (2016): HS-PS1-1
MA Curriculum Frameworks (2006): 3.1, 3.2
Mastery Objective(s): (Students will be able to...)
- Rank elements according to ionization energy based on their location on the periodic table.

Success Criteria:
- Rankings account for size and effective nuclear charge.

Tier 2 Vocabulary: ionization

Language Objectives:
- Explain why ionization energy increases as you go up and to the right on the periodic table.

Notes:

ionization energy: the amount of energy that it takes to remove an electron from an atom. (This makes it an ion. Ionization energy is literally the amount of energy it takes to make an atom into an ion.) Ionization energy is a measure of how tightly an element holds onto its electrons.

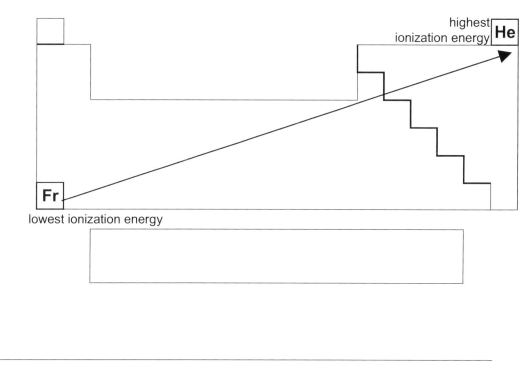

Use this space for summary and/or additional notes:

Ionization Energy

Big Ideas | **Details** | **Unit: Periodicity**

The more an element "wants" to gain electrons (meaning the more energetically favorable it is), the more tightly it will hold onto its own electrons, and the higher the ionization energies.

- Atoms of elements farther to the right hold on to electrons more tightly, because obtaining a full valent shell by gaining electrons is more stable than obtaining a full valent shell by losing them. Similarly, elements farther to the left are more stable if they attain a full valent shell by losing electrons than by gaining them.
- Atoms of elements higher up within a group hold on to electrons more tightly because they have fewer energy levels & sublevels to spread the electrons over. Also, because the atom has fewer levels, the electrons are closer to the nucleus, which means the force of attraction by the positive charge is stronger.
- Noble gases *and ions that have the same electron configuration as a noble gas* have the highest ionization energies, because they have the most stable electron configurations.
- Helium has the highest ionization energy. Francium has the lowest.

Ionization energy is calculated by measuring how much heat it takes to get an element to lose an electron and become a positive ion.

1st ionization energy: the amount of energy it takes to remove one electron from a neutral atom, to make a +1 ion.

2nd ionization energy: the amount of energy it takes to remove a second electron to make a +2 ion from a +1 ion.

3rd, 4th ionization energy, *etc.*: the amount of energy it takes to remove a third electron to make a +3 ion from a +2 ion, to remove a fourth electron to make a +4 ion, *etc.*

Use this space for summary and/or additional notes:

Ionization Energy

Unit: Periodicity

Big Ideas	Details

Consider the following table of ionization energies.

Element	1st Ionization Energy (kJ/mol) (neutral atom)	2nd Ionization Energy (kJ/mol) (+1 ion)	3rd Ionization Energy (kJ/mol) (+2 ion)
Ne	2081	3952	6122
Na	496	4562	6912
Mg	738	1451	7733

Notice that there is a large jump in the ionization energy once the atom or ion has a full valent shell (the same electron configuration as a noble gas, or a "noble gas core").

In the above table, neon (Ne) has the largest 1st ionization energy, because it is a noble gas and it already has a full valent shell.

Sodium (Na) has the smallest 1st ionization energy. This is because removing one electron will give it a "noble gas core". However, sodium has the largest 2nd ionization energy, because the +1 ion has a full valent shell, and is therefore more stable.

Magnesium (Mg) has the lowest 2nd ionization energy, because removing that second electron will give it a noble gas core. However, because the +2 ion has a noble gas core, Mg has the largest 3rd ionization energy.

First ionization energies for all elements are listed in your Chemistry Reference Tables in "**Error! Reference source not found.**," which begins on page **Error! Bookmark not defined.**.

Use this space for summary and/or additional notes:

Ionization Energy

Big Ideas	Details

Homework Problems

For each pair of elements:

- Answer the question about which element has the higher or lower ionization energy.
- State the direction(s) on the periodic table (up *vs.* down and/or left *vs.* right) that you based your choice on.
- Explain *why* moving that direction (up *vs.* down and/or left *vs.* right) caused the difference in ionization energy.

1. Which element has a *higher* first ionization energy: Na or Al ?

    Direction(s):

    Explanation:

2. Which element has a *higher* first ionization energy: Mg or Ca ?

    Direction(s):

    Explanation:

3. Which element has a *lower* second ionization energy: K or Ca ?

    Direction(s):

    Explanation:

4. Which element has a *lower* second ionization energy: Sr or Ba ?

    Direction(s):

    Explanation:

5. Which element has a *higher* first ionization energy: K or B ?

    Direction(s):

    Explanation:

6. Which element has a *lower* first ionization energy: Rb or P ?

    Direction(s):

    Explanation:

Use this space for summary and/or additional notes:

Electronegativity

Unit: Periodicity

MA Curriculum Frameworks (2016): HS-PS1-1

MA Curriculum Frameworks (2006): 3.4

Mastery Objective(s): (Students will be able to...)

- Rank elements according to electronegativity energy based on their location on the periodic table.

Success Criteria:

- Rankings account for electron configuration and size.

Tier 2 Vocabulary:

Language Objectives:

- Explain why electronegativity increases as you go up and to the right on the periodic table.

Notes:

electronegativity: the tendency of an atom to attract electrons.

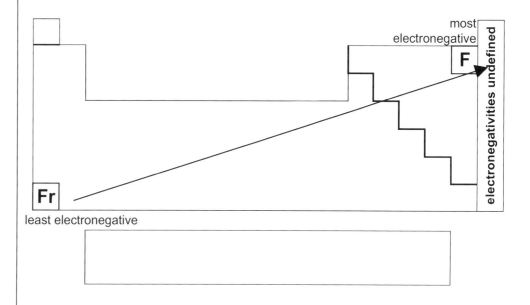

Pauling electronegativity scale: a set of numbers that describe an element's tendency to attract electrons. (Named for American chemist, biochemist, peace activist, author, and educator Linus Pauling.)

Use this space for summary and/or additional notes:

Electronegativity

Big Ideas | **Details** | Unit: Periodicity

Elements that "want" electrons (for which attracting electrons is energetically favorable) pull harder, which makes them more electronegative.

- Atoms of elements farther to the right on the periodic table (except for noble gases) pull harder, because it's easiest for them to get a full valence shell by gaining electrons. Elements to the left pull the least hard, because the positive ion (formed by losing electrons) is more stable than the negative ion.

- Atoms of elements higher up within a group or family pull harder because they have fewer energy levels & sublevels to spread the electrons over, which makes it harder for them to be stable with only their existing electrons.

- Noble gases do not have electronegativities, because they neither "pull" electrons, nor "push" them away.

- Fluorine is the most electronegative element (3.98 on the Pauling scale). Francium is the least electronegative (or most electropositive) element (0.7 on the Pauling scale).

Pauling originally assigned arbitrary electronegativity numbers that increased in steps of 0.5 across period 2 on the periodic table:

Li	Be	B	C	N	O	F	Ne
1.0	1.5	2.0	2.5	3.0	3.5	4.0	—

Electronegativity is currently calculated by comparing the amount of energy it takes to break bonds between atoms of different elements.

Now that we are able to make more precise measurements, the current values of the electronegativities for these elements are close to, but not exactly equal to Pauling's original numbers.

Electronegativities for all elements are listed in your Chemistry Reference Tables in "**Error! Reference source not found.**," which begins on page **Error! Bookmark not defined.**.

Use this space for summary and/or additional notes:

Electronegativity

Big Ideas | **Details** | **Unit: Periodicity**

Homework Problems

For each pair of elements:

- Answer the question about which element has the higher or lower electronegativity.
- State the direction(s) on the periodic table (up *vs.* down and/or left *vs.* right) that you based your choice on.
- Explain *why* moving that direction (up *vs.* down and/or left *vs.* right) caused the difference in electronegativity.

1. Which element is *more* electronegative: Li or K ?

    Direction(s):

    Explanation:

2. Which element is *more* electronegative: P or Cl ?

    Direction(s):

    Explanation:

3. Which element is *more* electronegative: Al or N ?

    Direction(s):

    Explanation:

4. Which element has a *lower* second ionization energy: Mg or Rb ?

    Direction(s):

    Explanation:

Use this space for summary and/or additional notes:

Big Ideas	Details	Atomic & Ionic Radius	Page: 270
			Unit: Periodicity

Atomic & Ionic Radius

Unit: Periodicity

MA Curriculum Frameworks (2016): HS-PS1-1

MA Curriculum Frameworks (2006): 3.4

Mastery Objective(s): (Students will be able to...)

- Rank elements according to atomic or ionic radius based on their charge, electron configuration and location on the periodic table.

Success Criteria:

- Rankings account for charge, electron configuration and size.

Tier 2 Vocabulary: radius

Language Objectives:

- Explain why atomic and ionic radius decrease as you go up and to the right on the periodic table.

Notes:

atomic radius: the average distance from the nucleus to the outermost electrons in an atom. The atomic radius is a measure of the "size" of the atom.

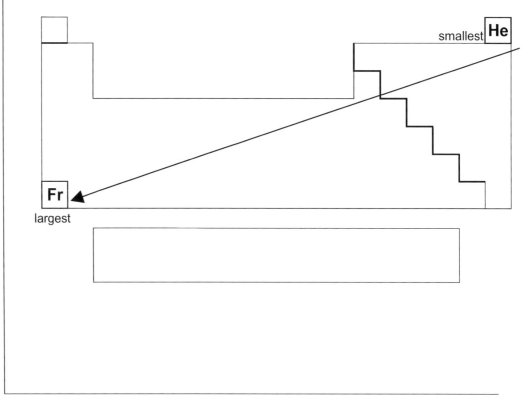

Use this space for summary and/or additional notes:

Atomic & Ionic Radius

Unit: Periodicity

<u>shielding</u>: when electrons in the lower energy levels (closer to the nucleus) shield (cancel) some of the nucleus's positive charge. This causes the outer electrons to be held less tightly by the nuclear charge, and the atom gets larger.

- Atoms of elements get larger as you move down a column, because each row adds a new energy level, and each new level is farther out from the nucleus than the previous level. Also, the inner electrons shield some of the charge in the nucleus, which means that the nucleus holds onto the outer electrons less tightly.

- Atoms of elements get smaller as you move to the right within the same row. This is because the amount of unshielded positive charge from the nucleus is greater, which pulls the electrons closer.

Covalent atomic radii for all elements are listed in your Chemistry Reference Tables in "**Error! Reference source not found.**," which begins on page **Error! Bookmark not defined.**.

Use this space for summary and/or additional notes:

Big Ideas	Details

Atomic & Ionic Radius

Unit: Periodicity

Ionic Radius

Because most of the space that an atom takes up is outside the nucleus (where the electrons are), changing the number of electrons (making an ion) changes the size of the atom.

- If you take away electrons, the ion gets smaller. This means ions with a positive charge are smaller than the neutral atom and also smaller than an atom of the neutral element with the same number of electrons. This is because the positive ions have more unshielded positive charge, which pulls the electrons closer.

- If you add electrons, the ion gets larger. This means ions with a negative charge are larger than the neutral atom and also larger than an atom of the neutral element with the same number of electrons.

Element	Covalent radius (of neutral atom) (Å)	Charge of ion	Radius of ion (Å)
O	0.75	−2	1.4
F	0.73	−1	1.33
Ne	0.72	0	—
Na	1.54	+1	1.02
Mg	1.36	+2	0.72
Al	1.18	+3	0.54

Use this space for summary and/or additional notes:

Atomic & Ionic Radius

Homework Problems

For each pair of elements:

- Answer the question about which element has the larger or smaller atomic or ionic radius.
- State the direction(s) on the periodic table (up *vs.* down and/or left *vs.* right) that you based your choice on.
- Explain *why* moving that direction (up *vs.* down and/or left *vs.* right) caused the difference in atomic radius.

1. Which atom is the *smallest*: Be, C, or F ?
   Direction(s):
   Explanation:

2. Which atom is the *smallest*: N, P, or Sb ?
   Direction(s):
   Explanation:

3. Which atom is the *largest*: Ba, Ga, or N ?
   Direction(s):
   Explanation:

4. Which ion is the *largest*: P^{3-}, S^{2-}, or Cl^- ?
   Direction(s):
   Explanation:

5. Which ion is the *smallest*: Na^+, Mg^{2+}, or Al^{3+} ?
   Direction(s):
   Explanation:

6. Which of the following is the *smallest*: O^{2-}, Ne, or Mg^{2+} ?
   Direction(s):
   Explanation:

Use this space for summary and/or additional notes:

Big Ideas	Details
	# Summary: Periodicity

Unit: Periodicity

List the main ideas of this chapter in phrase form:

Write an introductory sentence that categorizes these main ideas.

Turn the main ideas into sentences, using your own words. You may combine multiple main ideas into one sentence.

Add transition words to make your writing clearer and rewrite your summary below.

Use this space for summary and/or additional notes: |

Introduction: Nomenclature & Formulas

Unit: Nomenclature & Formulas

Topics covered in this chapter:

Bonding 277
Chemical Formulas 279
Balancing Charges 280
Polyatomic Ions 284
Naming Ionic Compounds 287
Naming Oxyanions 291
Naming Acids 295
Naming Molecular (Covalent) Compounds 297

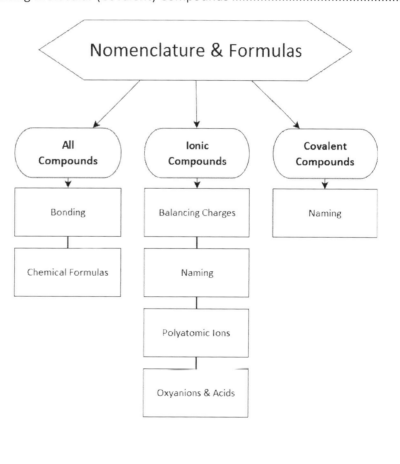

Use this space for summary and/or additional notes:

Introduction: Nomenclature & Formulas

Standards addressed in this chapter:

Massachusetts Curriculum Frameworks & Science Practices (2016):

HS-PS2-6 Communicate scientific and technical information about the molecular-level structures of polymers, ionic compounds, acids and bases, and metals to justify why these are useful in the functioning of designed materials.

Massachusetts Curriculum Frameworks (2006):

4.1 Explain how atoms combine to form compounds through both ionic and covalent bonding. Predict chemical formulas based on the number of valence electrons.

4.6 Name and write the chemical formulas for simple ionic and molecular compounds, including those that contain the polyatomic ions: ammonium, carbonate, hydroxide, nitrate, phosphate, and sulfate.

Use this space for summary and/or additional notes:

Bonding

Unit: Nomenclature & Formulas
MA Curriculum Frameworks (2016): HS-PS2-6
MA Curriculum Frameworks (2006): 4.1
Mastery Objective(s): (Students will be able to…)

- Explain how atoms bond together to form compounds.
- Identify different types of chemical bonds.

Success Criteria:

- Explanations account for sharing or transfer of electrons.

Tier 2 Vocabulary: bond

Language Objectives:

- Explain what happens with electrons in order to form chemical bonds.

Notes:

bonding: any joining together of atoms or molecules

chemical bond or intramolecular bond: a strong bond between atoms or individual ions, resulting from the sharing or transfer of electrons

intermolecular bond: a weak bond between molecules or ions, which holds the molecules of a liquid or solid together. (We will study these in more detail later in the section on "Intermolecular Forces" on page 340.)

ion: an atom or group of atoms that has a charge, because it has either gained or lost electrons.

Use this space for summary and/or additional notes:

Bonding

Types of Chemical Bonds

ionic bond: when a positive ion and a negative ion are held together by the electrical attraction of their charges.

- ionic bonds occur between ions, usually between a metal ion and a non-metal ion.
- the positive ion (cation) is always either the ion of a metal or a positive polyatomic ion.
- the negative ion (anion) is always either the ion of a nonmetal or a negative polyatomic ion.
- the difference between the electronegativity of the nonmetal and the electronegativity of the metal ($\Delta\chi$) is usually ≥ 1.7. (This will be addressed further in the section on "Intermolecular Forces" on page 340.)

covalent bond: when two atoms form a bond by sharing ("co-") their valence ("-valent") electrons.

- covalent bonds occur only between non-metals
- the electronegativity difference ($\Delta\chi$) between the two nonmetals is usually < 1.7

metallic bond: when atoms in a metal form a network of positive ions and loosely held electrons.

- metallic bonds, as the name suggests, occur only between metals
- metallic bonds are often described as a "sea of electrons" because the valence electrons can move easily from one atom to another. This is how metals conduct electricity—electricity is simply the flow of electrons.

Use this space for summary and/or additional notes:

Chemical Formulas

Big Ideas | Details | Page: 279
Unit: Nomenclature & Formulas

Chemical Formulas

Unit: Nomenclature & Formulas

MA Curriculum Frameworks (2016): HS-PS2-6

MA Curriculum Frameworks (2006): 4.1

Mastery Objective(s): (Students will be able to…)

- Write chemical formulas for ionic compounds when the number of atoms of each element is given.

Success Criteria:

- Elements are listed in the correct order.
- Subscripts give the correct number for each element.

Tier 2 Vocabulary: formula

Language Objectives:

- Explain what the subscripts mean in a chemical formula.

Notes:

<u>chemical formula</u>: a formula that describes a compound by listing how many of each element it's made of.

Some examples:

- Fe_2O_3 has 2 Fe (iron) atoms and 3 O (oxygen) atoms.
- $CaCl_2$ has 1 Ca (calcium) atom and 2 Cl (chlorine) atoms.
- $C_{21}H_{30}O_2$ has 21 C (carbon) atoms, 30 H (hydrogen) atoms, and 2 O (oxygen) atoms.

Elements in a chemical formula are listed with metals first, then non-metals, and almost always in order by increasing electronegativity: the <u>least</u> electronegative element is listed first, and the most electronegative one is listed last. (Exceptions are organic compounds and acids.)

(Note: the variable χ is usually used for electronegativity.)

For example: a compound made from Mg^{2+} ions (χ_{Mg} = 1.31) and Cl^- ions (χ_{Cl} = 3.16) would be $MgCl_2$, not Cl_2Mg.

Use this space for summary and/or additional notes:

Chemistry 1 | Mr. Bigler

Balancing Charges

Unit: Nomenclature & Formulas
MA Curriculum Frameworks (2016): HS-PS2-6
MA Curriculum Frameworks (2006): 4.1
Mastery Objective(s): (Students will be able to...)
- Write chemical formulas for ionic compounds.

Success Criteria:
- Subscripts are chosen so that positive and negative charges are balanced (equal).

Tier 2 Vocabulary: bond, charge

Language Objectives:
- Explain the process and necessity of balancing charges.

Notes:

If you have an ionic compound (a compound made of positive and negative ions), the positive and negative charges will attract each other (because opposite charges attract). This will continue to happen until the total amount of positive charge equals the total amount of negative charge, and there is no more attraction. When this happens in chemistry, we say that the charges are *balanced*.

Use this space for summary and/or additional notes:

Balancing Charges

For example, suppose we made a compound from Al^{3+} ions and O^{2-} ions.

1. If we start with an Al^{3+} ion, it is positive, so it will attract a negative O^{2-} ion. This gives:

      the total charge is now +1

2. Because the net (overall) charge is positive (+1), it will attract another negative O^{2-} ion, giving:

      total charge is now −1

3. Now the group is negative, so it will attract a positive Al^{3+} ion, giving us:

   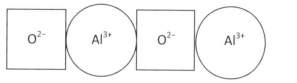   total: +2

4. Now the group is positive, so it will attract another negative O^{2-} ion, giving us:

      total: 0

5. Finally, all of the positive and negative charges have exactly balanced, and the compound has no net charge. Now it doesn't attract any more positive or negative ions.

6. To balance the charges, we needed 2 Al^{3+} ions and 3 O^{2-} ions, which means the formula for this compound is $(Al^{3+})_2(O^{2-})_3$ or simply Al_2O_3.

Use this space for summary and/or additional notes:

Balancing Charges

Shortcuts for Balancing Charges

Find the L.C.M.

In an ionic compound, *the total positive and total negative charge will always be equal*, and will be the least common multiple (L.C.M.) of the charges of the positive and negative ions.

In the compound made from aluminum and oxygen the charges of the ions are +3 (for Al), and −2 (for O). The LCM of 3 and 2 is 6, which means the total positive charge in the formula will be +6, and the total negative charge in the formula will be −6.

To get +6, we need 2 Al^{3+} ions, and to get −6 we need 3 O^{2-} ions. Thus, the formula is once again $(Al^{3+})_2(O^{2-})_3$ or simply Al_2O_3.

Cross the Charges and Reduce to Lowest Terms

Often, you can "cross the charges" to get the formula:

$$Al^{3+}_{2} \! \! \! \times \! \! \! O^{2-}_{3}$$

This always gives a correct ratio, although the ratio may not be in lowest terms. (For example, crossing the charges for Fe^{2+} and O^{2-} would give the compound Fe_2O_2, but the correct formula should be reduced to FeO.)

Use this space for summary and/or additional notes:

Balancing Charges

Unit: Nomenclature & Formulas

Homework Problems

In the chart below,
- Look up and add the appropriate charges to the cation and anion (if the charges are not already labeled).
- Balance the charges and write the formula of the resulting compound.

Cation	Anion	Chemical Formula
NH_4^+	$(PO_4)^{3-}$	$(NH_4)_3 PO_4$
Sr	S	
Na	Cl	
Ca	Br	
K	O	
Cu^+	Cl	
Cu^{2+}	Cl	
Mg	S	
Ba	P	
Cr^{6+}	O	

Use this space for summary and/or additional notes:

Polyatomic Ions

Unit: Nomenclature & Formulas

MA Curriculum Frameworks (2016): HS-PS2-6

MA Curriculum Frameworks (2006): 4.6

Mastery Objective(s): (Students will be able to…)

- Write chemical formulas that include polyatomic ions.

Success Criteria:

- Subscripts are chosen so that positive and negative charges are balanced (equal).
- Formulas for polyatomic ions are in parentheses if more than one is needed.

Tier 2 Vocabulary: bond, charge

Language Objectives:

- Explain the process and necessity of balancing charges.

Notes:

polyatomic ion: a group of atoms that are bonded to each other that behave chemically like a single ion. A polyatomic ion always has a specific name, chemical formula, and charge.

For example: the sulfate ion has the chemical formula SO_4^{2-}. It is made of one sulfur atom and 4 oxygen atoms. Chemically, it behaves like a single atom with a −2 charge.

The formula of a polyatomic ion never changes!

I.e., the sulfate ion is *always* SO_4^{2-}, and the 4 is an important part of the formula. If you wrote SO_2^{2-} instead, you would be talking about the hyposulfite ion instead of the sulfate ion—a different polyatomic ion with different chemical properties.

Use this space for summary and/or additional notes:

Polyatomic Ions

Unit: Nomenclature & Formulas

Polyatomic Ions in Chemical Formulas

If a compound contains a polyatomic ion, you write the formula for the polyatomic ion, *including the subscript numbers*, in the place where the ion goes. For example, a compound with Na^+ and SO_4^{2-} would simply be Na_2SO_4.

Balancing Charges with Polyatomic Ions

If you need more than one of a polyatomic ion in a chemical formula, put the entire polyatomic ion, *including any subscript numbers*, in parentheses, and put the number that tells how many ions you need outside the parentheses.

For example, to balance the compound made from Al^{3+} and SO_4^{2-}, you need 2 Al^{3+} ions and 3 SO_4^{2-} ions. The formula is:

$$Al_2(SO_4)_3$$

Note: there are positive and negative polyatomic ions. A compound can have either, neither, or both kinds. For example, if you had a compound made from the positive ion ammonium (NH_4^+) and the negative ion sulfate (SO_4^{2-}), the compound would have the formula:

$$(NH_4)_2SO_4$$

Determining the Number of Atoms in a Formula

The subscripts tell you how many you have of *whatever came immediately before the subscript*. If the thing before the subscript is an element, as in $CaCl_2$, the 2 tells us that we have 2 Cl atoms. There's no subscript after Ca, so this means we have only 1 Ca atom.

If the thing before the subscript is parentheses, as in $Al_3(SO_4)_2$, the 3 tells us that we have 3 Al atoms, the 2 outside the parentheses tells us that we have 2 entire SO_4 ions. This means we really have 2 atoms of S and 2 × 4 = 8 atoms of O.

Sample Problem:

How many hydrogen atoms are in the compound $(NH_4)_2HPO_4$?

We have 2 × 4 = 8 from the two NH_4 ions, plus 1 from the HPO_4 ion, giving us a total of 9 hydrogen atoms.

Use this space for summary and/or additional notes:

Polyatomic Ions

Table of Polyatomic Ions

ion	formula	ion	formula	ion	formula
americyl	AmO_2^{2+}	acetate	CH_3COO^-	tetraborate	$B_4O_7^{2-}$
carbonyl	CO_2^{2+}	amide	NH_2^-	carbide	C_2^{2-}
thiocarbonyl	CS_2^{2+}	hydroxylamide	$NHOH^-$	carbonate	CO_3^{2-}
chromyl	CrO_2^{2+}	azide	N_3^-	chromate	CrO_4^{2-}
neptunyl	NpO_2^{2-}	hydrazide	$N_2H_3^-$	dichromate	$Cr_2O_7^{2-}$
plutoryl	PuO_2^{2+}	bromate	BrO_3^-	imide	NH^{2-}
seleninyl	SeO^{2+}	chlorate	ClO_3^-	molybdate	MoO_4^{2-}
selenoyl	SeO_2^{2+}	cyanide	CN^-	peroxide	O_2^{2-}
thionyl/sulfinyl	SO^{2+}	cyanate	OCN^-	oxalate	$C_2O_4^{2-}$
sulfonyl/sulfuryl	SO_2^{2+}	thiocyanate	SCN^-	phthalate	$C_8H_4O_4^{2-}$
uranyl	UO^{2+}	selenocyanate	$SeCN^-$	selenite	SeO_4^{2-}
vanadyl	VO^{2+}	tellurocyanate	CH_3S^-	silicate	SiO_3^{2-}
ammonium	NH_4^+	hydroxide	OH^-	sulfate	SO_4^{2-}
hydronium	H_3O^+	iodate	IO_3^-	thiosulfate	$S_2O_3^{2-}$
iodyl	IO_2^+	methanolate	CH_3O^-	dithionate	$S_2O_4^{2-}$
nitrosyl	NO^+	methanethiolate	CH_3S^-	silicate	SiO_3^{2-}
thionitrosyl	NS^+	ethanolate	$C_2H_5O^-$	borate	BO_3^{3-}
phosphoryl	PO^+	permanganate	MnO_4^-	arsenate	AsO_4^{3-}
thiophosphoryl	PS^+	nitrate	NO_3^-	phosphate	PO_4^{3-}
phosphor	PO_2^+	superoxide	O_2^-	orthosilicate	SiO_4^{4-}

Use this space for summary and/or additional notes:

Big Ideas	Details
	Naming Ionic Compounds

Naming Ionic Compounds

Unit: Nomenclature & Formulas

MA Curriculum Frameworks (2016): HS-PS2-6

MA Curriculum Frameworks (2006): 4.6

Mastery Objective(s): (Students will be able to...)

- Write names for ionic compounds using the stock system, including Roman numerals where appropriate.
- Write formulas for ionic compounds based on their names.

Success Criteria:

- Compound names contain the correct cation (including a Roman numeral if necessary) and anion in the correct order (cation first, then anion).
- Chemical formulas have correctly balanced charges.
- Chemical formulas have polyatomic ions in parentheses when necessary.

Tier 2 Vocabulary: compound

Language Objectives:

- Explain when a cation does or does not need a Roman numeral.
- Explain the relationship between the charges of the ions and the subscripts in the formula.

Notes:

<u>ionic compound</u>: a compound made out of the ions of a metal and a nonmetal

<u>cation</u>: an ion with a positive charge, such as Na^+ or Ca^{2+}

<u>anion</u>: an ion with a negative charge, such as Cl^- or S^{2-}

Naming the Anion

- If an anion is a single element, the name of the ion is the name of the element with the ending changed to "ide". For example, the Cl^- ion is made from chlorine, so it is called "chloride". The O^{2-} ion is made from oxygen, so it is called oxide.

- If the anion is a polyatomic ion, its name is just the name of the polyatomic ion. For example, the NO_3^- ion is named "nitrate".

Use this space for summary and/or additional notes:

Naming Ionic Compounds

Naming the Cation

- If the cation is a single element that has only one possible charge, the name of the cation is the name of the element. For example, the K^+ ion is simply named "potassium", and the Ca^{2+} ion is simply named "calcium".

- If the element can have more than one possible charge, the name of the cation is the name of the element followed by a Roman numeral, indicating the charge, in parentheses. For example, chromium can make cations with three different charges:

Formula of Cation	Name of Cation
Cr^{2+}	chromium (II)
Cr^{3+}	chromium (III)
Cr^{6+}	chromium (VI)

- If the cation is a polyatomic ion, its name is the name of the polyatomic ion. For example, the NH_4^+ ion is named "ammonium".

Naming the Compound

stock system: a system of naming compounds by naming the ions that they're made of. The cation (positive ion) is always listed first and the anion (negative ion) is always listed last.

Examples:

Formula	Cation		Anion		Name of Compound
NaCl	Na^+	sodium	Cl^-	chloride	sodium chloride
$CaBr_2$	Ca^{2+}	calcium	Br^-	bromide	calcium bromide
Fe_2O_3	Fe^{3+}	iron (III)	O^{2-}	oxide	iron (III) oxide
FeO	Fe^{2+}	iron (II)	O^{2-}	oxide	iron (II) oxide
K_2SO_4	K^+	potassium	SO_4^{2-}	sulfate	potassium sulfate

Notice that the number of atoms in the chemical formula is not represented anywhere in the name of an ionic compound.

Use this space for summary and/or additional notes:

Naming Ionic Compounds

Big Ideas | **Details**

Unit: Nomenclature & Formulas

Writing Formulas from Names

The name of every ion must have enough information to describe its element(s) and its charge. This means that if you know the name of an ion, you know the formula and the charge.

To convert the name of a compound to its formula:

1. Write down the cation and anion, including their charges.

2. Add subscripts to balance the charges. (Don't forget to put polyatomic ions in parentheses when you need more than one!)

Sample Problems:

1. What is the chemical formula for calcium phosphate?

    Calcium is Ca^{2+} and phosphate is PO_4^{3-}. The L.C.M. of the charges is 6. This means we need 3 Ca^{2+} ions to get to +6, and 2 PO_4^{3-} ions to get to −6. Therefore, the formula must be:

    $$Ca_3(PO_4)_2$$

2. What is the formula for nickel (II) chloride?

    Nickel (II) is Ni^{2+} (remember that the Roman numeral tells us the charge) and chloride is Cl^-. The L.C.M. of the charges is 2, which means we need one Ni^{2+} ion to get to +2, and 2 Cl^- ions to get to −2. Therefore, the formula must be:

    $$NiCl_2$$

Use this space for summary and/or additional notes:

Naming Ionic Compounds

Unit: Nomenclature & Formulas

Homework Problems

In the chart below:
- Write the element symbol(s) for the cation (positive ion) and anion (negative ion).
- Determine the charges of both ions.
- Balance the charges and write the formula of the resulting compound.

Cation Name	Cation Formula	Anion Name	Anion Formula	Formula of Compound
ammonium	NH_4^+	phosphate	PO_4^{3-}	$(NH_4)_3PO_4$
				$SnCl_2$
				FeO
calcium		bromide		
potassium		oxide		
copper (I)		carbonate		
copper (II)		chloride		
magnesium		nitrate		
ammonium		hydroxide		
barium		phosphate		
chromium (VI)		sulfate		

Use this space for summary and/or additional notes:

Naming Oxyanions

Unit: Nomenclature & Formulas

MA Curriculum Frameworks (2016): HS-PS2-6

MA Curriculum Frameworks (2006): 4.6

Mastery Objective(s): (Students will be able to...)

- Write names for ionic compounds that contain oxyanions.
- Write chemical formulas for ionic compounds that contain oxyanions.

Success Criteria:

- Compound names contain the correct cation (including a Roman numeral if necessary) and the correct anion (with prefix and suffix that together determine the number of oxygen atoms).
- Chemical formulas have correctly balanced charges.
- Chemical formulas have polyatomic ions in parentheses when necessary.

Tier 2 Vocabulary: compound, anion

Language Objectives:

- Explain what the prefix and suffix tell about the number of oxygens in an anion.

Notes:

<u>oxyanion</u> (or oxoanion): polyatomic anion (negative polyatomic ion) that contains oxygen.

Examples: NO_3 (nitrate), PO_4 (phosphate), SO_3 (sulfite)

Use this space for summary and/or additional notes:

Naming Oxyanions

Unit: Nomenclature & Formulas

Big Ideas	Details

The names of oxyanions always end in either "ate" or "ite", depending on the *relative* number of oxygens.

Note that most of the "_____ate" ions have either three or four oxygens, but there is no way to predict how many oxygens any particular "_____ate" ion will have.

- The "_____ate" ion has its own specific formula and charge.
- The "_____ite" ion has one fewer oxygen than the "_____ate" ion, and the same charge.

Some examples:

"_____ate" ion	formula	"_____ite" ion	formula
sulfate	SO_4^{2-}	sulfite	SO_3^{2-}
chlorate	ClO_3^-	chlorite	ClO_2^-
nitrate	NO_3^-	nitrite	NO_2^-
phosphate	PO_4^{3-}	phosphite	PO_3^{3-}

If you have one fewer oxygen than the "_____ite" ion, add the prefix "hypo". Again, the charge stays the same. For example:

- chlorite = ClO_2^-
- hypochlorite = ClO^-

If you have one more oxygen than the "_____ate" ion, add the prefix "per". Again, the charge stays the same. For example:

- chlorate = ClO_3^-
- perchlorate = ClO_4^-

Use this space for summary and/or additional notes:

Naming Oxyanions

Unit: Nomenclature & Formulas

Homework Problems

Give the name for each of the following compounds. You may need to look up formulas for polyatomic ions in "Table K. Polyatomic Ions" in your Chemistry Reference Tables on page 552.

1. $Rb_2B_4O_7$
2. $Mo_2(MoO_3)_3$
3. $CoSO_2$
4. Ca_2SiO_4
5. $Cr(CO_2)_3$
6. Na_2SeO_2
7. $NaNO_3$
8. KNO_2

Give the chemical formula for each of the following compounds.

9. sodium nitrate
10. ammonium periodate
11. calcium hypochlorite
12. barium carbonite
13. potassium dichromate
14. magnesium thiosulfite
15. potassium hypobromite
16. sodium phthalate

Use this space for summary and/or additional notes:

Big Ideas	Details

Naming Acids

Unit: Nomenclature & Formulas

MA Curriculum Frameworks (2016): HS-PS2-6
MA Curriculum Frameworks (2006): 4.6

Mastery Objective(s): (Students will be able to...)
- Write names for inorganic acids.
- Write chemical formulas for inorganic acids.

Success Criteria:
- Compound names contain the name of the anion with the correct prefix and/or suffix, and the word "acid".
- Chemical formulas have correctly balanced charges.
- Chemical formulas have polyatomic ions in parentheses when necessary.

Tier 2 Vocabulary: acid, formula

Language Objectives:
- Explain what the prefix and suffix tell about the type of anion in an acid.

Notes:

<u>acid</u>: a chemical compound that creates hydrogen (H^+) ions in water.

Acids behave somewhat like ionic compounds in which the cation (positive ion) is H^+. (We will study acids and bases in detail later in the year.)

Because the cation is always H^+, the name of the acid is based on the name of the anion (negative ion).

Anion Ends With	Example	Acid Name	Example
_____ate	nitrate (NO_3^-)	_____ic acid	nitric acid (HNO_3)
_____ite	arsenite (AsO_3^{3-})	_____ous acid	arsenous acid (H_3AsO_3)
_____ide	chloride (Cl^-)	hydro_____ic acid	hydrochloric acid (HCl)

Any prefixes, such as "per-" and "hypo-", are kept:
- periodate is IO_4^- so the acid HIO_4 is periodic acid
- hypochlorite is ClO^- so the acid HClO is hypochlorous acid.

A stupid mnemonic that some students seem to like for remembering the pair of suffix changes is: "I *ate* something *ic*ky. It m*ite* be a hippopotam*ous*."

Use this space for summary and/or additional notes:

Naming Acids

Homework Problems

Fill in the chart below. Use the first row as an example.

Chemical Formula	Anion	Anion Name	Acid Name
HNO_3	NO_3^-	nitrate	nitric acid
H_2CO_3			
HBr			
			acetic acid
HNO_2			
			phosphoric acid
			sulfurous acid
			hydroïodic acid
HCl			
			chloric acid
$HClO_2$			

Use this space for summary and/or additional notes:

Naming Molecular (Covalent) Compounds

Unit: Nomenclature & Formulas

MA Curriculum Frameworks (2016): HS-PS2-6

MA Curriculum Frameworks (2006): 4.6

Mastery Objective(s): (Students will be able to...)

- Write names for molecular (covalent) inorganic compounds.
- Write chemical formulas for molecular (covalent) inorganic compounds.

Success Criteria:

- Compound names contain the name of each element with the appropriate prefix (except that a compound cannot start with "mono-").
- Chemical formulas have elements listed in the correct order (increasing electronegativity).
- Chemical formulas have the correct subscript for each element.

Tier 2 Vocabulary: molecule, compound

Language Objectives:

- Explain how prefixes are used with molecular compounds.

Notes:

molecular compound (also known as a covalent compound): a compound made of atoms joined by covalent bonds (shared electrons).

molecule: a set of atoms joined by covalent bonds.

Use this space for summary and/or additional notes:

Naming Molecular (Covalent) Compounds

Unit: Nomenclature & Formulas

Unlike ionic compounds, covalent compounds have names that give the chemical formula of the molecule. The numbers in the formula are Greek numbers made into prefixes.

No.	Prefix	No.	Prefix	No.	Prefix	No.	Prefix
1	mono- or hen-	10	deca-	100	hecta-	1000	kilia-
2	di- or do-	20	(i)cosa-	200	dicta-	2000	dilia-
3	tri-	30	triaconta-	300	tricta-	3000	trilia-
4	tetra-	40	tetraconta-	400	tetracta-	4000	tetralia-
5	penta-	50	pentaconta-	500	pentacta-	5000	pentalia-
6	hexa-	60	hexaconta-	600	hexacta-	6000	hexalia-
7	hepta-	70	heptaconta-	700	heptacta-	7000	heptalia-
8	octa-	80	octaconta-	800	octacta-	8000	octalia-
9	nona-	90	nonaconta-	900	nonacta-	9000	nonalia-

Rules:

1. The first atom in the formula has a number prefix only if the molecule contains more than one of that atom.

2. The last atom in the formula always has a number prefix, whether or not there are more than one of that atom.

3. For huge numbers of atoms, the number prefixes combine in *reverse* order of place value. (Yes, it's weird.)

   For example, the compound $C_{51}S_{22}$ (if it existed) would be called henpentacontacarbon docosasulfide.

Use this space for summary and/or additional notes:

Naming Molecular (Covalent) Compounds

Unit: Nomenclature & Formulas

Examples:

- CO_2 has 1 carbon atom ("carbon") and 2 oxygen atoms ("di-oxide"), so its name is <u>carbon dioxide</u>.

- P_2O_5 has 2 phosphorus atoms ("di-phosphorus") and 5 oxygen atoms ("pent-oxide"), so its name is <u>diphosphorus pentoxide</u>.

- N_2O has 2 nitrogen atoms ("di-nitrogen") and one oxygen atom ("mono-oxide," which we elide to "monoxide"), so its name is <u>dinitrogen monoxide</u>.

- H_2O has 2 hydrogen atoms and one oxygen atom, but it is always "water," not "dihydrogen monoxide[*]". The same goes for NH_3, which is always called "ammonia."

---

[*] Someone has created a humorous website, http://www.dhmo.org, which attempts to make people aware of the "dangers" of dihydrogen monoxide, or "DHMO."

Use this space for summary and/or additional notes:

Naming Molecular (Covalent) Compounds

Homework Problems

Give the name for each of the following covalent compounds, using the prefix system.

1. NF_3
2. NO
3. NO_2
4. B_2O_3
5. N_2O
6. N_2O_4
7. PCl_3
8. PCl_5
9. SF_6
10. CO_2

Give the chemical formula for each of the following compounds.

11. phosphorus triiodide
12. sulfur dichloride
13. xenon trioxide
14. dinitrogen tetrafluoride
15. sulfur tetrafluoride
16. boron trichloride
17. phosphorus pentafluoride
18. diphosphorus trioxide
19. dichlorine heptoxide
20. carbon tetrachloride

Use this space for summary and/or additional notes:

Big Ideas	Details
	# Summary: Nomenclature & Formulas **Unit:** Nomenclature & Formulas List the main ideas of this chapter in phrase form: Write an introductory sentence that categorizes these main ideas. Turn the main ideas into sentences, using your own words. You may combine multiple main ideas into one sentence. Add transition words to make your writing clearer and rewrite your summary below.

Use this space for summary and/or additional notes:

Introduction: Bonding & Molecular Geometry

Unit: Covalent Bonding & Molecular Geometry

Topics covered in this chapter:

Covalent Bonding & Lewis Structures 305

Charged Atoms in Lewis Structures 311

VSEPR Theory 315

Orbital Hybridization 321

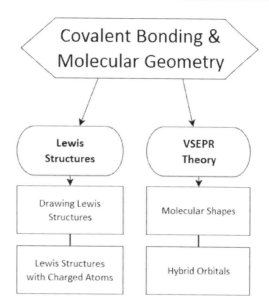

Standards addressed in this chapter:

Massachusetts Curriculum Frameworks & Science Practices (2016):

HS-PS1-2 Use the periodic table model to predict and design simple reactions that result in two main classes of binary compounds, ionic and molecular. Develop an explanation based on given observational data and the electronegativity model about the relative strengths of ionic or covalent bonds.

Use this space for summary and/or additional notes:

Introduction: Bonding & Molecular Geometry

Unit: Covalent Bonding & Molecular Geometry

Massachusetts Curriculum Frameworks (2006):

4.1 Explain how atoms combine to form compounds through both ionic and covalent bonding. Predict chemical formulas based on the number of valence electrons.

4.2 Draw Lewis dot structures for simple molecules and ionic compounds.

4.3 Use electronegativity to explain the difference between polar and nonpolar covalent bonds.

4.4 Use valence-shell electron-pair repulsion theory (VSEPR) to predict the molecular geometry (linear, trigonal planar, and tetrahedral) of simple molecules.

Use this space for summary and/or additional notes:

Big Ideas | Details

Covalent Bonding & Lewis Structures

Page: 305

Unit: Covalent Bonding & Molecular Geometry

Covalent Bonding & Lewis Structures

Unit: Covalent Bonding & Molecular Geometry

MA Curriculum Frameworks (2016): HS-PS1-2

MA Curriculum Frameworks (2006): 4.1, 4.2

Mastery Objective(s): (Students will be able to...)

- Draw Lewis structures for diatomic molecules, showing bonds and lone pairs.

Success Criteria:

- Lewis structures show the correct number of bonds.
- Lewis structures show the correct number of unpaired electrons, drawn in pairs.

Tier 2 Vocabulary: bond, lone pair

Language Objectives:

- Explain how a Lewis structure represents the electrons in a molecular compound.

Notes:

<u>covalent bond</u>: a chemical bond consisting of one or more pair(s) of shared electrons.

<u>covalent compound</u> (also known as a <u>molecular compound</u>): a compound made of atoms joined by covalent bonds.

- covalent bonding occurs between non-metals
- electrons are shared in pairs—each pair usually contains one electron that came from each atom.

For example, a chlorine atom has seven valence electrons. The Lewis dot structure for chlorine is:

Note that the chlorine atom has seven valence electrons.

Use this space for summary and/or additional notes:

Covalent Bonding & Lewis Structures

In a Lewis dot structure, we draw the valence electrons in pairs, representing the orbitals that contain them.

electron cloud: an orbital (a region within the atom) that contains either one or more of an atom's valence electrons or the shared electrons in a covalent bond.

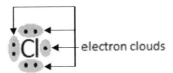

Because the chlorine atom contains seven valence electrons, it needs one more to fill its valent shell. If it is able to receive an electron from another atom (such as sodium), it can from a Cl⁻ (chloride) ion, which can then become part of an inorganic compound, such as NaCl (sodium chloride).

However, if chlorine is only in the presence of other atoms that are also trying to gain electrons, it can form a covalent bond, in which the electrons are shared.

You can think of this as if the chlorine atom were "borrowing" an electron to fill its valent shell. In order to "borrow" an electron, the chlorine atom needs to share an electron of its own. This shared pair of electrons becomes an electron cloud that is shared between the two chlorine atoms, which we call a covalent bond:

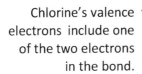 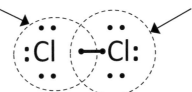

Chlorine's valence electrons include one of the two electrons in the bond.

When a Cl atom has *both* shared electrons, it has a full valent shell.

When each of the Cl atoms has both of the shared electrons, it has a full valent shell. This means each Cl atom gets a full valent shell some of the time.

This drawing is called a "Lewis structure," named after its inventor, American chemist G.N. Lewis. In a Lewis structure, unshared electrons are shown as dots (just like the Lewis dot diagram), and bonds are shown as line segments:

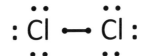

(Usually, the dots at the ends of the line segments are omitted. They are shown here as a reminder that the line representing the bond contains two electrons.)

Use this space for summary and/or additional notes:

For another example, oxygen has six unpaired electrons, which means an oxygen atom needs two more electrons to fill its valent shell. This means that each oxygen atom in O₂ will share two electrons:

Just like the example with Cl₂, each oxygen atom has a full valent shell whenever it has all four of the shared electrons.

One pair of shared electrons is called a <u>single bond</u>, such as the bond in Cl₂. The O₂ molecule has two pairs of shared electrons, which is called a <u>double bond</u>. Three pairs of shared electrons would be a <u>triple bond</u>. Note that a double bond is a large electron cloud with four electrons, and a triple bond is an even larger electron cloud with six electrons.

An atom can make bonds to more than one other atom. For example, oxygen needs two additional electrons. IT doesn't matter where the oxygen atom gets these electrons from—oxygen can just as easily share one electron with each of two different atoms. An example is the H₂O molecule:

This reasoning can be extended to any non-metal. The rule of thumb is that atoms have to "share one to get one". This means that the number of bonds to any atom will be equal to the number of electrons the atom needs in order to fill its valent shell.

Because of space constraints and the shapes of the *s* and *p* orbitals involved in forming bonds, it is not possible for atoms to make more than a triple bond to a single atom. Also, in a first-year high school chemistry course, we will only consider structures made by electrons in *s* and *p* orbitals. This means the structures we will consider in this course will have no more than four total electron clouds. Structures with expanded octets are studied in AP® Chemistry.

Use this space for summary and/or additional notes:

Common Atoms in Lewis Structures

Assuming that each atom (other than hydrogen) needs a full octet (8 electrons), including unshared pairs of electrons ("lone pairs") plus the electrons in bonds. Here are the most common combinations.

The halogens (F, Cl, Br, and I) have 1 bond and 3 lone pairs; hydrogen (H) has 1 bond and no lone pairs:

Atoms in the oxygen group (O, S, Se, and Te) always have 2 bonds and 2 lone pairs. The two possible combinations are:

Atoms in the nitrogen group (N, P, and As) always have 3 bonds and one lone pair. There are three possible combinations:

Atoms in the carbon group (C and Si) always have 4 bonds and no lone pairs, There are four possible combinations:

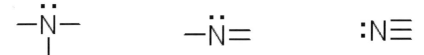

Use this space for summary and/or additional notes:

Covalent Bonding & Lewis Structures

Unit: Covalent Bonding & Molecular Geometry

Rules for Drawing Lewis Structures

1. Neutral atoms will make the same number of bonds as the number of additional electrons they need to fill their valent shell.
2. No pair of atoms can have more than a triple bond between them.
3. Atoms that make only one bond (*i.e.*, that need only one more electron) will almost always be on the outside (*i.e.*, not in the center).
4. The least electronegative atom will almost always be in the center.
5. The total number of bonds in a Lewis structure will usually be:

$$\frac{\text{\# electrons wanted} - \text{actual \# electrons}}{2}$$

If you cannot draw a correct Lewis structure using neutral atoms, try the following. (These are explained in the section titled "Charged Atoms in Lewis Structures" on page 311.)

6. Try moving one electron (at a time) from the central atom to the most electronegative atom that doesn't yet have a negative charge.
7. If the structure is for a charged ion, add or remove the appropriate number of electrons from your structure.
    a. If you need to add electrons, add them to the most electronegative atom first.
    b. If you need to remove electrons, remove them from the least electronegative atom first.
8. If you can draw more than one valid structure, the one with the lowest maximum charge on an atom is preferred. If you can draw more than one such structure, the structure with the fewest number of charged atoms is preferred.

Use this space for summary and/or additional notes:

Covalent Bonding & Lewis Structures

Unit: Covalent Bonding & Molecular Geometry

Homework Problems

Draw a correct Lewis structure for each of the following compounds.

1. PI_3
2. N_2
3. H_2O
4. PBr_3
5. $SiCl_4$
6. HCl
7. FCN
8. HNO
9. PN
10. CH_3NH_2
11. CH_2O
12. CO_2
13. $CBrClFl$
14. NF_3
15. N_2H_2
16. IBr
17. CH_3OH
18. C_2H_4
19. OF_2
20. $COCl_2$

Use this space for summary and/or additional notes:

Charged Atoms in Lewis Structures

Unit: Covalent Bonding & Molecular Geometry
MA Curriculum Frameworks (2016): HS-PS1-2
MA Curriculum Frameworks (2006): 4.2
Mastery Objective(s): (Students will be able to…)

- Draw Lewis structures in which one or more atoms has a formal charge.

Success Criteria:

- Lewis structures show the correct number of bonds.
- Lewis structures show the correct number of unpaired electrons in the correct locations.
- Individual charges are assigned correctly. (Positive charges on least electronegative atom, negative charges on most electronegative atom, *etc.*)
- Total charge adds up to the correct value.

Tier 2 Vocabulary: bond, charge

Language Objectives:

- Explain how charges are assigned to Lewis structures.

Notes:

Lewis structures show the shape of a molecule and how the atoms share electrons with each other. If a molecule exists, that means it must have a Lewis structure.

If you can't find a way to draw the Lewis structure for a molecule using neutral atoms, you may need to take electrons away from one atom and distribute them to other atoms, creating atoms with charges.

For example, consider the sulfur dioxide (SO_2) molecule. Sulfur and oxygen both have 6 valence electrons and need two bonds. If you draw the following:

$$\cdot \ddot{\underset{..}{O}} - \ddot{\underset{..}{S}} - \ddot{\underset{..}{O}} \cdot$$

sulfur has enough bonds, but each oxygen needs one more electron.

Use this space for summary and/or additional notes:

Charged Atoms in Lewis Structures

Unit: Covalent Bonding & Molecular Geometry

If you move one electron from sulfur to one of the oxygens, you have the following situation:

$$:\!\ddot{\underset{..}{O}}\!{}^{\ominus}\!-\!\ddot{S}{}^{\oplus}\!-\!\ddot{\underset{..}{O}}\!:$$

(Notice that when an atom in a Lewis structure has a charge, the charge is written next to the atom and circled.)

One way to think of charged atoms in Lewis structures is to *temporarily* substitute the charged atom with a neutral atom that has the same number of valence electrons, draw the Lewis structure, and then switch the atoms back.

When we take an electron away from S to make it S⁺, the S⁺ atom has the same number of valence electrons as P, so we could temporarily substitute P for S. Similarly, when we add an electron to O to make it O⁻, the O⁻ atom has the same number of electrons as F, so we could temporarily substitute F for O. If we made both of these substitutions, the compound would be OPF. We would place a double bond between P and O, to give the following Lewis structure:

$$:\!\ddot{\underset{..}{F}}\!-\!\ddot{P}\!=\!\ddot{\underset{..}{O}}$$

Now, we switch the F back to O⁻ and the P back to S⁺, which gives the correct Lewis structure for SO₂:

$$:\!\ddot{\underset{..}{O}}\!{}^{\ominus}\!-\!\ddot{S}{}^{\oplus}\!=\!\ddot{\underset{..}{O}}$$

If you are drawing a Lewis structure for a charged polyatomic ion, you need to show the overall charge as well as the charges on the individual atoms. To do this, put the entire structure in square brackets and write the overall charge outside the brackets, as in the example to the right:

This means that when we write the formula CaCO₃ for the ionic compound calcium carbonate, the actual arrangement of the atoms is:

Ca^{2+}

Use this space for summary and/or additional notes:

Charged Atoms in Lewis Structures

Unit: Covalent Bonding & Molecular Geometry

Homework Problems

Draw a correct Lewis structure for each of the following compounds.

1. PF_3
2. O_2
3. H_2S
4. $AsCl_3$
5. CBr_4
6. HI
7. $HSCN$
8. SCl_2
9. NH_2SeF
10. CH_3OH
11. SiO_2
12. CH_3COOH
13. POF
14. SO_4^{2-}
15. NO_3^-
16. PO_4^{3-}

Use this space for summary and/or additional notes:

Big Ideas	Details	VSEPR Theory — Unit: Covalent Bonding & Molecular Geometry

VSEPR Theory

Unit: Covalent Bonding & Molecular Geometry
MA Curriculum Frameworks (2016): HS-PS1-2
MA Curriculum Frameworks (2006): 4.4
Mastery Objective(s): (Students will be able to...)
- Identify the VSEPR shapes and bond angles for simple molecules (one central atom).

Success Criteria:
- VSEPR shapes show the correct number of lone pairs in the correct locations.
- VSEPR shapes have the correct bond angles.

Tier 2 Vocabulary: bond, cloud

Language Objectives:
- Explain how repulsion between electron clouds results in VSEPR shapes.

Notes:

<u>Valence Shell Electron Pair Repulsion (VSEPR*) theory</u>: a theory that the shape of a molecule is determined by the repulsion between electrons in the bonds and unshared pairs of the atoms.

The Lewis structure of a molecule represents the structure in 2 dimensions. The VSEPR shape is the 3-dimensional equivalent.

The VSEPR shapes are determined by the following constraints:

Electrons are all negatively charged, so they repel each other. Valence electrons exist in electron clouds, which can be either:

- unshared electrons (in pairs), attached to only one atom
- as part of a covalent bond (shared pair of electrons) between two atoms

The VSEPR shape of the molecule is the shape that occurs when all of these "clouds" of electrons are as far apart as possible.

---

* VSEPR is pronounced as if it were written "vesper".

Use this space for summary and/or additional notes:

VSEPR Theory

Unit: Covalent Bonding & Molecular Geometry

For example, in CH_4, the electron clouds around carbon are the four bonds to the hydrogen atoms. These electrons repel, which means they get as far apart as possible. In the Lewis structure, we draw the bonds at 90° angles, which is as far apart as possible in a 2-dimensional drawing:

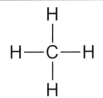

However, the molecule is really 3-dimensional. This means the bonds are actually equally spaced around a *sphere*. This would result in a 3-dimensional molecule, with the hydrogens at 109.5° angles around the carbon atom:

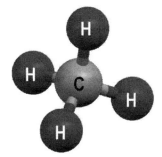

If we described this molecule as a geometric shape, it would be a regular (all edges and angles equal) tetrahedron, with the carbon atom in the center and hydrogen atoms at the vertices:

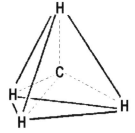

This means that, according to VSEPR theory, CH_4 is a "tetrahedral" molecule.

Use this space for summary and/or additional notes:

VSEPR Theory

Unit: Covalent Bonding & Molecular Geometry

Now, consider the NH₃ molecule. The Lewis structure looks like this:

The "lone pair" of electrons (above the N atom) and the three bonds all repel each other.

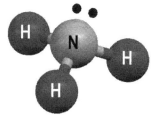

This gives *four* electron clouds, just like CH₄. However, because the lone pair is closer to the nucleus, it repels the other electrons more strongly than bond electrons. This compresses the bond angles slightly, to about 107.5°.

This time, the shape of the molecule is a triangular pyramid, but not a regular tetrahedron. This means the VSEPR shape of NH₃ is "trigonal pyramidal". The shape at right shows the pyramid (from the N atom downward), plus the "invisible" lone pair of electrons above.

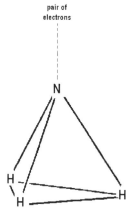

H₂O has two bonds to hydrogen atoms, and two lone pairs of electrons.

The VSEPR shape of the H₂O molecule is therefore "bent". Because the lone pairs are closer to the nucleus, they repel a little more strongly than bond electrons, and the bond angle compresses to 104.5°.

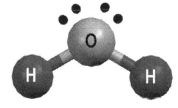

Use this space for summary and/or additional notes:

Now, suppose we have a molecule with a double bond, such as CH_2O. The Lewis structure is:

The electrons around the carbon atom are in <u>three</u> clouds, two smaller clouds for the C-H single bonds, and one larger cloud for the C=O double bond.

If these bonds got as far apart as possible in 3-dimensional space, they would be the points of a triangle, all in the same plane. This means that CH_2O is a "trigonal planar" molecule:

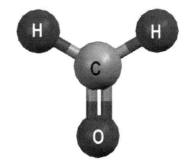

Finally, the CO_2 molecule has the following Lewis structure:

$\ddot{O}=C=\ddot{O}$

It has two large electron clouds for the C=O double bonds. The farthest these clouds can be from each other is 180° apart. This means the molecule forms a straight line, and its VSEPR shape is "linear":

The VSEPR shapes in this document are summarized in the table on the following page.

Use this space for summary and/or additional notes:

VSEPR Theory

Unit: Covalent Bonding & Molecular Geometry

Table of VSEPR Shapes

Electron Clouds	Bond Atoms	Lone Pairs	Hybridization	Bond Angle	Picture	VSEPR Shape
4	4	0	sp^3	109.5°		tetrahedral
4	3	1	sp^3	107.5°		trigonal pyramidal
4	2	2	sp^3	104.5°		bent
3	3	0	sp^2	120°		trigonal planar
3	2	1	sp^2	118°		bent
2	2	0	sp	180°		linear

Use this space for summary and/or additional notes:

VSEPR Theory

Unit: Covalent Bonding & Molecular Geometry

Homework Problems

For each of the following molecules, draw the Lewis structure. Then build a model of the molecule, and use your model to determine the shape of the electron clouds, and the shape of the molecule.

Formula	Lewis Structure	# of Electron Clouds around Central Atom	VSEPR shape
CHF_3	H—C(—F)(—F)(—F) with lone pairs on F	4	tetrahedral
NI_3			
H_2S			
CO_2			
AsH_3			
$COCl_2$			
PCl_3			

Use this space for summary and/or additional notes:

Orbital Hybridization

Unit: Covalent Bonding & Molecular Geometry
MA Curriculum Frameworks (2016): HS-PS1-2
MA Curriculum Frameworks (2006): N/A
Mastery Objective(s): (Students will be able to...)

- Determine the hybridization of the central atom in simple molecules.

Success Criteria:

- VSEPR shapes show the correct number of lone pairs in the correct locations and correct bond angles.
- Hybridization is correct (sp, sp^2 or sp^3).

Tier 2 Vocabulary: hybrid
Language Objectives:

- Explain how electron clouds change shape.

Notes:

<u>orbital</u>: the name for one of the spaces around an atom where electrons are.

<u>hybrid orbital</u>: an orbital whose shape is a hybrid of the shapes of different types of orbitals (such as a cross between an s-orbital and a p-orbital).

It is tempting to think of electrons as well-behaved particles that stay within the rigid boundaries defined by their energy levels. However, electrons are actually tiny charged particles moving randomly at speeds close to the speed of light. Because of their energies and the energies of the nuclei and the other electrons around them, they bounce around within a specific area. If that area is the shared electrons in a covalent bond, the region has a different shape than the electrons of an unbonded atom.

When atoms form covalent bonds, the electrons occupy the space between the two atoms. The space where the bonding electrons are is still called an orbital, even though its shape is now *different* from the shapes of the orbitals in the s, p, d, or f sub-levels of a single atom.

Use this space for summary and/or additional notes:

Orbital Hybridization

Unit: Covalent Bonding & Molecular Geometry

Recall that molecules with four electron clouds (tetrahedral, trigonal pyramidal, or bent with single bonds, like H_2O), are based on a tetrahedral VSEPR shape:

The shape of the orbitals surrounding the central atom is the shape determined by the four electron clouds. It looks like the following:

If we wanted to create four orbitals like this one by reshaping the *s* and *p* orbitals of an atom's valent shell, we would need to start with one *s* and three *p* orbitals. We therefore call this bonding orbital an **sp^3 hybrid orbital**, because it looks like a hybrid made from the one *s* and three *p* orbitals.

Similarly, molecules with three electron clouds are based on the trigonal planar VSEPR shape:

This hybrid orbital would come from one s and two p orbitals, and would be called an **sp^2 hybrid orbital**:

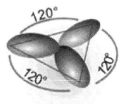

Finally, the hybrid orbital from one *s* and one *p* orbital is indeed called an **sp hybrid orbital**.

Use this space for summary and/or additional notes:

Orbital Hybridization

Summary of VSEPR Shapes for Hybrid Orbitals

Hybridization	VSEPR Shape(s)	Bond Angles
sp^3	tetrahedral	109.5°
	trigonal pyramidal,	107.5°
	bent	104.5°
sp^2	trigonal planar	120°
	bent	118°
sp	linear	180°

Use this space for summary and/or additional notes:

Orbital Hybridization

Page: 324

Unit: Covalent Bonding & Molecular Geometry

Homework Problems

For each of the following molecules, draw the Lewis structure. Then build a model of the molecule, and use your model to determine the shape of the electron clouds, and the shape of the molecule.

Formula	Hybrid-ization	Lewis Structure	# of Electron Clouds around Central Atom	VSEPR shape
CHF_3	sp^3	H—C(—F:)(—F:)(—F:) with lone pairs on F	4	tetrahedral
SCl_2				
SiO_2				
PH_3				
CH_2O				

Use this space for summary and/or additional notes:

Orbital Hybridization

Page: 325
Unit: Covalent Bonding & Molecular Geometry

Big Ideas | Details

Formula	Hybrid-ization	Lewis Structure	# of Electron Clouds around Central Atom	VSEPR shape
C₂H₂				
HCN				
NO₃⁻				
BF₃				

Use this space for summary and/or additional notes:

Summary: Bonding & Molecular Geometry

Unit: Bonding & Molecular Geometry

List the main ideas of this chapter in phrase form:

Write an introductory sentence that categorizes these main ideas.

Turn the main ideas into sentences, using your own words. You may combine multiple main ideas into one sentence.

Add transition words to make your writing clearer and rewrite your summary below.

Use this space for summary and/or additional notes:

Introduction: Intermolecular Forces

Unit: Intermolecular Forces

Topics covered in this chapter:

Polar Bonds 329

Polar Molecules 335

Intermolecular Forces 340

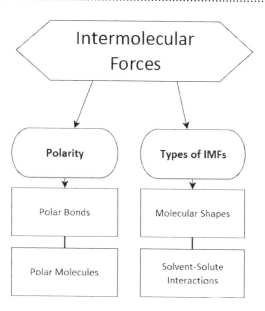

Use this space for summary and/or additional notes:

Introduction: Intermolecular Forces

Standards addressed in this chapter:

Massachusetts Curriculum Frameworks & Science Practices (2016):

HS-PS1-2 Use the periodic table model to predict and design simple reactions that result in two main classes of binary compounds, ionic and molecular. Develop an explanation based on given observational data and the electronegativity model about the relative strengths of ionic or covalent bonds.

HS-PS1-3 Cite evidence to relate physical properties of substances at the bulk scale to spatial arrangements, movement, and strength of electrostatic forces among ions, small molecules, or regions of large molecules in the substances. Make arguments to account for how compositional and structural differences in molecules result in different types of intermolecular or intramolecular interactions.

HS-PS2-7(MA) Construct a model to explain how ions dissolve in polar solvents (particularly water). Analyze and compare solubility and conductivity data to determine the extent to which different ionic species dissolve.

Massachusetts Curriculum Frameworks (2006):

4.3 Use electronegativity to explain the difference between polar and nonpolar covalent bonds.

4.5 Identify how hydrogen bonding in water affects a variety of physical, chemical, and biological phenomena (e.g., surface tension, capillary action, density, boiling point).

Use this space for summary and/or additional notes:

Polar Bonds

Unit: Intermolecular Forces

MA Curriculum Frameworks (2016): HS-PS1-2, HS-PS1-3

MA Curriculum Frameworks (2006): N/A

Mastery Objective(s): (Students will be able to...)

- Calculate the electronegativity difference between atoms in a bond.
- Identify polar bonds based on electronegativity differences.

Success Criteria:

- Bonds are correctly identified as polar or non-polar based on electronegativity difference.

Tier 2 Vocabulary: polar

Language Objectives:

- Explain how electrons are distributed unevenly in polar bonds.

Notes:

polar: anything with two sides that are opposite with respect to something. For example, a battery is polar because it has a positive and negative end.

polar bond: a covalent bond that has opposite partial charges on each side (one side partially positive and one side partially negative), because of unequal sharing of electrons.

Use this space for summary and/or additional notes:

Polar Bonds

For example: The bond between H and Cl in the H-Cl molecule is a polar bond.

The bond is polar because hydrogen and chlorine share a pair of electrons, but the sharing is not equal. Chlorine has an electronegativity of 3.16, but hydrogen has an electronegativity of only 2.2. This means the electrons spend more time with chlorine than with hydrogen.

One way to show a polar bond is by using a wedge-shaped bond, which is wider on the side where the electrons spend the most time. The HCl molecule would look like this:

H◀Cl

The wedge is narrower on the H side and wider on the Cl side because the chlorine atom has the electrons more of the time.

It is also common to label atoms in the structure with partial charges. The lower-case Greek letter "delta" (δ) is used to mean "partial". A partially positive charge would be shown as δ+ and a partially negative charge would be shown as δ−, as in the following example:

In the above example, hydrogen has a partial positive charge, and chlorine has a partial negative charge.

Use this space for summary and/or additional notes:

Bond Type and Bond Character

bond type: whether we define a bond as ionic or covalent. This depends on several factors, but for the purposes of this class, it just depends on which types of elements the bond is joining.

- covalent bond: between two nonmetals
- ionic bond: between two ions (usually a metal & nonmetal)
- metallic bond: between two metals

bond character: what a bond *acts like*, based on how equally the electrons are shared. This depends on the electronegativities (χ) of the two elements. In simplest terms, the larger the difference between the two electronegativities ($\Delta\chi$) the greater the extent to which the electrons are associated with one element more than the other.

American chemist Linus Pauling developed the following formula for calculating bond character based on the electronegativity difference between the elements sharing the bond:

$$\% \text{ ionic character} = 1 - e^{-\frac{1}{4}(\chi_A - \chi_B)^2}$$

The following is a graph of Pauling's formula:

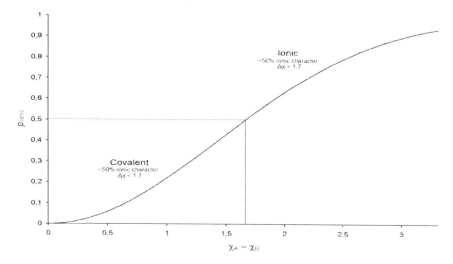

Notice that an ionic character of 0.5 (50 %) would result from an electronegativity difference of approximately 1.7. Therefore, we can define a bond as being primarily ionic in character if the electronegativity difference between the two atoms in the bond is 1.7 or more.

A bond that is less than 50 % ionic is said to be primarily covalent in character.

Use this space for summary and/or additional notes:

Polar Bonds

Big Ideas	Details

However, even covalent bonds can have measurably unequal sharing. it takes a relatively small difference to create a measurable bond polarity. The lower boundary is defined by the polarity of a C—H bond, which has no observable polarity under most conditions. The difference between the Pauling electronegativity of carbon (2.55) and hydrogen (2.20) is 0.35, so $\Delta\chi = 0.35$ is chosen to be the maximum electronegativity difference for a nonpolar bond. This represents about 3 % ionic character.

These values are summarized in the following table:

Electronegativity Difference ($\Delta\chi$)	% Ionic Character	Bond Character
0.35 or less (C—H)	< 3 %	nonpolar covalent
between 0.35 and 1.7	3 %–50 %	polar covalent
1.7 or more	> 50 %	ionic

Use this space for summary and/or additional notes:

Bond Triangle

As it turns out, looking only at the difference between the electronegativities of the elements in a chemical bond is not always a good predictor of bond character. For example, metallic bonds behave neither like ionic nor covalent bonds, and electronegativity difference could not predict metallic bond character. The idea of a triangular region, with covalent, ionic, and metallic bonds in each of the corners has been around since the 1920s. The modern bond triangle was developed by two chemists: the Dutch chemist Anton Eduard van Arkel published a bond triangle in 1941 that placed compounds with in specific positions around the triangle. In 1947, the Dutch chemist J. A. A. Ketelaar published a significant paper expanding on van Arkel's ideas. Bond triangles are often called van Arkel-Ketelaar triangles after these two chemists.

American chemist William Jensen quantified the van Arkel-Ketelaar triangle in a 1995 paper, in which he plotted the average of the two electronegativities on the *x*-axis, and the difference between the two electronegativities on the *y*-axis.

The following bond triangle, which gives reasonably good predictions of the nature of various chemical bonds based on each atom's electronegativity, was published by the chemistry department at Purdue University:.

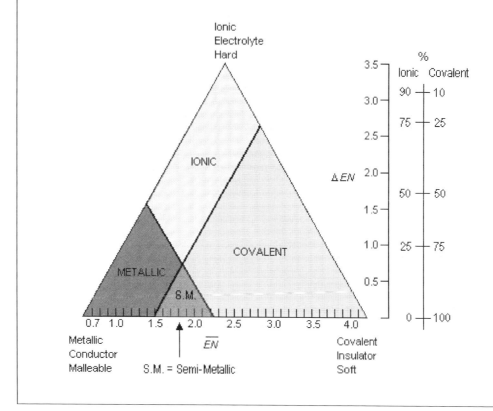

Use this space for summary and/or additional notes:

Polar Bonds

Homework Problems

Complete the table. You will need to look up electronegativity values (χ) from **"Error! Reference source not found."** Error! Bookmark not defined. of your Chemistry Reference Tables.

Elements	Bond Type	χ1	χ2	Δχ	Bond Character
Pb–S	ionic	2.33	2.58	0.25	nonpolar covalent
Ag–Cl					
Cu–C					
C–N					
C–I					
H–O					
Al–Cl					
K–F					
N–H					
N–O					
C–S					
Ba–Cl					
S–O					
Si–H					

Use this space for summary and/or additional notes:

Polar Molecules

Unit: Intermolecular Forces
MA Curriculum Frameworks (2016): HS-PS1-2, HS-PS1-3
MA Curriculum Frameworks (2006): N/A
Mastery Objective(s): (Students will be able to...)

- Draw polarity arrows indicating polarity of a molecule.

Success Criteria:

- Arrow is in the correct direction and points from the δ+ atom to the δ− atom.

Tier 2 Vocabulary: polar

Language Objectives:

- Explain how electrons are distributed unevenly in polar molecules.

Notes:

<u>polar bond</u>: a covalent bond that has opposite partial charges on each end (one end partially positive and one end partially negative), because of unequal sharing of electrons.

<u>polar molecule</u>: a molecule that can be oriented so that it has opposite charges on opposite sides.

In order to be polar, a molecule must have both:

1. one or more polar bonds
2. an "axis of asymmetry," meaning a way to orient the molecule so that there is more partial positive charge on one side (relative to the central atom), and more partial negative charge on the opposite side.

Use this space for summary and/or additional notes:

Polar Molecules

For example, the CH₃Cl molecule is polar, because the C-Cl bond is polar (Δ𝜒 = 0.61), and because you can view the central atom (carbon) so that the negative charge (towards chlorine) is on one side:

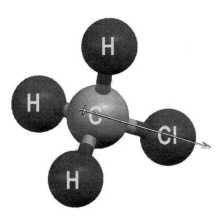

Notice the polarity arrow running from C to Cl in the molecule. The arrow shows the direction of polarization (pointing towards the more electronegative atom), and the "+" at the tail of the arrow indicates the end that has the partial positive (δ+) charge. Polarity arrows are often used with Lewis structures:

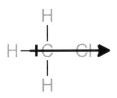

Use this space for summary and/or additional notes:

Polar Molecules

Because lone pairs of electrons change the symmetry of the molecule, trigonal pyramidal and bent molecules that contain polar bonds will usually be polar molecules. For example, NH_3 is polarized towards the nitrogen atom:

and water is polarized towards the oxygen atom:

However, if a molecule has multiple polar bonds that are pulling equally in opposite directions, then the forces cancel out and the molecule is not polar. An example of a *nonpolar molecule* that does have polar bonds is CCl_4.

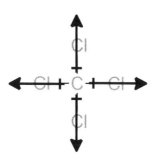

Each of the C-Cl bonds is polar, but the forces all cancel out, so there is no net force in any direction.

Use this space for summary and/or additional notes:

Polar Molecules

Dipole Moment

The polarity of a molecule can be expressed quantitatively as its dipole moment.

moment: in physics, the degree to which mass is spread out from the center of (potential) rotation. For example, an object's moment of inertia measures how much the object resists forces that would cause it to rotate.

dipole moment: a measure of how strongly a dipole will react to an external field. The dipole moment is expressed as the moment of inertia caused by a pair of charges (+q and −q) separated by a distance d.

The dipole moment (μ) is expressed by the formula:

$$\mu = qd$$

The unit for dipole moment is the debye (D).

Stronger charges (or partial charges) and/or greater distance between those charges will result in a larger dipole moment (and therefore a more polar molecule).

Use this space for summary and/or additional notes:

Polar Molecules

Unit: Intermolecular Forces

Homework Problems

For each of the following Lewis structures:
- Draw polarity arrows next to each polar bond.
- If the molecule is polar, draw a large polarity arrow for the molecule.
- If the molecule is non-polar, write "non-polar" next to it.

1. H—P(..)—H with H below (PH₃)

2. :N≡N:

3. H—O(..)(..)—H

4. H₂C=CH₂ (ethene)

5. CF₃I (F, F, I around central C with F on top)

6. :F—C≡P:

7. H—N=O (with lone pair on N, lone pairs on O)

8. SiCl₄ (Cl around central Si)

9. H—N(..)H with H below (NH₃)

10. H₂C=O (formaldehyde)

Use this space for summary and/or additional notes:

Intermolecular Forces

Unit: Intermolecular Forces
MA Curriculum Frameworks (2016): HS-PS1-2, HS-PS1-3, HS-PS2-7(MA)
MA Curriculum Frameworks (2006): 4.3, 4.5
Mastery Objective(s): (Students will be able to...)
- Rank attractions from strongest to weakest based on the type of intermolecular force.

Success Criteria:
- Attractions are correctly identified and correctly ranked.

Tier 2 Vocabulary: polar

Language Objectives:
- Explain the different types of intermolecular forces and their relative strengths.

Notes:

intramolecular forces: forces within a molecule (chemical bonds)

intermolecular forces ("IMFs"): forces between molecules (solids & liquids). Weaker than intramolecular forces.

Note that both intramolecular forces (chemical bonds) and IMFs form because the process of these atoms or molecules coming together releases energy. If you wanted to separate the atoms/molecules/particles, you would need to add enough energy to make up for the amount of energy that was released.

Forming bonds always releases energy.
Breaking bonds always requires energy.

soluble: when the attraction between solvent molecules and solute molecules or ions is strong enough to keep the solute distributed throughout the solvent.

miscible: when two or more liquids are soluble in each other.

Recall the 3 types of compounds:

ionic: compound made of ions (usually metal + nonmetal), which have charges with integer values (±1 or more)

covalent: compound made by sharing of electrons (usually all nonmetals),

metallic: compound made of metal atoms with delocalized electrons

Use this space for summary and/or additional notes:

Intermolecular Forces

Types of intermolecular forces (IMF), strongest to weakest:

The stronger the IMFs, the <u>higher</u> the melting and boiling point of the compound, because you have to overcome the IMF in order to separate the molecules going from solid → liquid or liquid → gas.

<u>ion-ion</u>: force of attraction between ions. The strength of the force is based on Coulomb's Law:

$$F = \frac{k q_1 q_2}{d^2}$$

where k is a constant, q_1 and q_2 are the strengths of the two charges, and d is the distance between them. Bigger charges (larger values of q) mean stronger forces. (*E.g.,* the attraction between a +2 ion and a −3 ion will produce a force that's six times as strong as the attraction between a +1 ion and a −1 ion.) If charges are the same, smaller molecules (smaller value of d) have stronger forces.

<u>metallic bonds</u>: metal atoms that delocalize their electrons and are held together by the "sea" of electrons surrounding them.

<u>dipole-dipole</u>: the force of attraction between two polar molecules (dipoles). Recall that the strength of attraction is based on the <u>dipole moment</u> (μ) of the molecule, given by the formula:

$$\mu = qd$$

The partial charge (q) is produced by the electronegativity difference ($\Delta \chi$) between the two atoms of a polar bond.

<u>hydrogen bonds</u>: the strongest type of dipole-dipole forces. Occurs in molecules that contain hydrogen (χ = 2.20) and an element with an electronegativity larger than 3.0 (F, O, Cl, or N). The hydrogen bonds that hold water molecules together are what give water its unusual properties:

- Water is more dense as a liquid than as a solid.
- Water has an unusually high heat capacity (specific heat).
- Water has a relatively high melting and boiling point. (Almost all covalent compounds with a molecular weight as light as 18 amu are gases.)
- Water exhibits high surface tension and capillary action.
- Water is known as the "universal solvent".

Use this space for summary and/or additional notes:

Intermolecular Forces

Big Ideas | **Details**

London dispersion forces (induced dipoles): random movement of electrons causes temporary dipoles to form within molecules, causing very weak attraction. (Named after the chemist Fritz London.) All molecules have these dispersion forces, but they can only be observed when there aren't any stronger forces present. Because dispersion forces are attractions between electrons, molecules with more electrons (generally those with higher molecular mass) have stronger dispersion forces.

There can also be IMFs between different types of molecules:

ion-dipole attraction: attraction between an ion (full positive/negative charge) and a dipole (partial positive/negative charge). This attraction is stronger than dipole-dipole forces but weaker than ion-ion forces.

induced dipole: attraction between a permanent dipole and a compound that experiences only dispersion forces. The permanent dipole induces (creates) a temporary dipole in the other compound.

Intermolecular Force	Type of Compound	Strength
ion-ion	ionic (metal + nonmetal)	strongest
metal-metal	metallic (all metals)	↕
hydrogen bonds (strong dipole-dipole)	H with F, O, Cl, or N	
dipole-dipole forces (other than hydrogen bonds)	polar covalent (all nonmetals)	
dispersion	nonpolar covalent (all nonmetals)	weakest

Use this space for summary and/or additional notes:

Polar *vs.* Nonpolar Solvents

Because the molecules polar liquids (especially those with hydrogen bonds) attract each other, polar liquids will:

- dissolve other polar liquids (The two liquids are said to be *miscible*.)
- dissolve ions (from ionic compounds that can separate). *E.g.,* NaCl ions will dissolve in H_2O.

In general, polar liquids will not dissolve nonpolar liquids or other uncharged molecules.

In general, polar liquids can dissolve ionic compounds that have relatively small ion-ion forces (such as ions with +1 or −1 charges). In general, most polar liquids cannot dissolve most ionic compounds in which the charges of all of the ions are ±2 or higher. (However, note that this is a rule of thumb and there are *many* exceptions!)

Because polar liquids form dipole-dipole bonds, they will squeeze out nonpolar liquids. If you mix a nonpolar liquid and a polar liquid (such as oil and water), the liquids will form two separate phases. The popular expression to describe this phenomenon is "like dissolves like."

Use this space for summary and/or additional notes:

Big Ideas	Details	Intermolecular Forces	Page: 344
			Unit: Intermolecular Forces

Homework Problems

For each pair of compounds:
- Write down the strongest intermolecular force that occurs in each of the two compounds.
- Circle the compound of each pair that would have the higher melting and boiling point (stronger intermolecular forces).
- Give the reason(s) for your choice.

1. NaCl   HCl

2. CaS   KCl

3. H_2O N_2O

4. CO CO_2

5. HBr   HF

6. CH_4 C_8H_{18}

Use this space for summary and/or additional notes:

Chemistry 1   Mr. Bigler

Big Ideas	Details
	## Summary: Intermolecular Forces
Unit: Intermolecular Forces

List the main ideas of this chapter in phrase form:

Write an introductory sentence that categorizes these main ideas.

Turn the main ideas into sentences, using your own words. You may combine multiple main ideas into one sentence.

Add transition words to make your writing clearer and rewrite your summary below.

 |

Use this space for summary and/or additional notes:

Introduction: The Mole

Unit: The Mole

Topics covered in this chapter:

Moles 349
Percent Composition & Empirical Formula 357
Hydrates 363

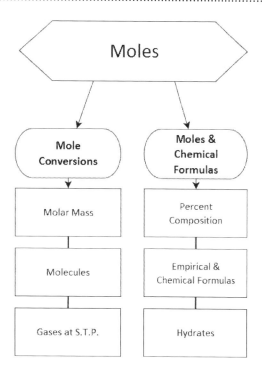

Standards addressed in this chapter:

Massachusetts Curriculum Frameworks & Science Practices (2016):

HS-PS1-7 Use mathematical representations and provide experimental evidence to support the claim that atoms, and therefore mass, are conserved during a chemical reaction. Use the mole concept and proportional relationships to evaluate the quantities (masses or moles) of specific reactants needed in order to obtain a specific amount of product.

Use this space for summary and/or additional notes:

Introduction: The Mole

Unit: The Mole

Big Ideas	Details
	Massachusetts Curriculum Frameworks (2006):
	5.3 Use the mole concept to determine number of particles and molar mass for elements and compounds.
	5.4 Determine percent compositions, empirical formulas, and molecular formulas.

Use this space for summary and/or additional notes:

Big Ideas	Details	Moles	Page: 349
			Unit: Moles

Moles

Unit: Moles

MA Curriculum Frameworks (2016): HS-PS1-2, HS-PS1-3

MA Curriculum Frameworks (2006): 5.3

Mastery Objective(s): (Students will be able to…)

- Determine the molar mass of a compound.
- Convert between moles, mass, volume, and molecules/atoms.

Success Criteria:

- Conversions are set up properly so undesired units are canceled and desired units appear in the correct place.
- Algebra and rounding to appropriate number of significant figures is correct.

Tier 2 Vocabulary: mole, molar

Language Objectives:

- Explain the concept of a quantitative collective noun (such as "dozen") and apply it to the mole concept.

Notes:

mole: (working definition) the amount a compound that is the same number of grams* as the compound's formula or molecular mass in amu.

mole: (formal definition) the amount of matter that contains the same number of objects (atoms, molecules, *etc.*) as the number of atoms in <u>exactly</u> 12 g of ^{12}C. When the British Imperial system was more commonly used, this quantity was often called a gram-mole.

Avogadro's constant: 1 mole = 6.022×10^{23} atoms, molecules, *etc.* (*Memorize this number!*)

molar volume: the space occupied by 1 mole of ANY gas. At S.T.P. (0 °C = 273 K) and 1 bar of pressure), the molar volume is 22.7 L. (*Memorize this number!*)

molar mass (m.m.): the mass (in grams) of 1 mole of a substance. For atoms, this is the same number as the atomic mass on the periodic table, but with the unit "grams". For compounds, add up the mass of each atom in the compound.

---

* Technically, this is only true for moles in the S.I. system, which used to be called "gram-moles". In the British Imperial system, a "pound-mole" is the amount of a compound that is the same number of *pounds* as its formula mass. Chemists use the term "ton-mole" the same way other people might use "@#%$-ton".

Use this space for summary and/or additional notes:

How Moles are Used in Chemistry

Moles are used as a way to make sure you have the desired number of atoms or molecules for a chemical reaction.

For example, consider the reaction:

$$2\ Fe + 3\ I_2 \rightarrow 2\ FeI_3$$

Suppose you wanted to perform this reaction in a lab with exact amounts of your starting materials (for example, let's say it's because you didn't want to have anything left over). You would need a ratio of 2 atoms of Fe for every 3 molecules of I_2. The problem is that all you have to work with is a jar of iron, a jar of iodine, and a balance.

To solve this problem, we define a mole as being the same number of grams as the average atomic mass. This means that a mole is always the same number of atoms or molecules (Avogadro's constant), so if we start with exactly 2 *moles* of Fe and exactly 3 *moles* of I_2, we will end up making exactly 2 *moles* of FeI_3.

You can think of a mole the same way you think of a dozen (or a 12-pack). If you have 3 dozen eggs and 6 dozen slices of toast, you could serve $\frac{1}{12}$ dozen eggs (one egg) and $\frac{2}{12}$ dozen slices of toast (2 slices) to each person. Similarly, if you have 2 moles of Fe (12.04×10^{23} atoms), and 3 moles of I_2 (18.06×10^{23} molecules), you can react them to make 2 moles of FeI_3 (12.04×10^{23} molecules).

In the chapter on "Stoichiometry," starting on page 443, we will use the coëfficients in chemical equations to calculate the number of moles of reactants used and the moles of products produced in a reaction.

Use this space for summary and/or additional notes:

Moles

Big Ideas | Details

Molar Mass

The molar mass is the mass (in grams) of one mole of a chemical. Molar mass is used to convert between grams and moles. (Recall the "Conversions (Factor-Label Method)" section on page 91.)

The molar mass of an atom is its (average) atomic mass, but expressed in grams instead of atomic mass units (amu):

    1 atom of Fe has a mass of 55.8 amu (from the periodic table), so
    1 mole of Fe has a mass of 55.8 g

The molar mass of a compound is the sum of the (average) atomic masses of the elements in the compound, but again expressed in grams instead of amu:

    1 molecule of I_2 has 2 atoms of I. One atom of I has a mass of 126.9 amu, so
    1 molecule of I_2 has a mass of 2 × 126.9 = 253.8 amu. This means that
    1 mole of I_2 has a mass of 2 × 126.9 = 253.8 g

Similarly:

    1 mole of FeI_3 contains 1 mole of Fe atoms and 3 moles of I atoms, so we add up all of the atoms in the molecule:

        1 mole Fe = 1 × 55.8 =   55.8 g
   + <u>3 moles I = 3 × 126.9 =   380.7 g</u>
        1 mole FeI_3 =   436.5 g

Use this space for summary and/or additional notes:

Moles

Big Ideas | **Details**

Homework Problems

Calculate the mass in grams of one mole of each of the following compounds.

1. HCl

   Answer: 36.46 $\frac{g}{mol}$

2. Fe_2O_3

   Answer: 36.46 $\frac{g}{mol}$

3. $LiAlH_4$

   Answer: 36.46 $\frac{g}{mol}$

4. $C_6H_{12}O_6$

   Answer: 36.46 $\frac{g}{mol}$

5. $Ca_3(PO_4)_2$

   Answer: 36.46 $\frac{g}{mol}$

Use this space for summary and/or additional notes:

Moles

Big Ideas	Details
	6. $UOCl_2$
	Answer: $36.46 \frac{g}{mol}$
	7. $NiCl_2$
	Answer: $36.46 \frac{g}{mol}$
	8. $(NH_4)_2SO_4$
	Answer: $36.46 \frac{g}{mol}$
	9. $AgNO_3$
	Answer: $36.46 \frac{g}{mol}$
	10. CH_3COOH
	Answer: $36.46 \frac{g}{mol}$

Use this space for summary and/or additional notes:

Mole Conversions

1 mol = ___ grams   (add up the mass of the formula)

1 mol = 6.02 x 10^{23} atoms, molecules, *etc.*

n mol = $\dfrac{PV}{RT}$ (1 mol of gas @ S.T.P.* = 22.7 L)

These conversions work just like the ones from earlier in the year.

Sample Problems:

1. 2.5 mol of NH$_3$ gas occupies what volume at S.T.P.?

$$\dfrac{2.5 \text{ mol NH}_3}{1} \times \dfrac{22.7 \text{ L NH}_3}{1 \text{ mol NH}_3} = 57 \text{ mol NH}_3$$

2. What is the mass of 4.1 mol NH$_3$ gas?

    The molar mass of 1 mol NH$_3$ is (1 × 14) + (3 × 1) = 17 g NH$_3$.

$$\dfrac{4.1 \text{ mol NH}_3}{1} \times \dfrac{17 \text{ g NH}_3}{1 \text{ mol NH}_3} = 70. \text{ g NH}_3$$

3. How many molecules are there in 0.75 mol of NH$_3$ gas?

$$\dfrac{0.75 \text{ mol NH}_3}{1} \times \dfrac{6.02 \times 10^{23} \text{ molecules NH}_3}{1 \text{ mol NH}_3} = 4.5 \times 10^{23} \text{ molecules NH}_3$$

4. What is the volume of 25.5 g of NH$_3$ gas at S.T.P.?

    The molar mass of 1 mol NH$_3$ = (1 × 14) + (3 × 1) = 17 g NH$_3$.

$$\dfrac{25.5 \text{ g NH}_3}{1} \times \dfrac{1 \text{ mol NH}_3}{17 \text{ g NH}_3} \times \dfrac{22.7 \text{ L NH}_3}{1 \text{ mol NH}_3} = 34.1 \text{ L NH}_3$$

Note: this chapter is a good time to start including the chemical formula as part of the units. This will be extremely useful when we study stoichiometry.

---

* S.T.P. = "Standard Temperature and Pressure". Since 1980, the official IUPAC definition of S.T.P. has been 0 °C and 100 kPa. Some texts (and the MA DESE) stubbornly insist on using the old definition of 0 °C and 1 atm. This would make the molar volume of an ideal gas 22.4 L instead of 22.7 L.

Use this space for summary and/or additional notes:

Moles

Homework Problems

1. How many moles are 65.0 grams of zinc?

   Answer: 0.99 mol Zn

2. How many moles are 1 250.5 g of lead (II) nitrate ($Pb(NO_3)_2$)?

   Answer: 3.775 mol lead (II) nitrate

3. How many moles are 2 500 g of tin (IV) chlorate ($Sn(ClO_3)_4$)?

   Answer: 5.5 mol tin (IV) chlorate

4. How many moles are 125.0 g of silver nitrate ($AgNO_3$)?

   Answer: 0.7357 mol silver nitrate

5. How many nitrogen atoms are there in 62.5 g of dinitrogen pentoxide?

   Answer: 0.579 mol dinitrogen pentoxide

6. How many oxygen atoms are there in 380 g of copper (II) phosphate?

   Answer: 1.00 mol copper (II) phosphate

7. How many hydrogen atoms in 454 g of aluminum hydroxide?

   Answer: 5.82 mol aluminum hydroxide

Use this space for summary and/or additional notes:

Moles

Page: 356
Unit: Moles

8. What is the mass (in grams) of 2.35 mol of S_2N_3?

   Answer: 249 g S_2N_3

9. What is the mass (in grams) of 0.25 mol of silver acetate?

   Answer: 42 g silver acetate

10. What is the mass (in grams) of a 2.00 kg bag of table sugar ($C_{12}H_{22}O_{12}$)?

    Answer: 2 000 g ☺

11. How many moles are in 123.5 L of oxygen gas at S.T.P.?

    Answer: 5.44 mol oxygen

12. How many moles are in a 40. gallon drum of chlorine gas at S.T.P.?
    (1 gal = 3.78 L)

    Answer: 6.7 mol chlorine gas

13. What is the volume (in liters) of 3.5 mol of argon gas at 1.1 atm and 20 °C?
    (Hint: this is not at S.T.P., so you need to use $PV = nRT$.)

    Answer: 76.5 L argon gas

14. What is the volume (in liters) of 4.90×10^{25} molecules of N_2 gas at S.T.P.?

    Answer: 1 850 L N_2 gas

Use this space for summary and/or additional notes:

Percent Composition & Empirical Formula

Unit: Moles

MA Curriculum Frameworks (2016): HS-PS1-2, HS-PS1-3

MA Curriculum Frameworks (2006): 5.4

Mastery Objective(s): (Students will be able to...)

- Determine the percentage of each element in a compound given its chemical formula.
- Determine the empirical and molecular formulas of a compound from percent composition data.

Success Criteria:

- Percentages are calculated correctly.
- Empirical and molecular formulas are calculated correctly.
- Subscripts in empirical and molecular formulas are whole numbers.
- Ratio of subscripts in empirical formulas is in lowest terms.
- Algebra and rounding to appropriate number of significant figures is correct.

Tier 2 Vocabulary: mole, composition

Language Objectives:

- Accurately describe the process for converting percentages to moles.

Notes:

<u>percent composition</u>: the percentage by mass of each element in a compound.

<u>molecular formula</u> (chemical formula): a formula that gives the numbers and types of atoms in a molecule.

<u>empirical formula</u>: a chemical formula with the subscripts reduced to lowest terms. *E.g.*, the empirical formula for C_2H_4 would be CH_2. The empirical formula for C_8H_{16} would *also* be CH_2. (You may remember that we <u>*always*</u> use empirical formulas for ionic compounds.)

<u>formula mass</u>: the mass in grams represented by a chemical formula. Sometimes called <u>molecular mass</u>, <u>formula weight</u> or <u>molecular weight</u>. (This is the same number as the molar mass, but with units of atomic mass units (amu) instead of grams.)

Use this space for summary and/or additional notes:

Percent Composition & Empirical Formula

Determining Percent Composition

To determine the percent by mass of each element in a compound:

1. Determine the atomic mass of the element of interest
2. Determine the formula mass of the entire compound.
3. $\text{percent composition} = \dfrac{\text{atomic mass of element of interest}}{\text{formula mass of entire compound}} \times 100$

Sample Problem:

Q: What is the percentage of carbon in the compound $C_6H_{12}O_6$?

A: Mass of C_6 = 6 × 12.01 = 72.06

Mass of $C_6H_{12}O_6$:

C_6 = 6 × 12.01 = 72.06
H_{12} = 12 × 1.008 = 12.096
+ O_6 = 6 × 16.00 = 96.00
             180.156

$\dfrac{\text{mass of } C_6}{\text{mass of } C_6H_{12}O_6} = \dfrac{72.06}{180.156} = 0.400 \times 100 = 40.0\,\%$

Determining Empirical and Molecular Formulas

The lowest-terms ratio of the atoms in a chemical formula is the empirical formula.

The ratio of atoms is the same as the ratio of moles, which means you can find the empirical formula of a compound by determining the ratio of moles, and converting that ratio to whole numbers:

1. Find the molar mass (in grams) of each element in the compound.
2. Convert the grams to moles for each element.
3. Convert the number of moles of each element to whole-number subscripts. The easiest way to do this is by dividing them all by whichever number is the smallest.
4. If a subscript is within ±5 % of a whole number after dividing, you can round it off. If a subscript is not within ±5 % of a whole number, multiply *all* of the subscripts by the smallest number that would cause *all* of the subscripts to be whole numbers (within ±5 %).

Use this space for summary and/or additional notes:

Percent Composition & Empirical Formula

Big Ideas | **Details** | Page: 359
Unit: Moles

Sample Problem:

A sample of a chemical compound contains 8.56 g of carbon and 1.44 g of hydrogen. What is the empirical formula of this compound?

1. Masses are C: 8.56 g and H: 1.44 g. Write the formula as $C_{8.56\,g}H_{1.44\,g}$

2. Convert grams to moles:

   C: $\dfrac{8.56\,g}{1} \times \dfrac{1\,mol}{12.011\,g} = 0.713\,mol$

   H: $\dfrac{1.44\,g}{1} \times \dfrac{1\,mol}{1.008\,g} = 1.429\,mol$

   The formula for this compound is therefore a whole-number ratio that equals $C_{0.713}H_{1.429}$

3. Convert the subscripts to simple whole numbers. The easiest way to do this is to divide them all by the smallest one and see what happens.

   $C_{\frac{0.713}{0.713}}H_{\frac{1.429}{0.713}} = C_1H_{2.004}$

4. Round the empirical formula off. $CH_{2.004}$ becomes CH_2. (You can—*and should*—round, as long as you are within ±5 %.)

Hints:

If the problem gives percentages instead of actual mass, just pretend the percentages are out of 100 g total. *E.g.*, if you had a compound containing 25.3 % nitrogen, you would use 25.3 g of nitrogen in your calculations.

If you have something like $NO_{2.5}$, you can't round 2.5 off to 2 or 3. Instead, you need to multiply both subscripts by 2, which gives you N_2O_5. (This means it's important to be able to recognize decimal equivalents for simple fractions, such as $0.50 = \frac{1}{2}$, $0.33 = \frac{1}{3}$, $0.25 = \frac{1}{4}$, $0.20 = \frac{1}{5}$, *etc.*)

Use this space for summary and/or additional notes:

Percent Composition & Empirical Formula

Empirical Formula *vs.* Molecular (actual) Formula

If you know the molar mass of the compound, you can use it to get from the empirical formula to the molecular formula.

For example, suppose you were told that the actual molar mass of the hydrocarbon from the example above is $42.08 \frac{g}{mol}$.

The empirical formula mass (*i.e.*, the molar mass of the empirical formula CH_2) is $(1 \times 12.011) + (2 \times 1.008) = 14.027$.

The actual molar mass of 42.08 is 3 times as much, *i.e.*, $\frac{42.08}{14.027} = 3.00$.

This means the molecule contains exactly 3 of the empirical formula units, so we need to multiply all of the subscripts by 3 to get the molecular formula:

$CH_2 \times 3 = \boxed{C_3H_6}$.

Use this space for summary and/or additional notes:

Percent Composition & Empirical Formula

Homework Problems

1. A 5.00 g sample of a compound was found to contain 1.93 g carbon, 0.49 g hydrogen and 2.58 g sulfur. What is the empirical formula of the compound?

    Answer: C_2H_8S

2. What is the percentage composition of each element in the compound tetrahydrocannabinol (THC), which has the formula $C_{21}H_{30}O_2$?

    Answers: C: 80.2 %; H: 9.6 %; O: 10.2 %

3. A sample of a compound was found to contain 42.56 g of palladium (Pd) and 0.80 g of hydrogen. If the molar mass of the compound is 2 216.8 $\frac{g}{mol}$, what is the molecular formula of the compound?

    Answer: Pd_2H_4

4. Find the empirical formula of a compound that contains 30.45% nitrogen and 69.55% oxygen.

    Answer: NO_2

Use this space for summary and/or additional notes:

Percent Composition & Empirical Formula

Page: 362
Unit: Moles

5. Find the percentage of boron in the compound boron triiodide (BI_3).

   Answer: 2.76 % boron

6. A compound containing only carbon and hydrogen has a molecular mass of 114.26 amu. If one mole of the compound contains 18.17 g of hydrogen, what is its molecular formula?

   Answer: C_8H_{18}

7. Find the molecular formula of a compound that contains 56.36 g of oxygen and 43.64 g of phosphorus. The molecular mass of the compound is 283.9 amu.

   Answer: P_4O_{10}

8. The compound caffeine has a molecular weight of 194.1926 amu. It contains 49.5% carbon, 5.2% hydrogen, 28.9% nitrogen, and 16.5% oxygen. What is its empirical formula? What is its molecular formula?

   Answers: empirical: $C_4H_5N_2O$;   molecular: $C_8H_{10}N_4O_2$

Use this space for summary and/or additional notes:

Hydrates

Unit: Moles

MA Curriculum Frameworks (2016): HS-PS1-2, HS-PS1-3

MA Curriculum Frameworks (2006): 5.4

Mastery Objective(s): (Students will be able to...)

- Determine the number of water molecules in a hydrate

Success Criteria:

- Empirical formula is calculated correctly (if necessary).
- Number of water molecules in each formula unit is calculated correctly.
- Formula of hydrate is written correctly, with the empirical first, then a dot, then the number of H_2O molecules.
- Algebra and rounding to appropriate number of significant figures is correct.

Tier 2 Vocabulary: hydrate

Language Objectives:

- Explain the concept of a hygroscopic compound.
- Explain the process of determining the amount of water in a hydrate, both numerically and experimentally.

Notes:

hydrate: an ionic solid that has H_2O molecules loosely bound to its crystals.

water of hydration: the water molecules that are bound into a hydrate.

anhydrous: a compound that has had its water of hydration removed, usually by heating.

hygroscopic: a compound that can absorb water from the air. In a humid environment, an anhydrous compound will absorb water until it becomes the hydrate.

Use this space for summary and/or additional notes:

Hydrates

Naming of Hydrates

The name of a hydrate is the name of the compound followed by a number prefix and the word "hydrate".

The number prefix (the same ones we used for molecular compounds; found in "Table J. Number Prefixes" on page 552 of your Chemistry Reference Tables) indicates the number of H_2O molecules in the hydrate. For example, the compound nickel (II) chloride forms a hydrate that contains six water molecules. Its name is therefore:

nickel (II) chloride hexahydrate

Chemical Formula of a Hydrate

The chemical formula of a hydrate is the chemical formula of the compound followed by a dot and the number of H_2O molecules bound to it. For example, the chemical formula of nickel (II) chloride hexahydrate is:

$$NiCl_2 \cdot 6\ H_2O$$

Molar Mass of a Hydrate

The molar mass of a hydrate includes the mass of the water of hydration. This means a hydrate will have a larger molar mass than the anhydrous compound. For example, the molar mass of $NiCl_2$ is 129.60 g. The molar mass of H_2O is 18.015 g. The molar mass of 6 H_2O molecules is 6 × 18.015 = 108.09 g.

Therefore, the molar mass of $NiCl_2 \cdot 6\ H_2O$ is 129.60 + 108.09 = 237.69 g

Experimentally Determining the Water of Hydration

You can figure out the formula of a hydrate by weighing it, heating it to remove the water of hydration, and figuring out how many moles of water were removed for every mole of the compound.

Use this space for summary and/or additional notes:

Hydrates

Sample Problem:

Q: Sodium sulfate forms a hydrate. We want to find the chemical formula of the hydrate. Suppose you weighed out 32.22 g of the hydrate. After heating it to remove all of the water, the final mass was 14.20 g.

A: The formula of anhydrous sodium sulfate is Na_2SO_4, which has a molar mass of 142.05 g. We have:

$$\frac{14.20 \text{ g } Na_2SO_4}{1} \times \frac{1 \text{ mol } Na_2SO_4}{142.05 \text{ g } Na_2SO_4} = 0.10 \text{ mol } Na_2SO_4$$

The amount of water removed was 32.22 − 14.20 = 18.02 g.

This 18.02 g of H_2O is:

$$\frac{18.02 \text{ g } H_2O}{1} \times \frac{1 \text{ mol } H_2O}{18.0152 \text{ g } H_2O} = 1.000 \text{ mol } H_2O$$

Our sample had 1 mole of H_2O and 0.1 mole of Na_2SO_4. This is 10 times as much H_2O as Na_2SO_4:

$$\frac{1 \text{ mol } H_2O}{0.1 \text{ mol } Na_2SO_4} = \frac{10 \text{ H}_2O}{1 \text{ Na}_2SO_4}$$

Therefore, the formula must be:

$$Na_2SO_4 \cdot 10 \text{ H}_2O$$

Use this space for summary and/or additional notes:

Hydrates

Homework Problems

1. What is the chemical formula of iron (III) chloride hexahydrate?

2. Give the stock name and molar mass of the compound $CoSO_4 \cdot 7H_2O$?

3. If 10.0 g of $Na_2CrO_4 \cdot 4H_2O$ is heated to constant mass (*i.e.*, until all of the water of hydration is removed), what will the final mass be?

    a. Find the molar mass of $Na_2CrO_4 \cdot 4H_2O$.

    b. Find the number of moles of $Na_2CrO_4 \cdot 4H2O$ in 10.0 g.

    c. Find the molar mass of anhydrous Na_2CrO_4.

    d. Convert the number of moles (*which you found in part b*) to grams, using the molar mass of the anhydrous compound (*which you found in part c*).

Use this space for summary and/or additional notes:

Hydrates

3. 14.70 g of a hydrate of $CaCl_2$ is heated to dryness. The anhydrous sample has a mass of 11.10 g after evaporating the H_2O. Use the following steps to determine the chemical formula of the hydrate.

   a. Find the moles of anhydrous compound left at the end.

   b. Find the moles of water evaporated. (You'll need to find the grams of water evaporated, and then convert to moles.)

   c. Find the ratio of the moles of water evaporated to the moles of the anhydrous compound. This will be the number of water molecules in the hydrate.

   d. Write the formula for the hydrate (*using the number of water molecules that you found in part c*)?

Use this space for summary and/or additional notes:

Big Ideas | Details | Summary: The Mole | Unit: The Mole

Summary: The Mole

Unit: The Mole

List the main ideas of this chapter in phrase form:

Write an introductory sentence that categorizes these main ideas.

Turn the main ideas into sentences, using your own words. You may combine multiple main ideas into one sentence.

Add transition words to make your writing clearer and rewrite your summary below.

Use this space for summary and/or additional notes:

Introduction: Solutions

Unit: Solutions

Topics covered in this chapter:

Solutions & Dissolution 371
Solubility 378
Concentration (Molarity) 385
Colligative Properties 390

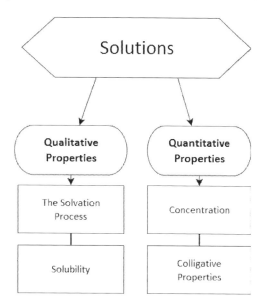

Standards addressed in this chapter:

Massachusetts Curriculum Frameworks & Science Practices (2016):

HS-PS1-11(MA) Design strategies to identify and separate the components of a mixture based on relevant chemical and physical properties.

HS-PS2-7(MA) Construct a model to explain how ions dissolve in polar solvents (particularly water). Analyze and compare solubility and conductivity data to determine the extent to which different ionic species dissolve.

Use this space for summary and/or additional notes:

Introduction: Solutions

Massachusetts Curriculum Frameworks (2006):

7.1 Describe the process by which solutes dissolve in solvents.

7.2 Calculate concentration in terms of molarity. Use molarity to perform solution dilution and solution stoichiometry.

7.3 Identify and explain the factors that affect the rate of dissolving (e.g., temperature, concentration, surface area, pressure, mixing).

7.4 Compare and contrast qualitatively the properties of solutions and pure solvents (colligative properties such as boiling point and freezing point).

Use this space for summary and/or additional notes:

Big Ideas	Details
	# Solutions & Dissolution

Unit: Solutions

MA Curriculum Frameworks (2016): HS-PS2-7(MA)

MA Curriculum Frameworks (2006): 7.1, 7.3

Mastery Objective(s): (Students will be able to…)
- Describe how a solution forms.
- Explain the effect of temperature changes on solubility.

Success Criteria:
- Descriptions account for solvent-solute interactions.
- Descriptions account for intermolecular forces.
- Explanations of the effect of temperature are consistent with solubility curves.

Tier 2 Vocabulary: solution

Language Objectives:
- Explain how solutes dissolve in solvents.

Notes:

<u>solute</u>: a substance that is broken down and dissolved into another substance. Solutes can be solids, liquids, or gases.

<u>solvent</u>: a substance that contains a solute. Solvents can be solids or liquids.

<u>solution</u>: a mixture that consists of a solute dissolved in a solvent.

<u>dissolution</u> or <u>solvation</u>: the process of a solute dissolving in a solvent.

<u>solubility</u>: the amount of a solute that can dissolve in a solvent. Often expressed in $\frac{mol}{L}$ or $\frac{g}{L}$.

<u>soluble</u>: when a solute can dissolve in a solvent.

<u>insoluble</u>: when a solute cannot dissolve in a solvent. Common threshold values are that solutes with solubilities of less than $1\frac{g}{L}$ or less than $0.01\frac{mol}{L}$ in a given solvent are considered insoluble.

<u>miscible</u>: when two liquids can dissolve in (mix freely with) each other

<u>dissociation</u>: when ions split apart in a solution. *E.g.,* when NaCl dissolves, the Na^+ and Cl^- ions separate and dissolve separately. |

Use this space for summary and/or additional notes:

Solutions & Dissolution

electrolyte: a solution that conducts electricity. Electrolytes are generally made when ionic compounds (salts) dissociate and dissolve, and the ions conduct electrons (electricity) through the solution.

saturated solution: a solution that holds as much solute as the solvent is capable of dissolving at a given temperature.

unsaturated solution: a solution that contains less solute than is capable of dissolving in a solvent.

supersaturated solution: a solution that temporarily contains more solute than is capable of remaining dissolved in a solvent. Supersaturated solutions are unstable.

A solution forms when solute molecules are dissolved in solvent molecules. This process involves the following steps:

1. Solvent molecules are attracted to the surface of the solute.

2. Intermolecular bonds (*e.g.,* ion-dipole bonds, hydrogen bonds, *etc.*) between solvent and solute particles pull the solute particles (ions, molecules, *etc.*) apart and into the solvent.

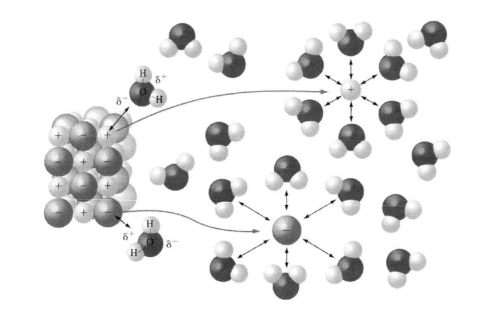

Use this space for summary and/or additional notes:

Enthalpy (Heat) of Solution

If a solute dissolves in a solvent, it is *always* the case that more energy had to be released when the solvent-solute intermolecular bonds are formed than it took to pull the solute particles apart. This means that the combined intermolecular forces between the solvent and solute particles are stronger than the intermolecular forces that had held the particles together in the solute.

If a solute does not dissolve, this means it would have taken more energy to pull the solute particles apart than the amount that would have been released by forming the solvent-solute intermolecular bonds. This means that the combined intermolecular forces between the solute particles are stronger than the combined intermolecular forces between solvent and solute particles.

This energy can exist in two forms: enthalpy (heat) and entropy (how much the energy is spread out among the particles). Enthalpy and entropy are discussed in more detail in the chapter on "Thermochemistry (Heat)," starting on page 469.

If the solution gets hotter as the solute dissolves, this means energy was *released* in the form of enthalpy (heat).

If the solution gets colder as the solute dissolves, this means heat energy was *absorbed*. However, it still must be true that energy had to be released when the solute dissolved. (Otherwise it would not have done so.) This means that entropy must have increased, and that more energy was released in the form of entropy than was absorbed in the form of enthalpy (heat).

Use this space for summary and/or additional notes:

For example, if you mix a strong acid with sodium hydroxide (a strong base), the solution gets very hot. (In fact, it can get hot enough to boil!) However, if you mix a strong acid with sodium carbonate ("soda ash") or sodium hydrogen carbonate (baking soda), the solution gets cold, because it releases CO_2 gas. As the gas is released, its heat energy spreads out into the surroundings (the room), which is a large increase in entropy. This increase in entropy releases so much energy that it takes thermal energy (heat) away from the solution, cooling it off. This is why baking soda is a good choice for neutralizing strong acids, whereas sodium hydroxide would be a poor choice.

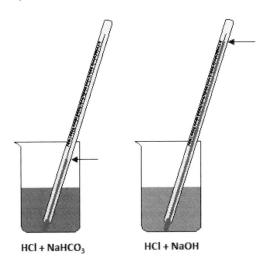

Use this space for summary and/or additional notes:

Solutions & Dissolution

Polar *vs.* Non-Polar Solvents

Whether a solute will dissolve in a solvent depends on the intermolecular forces between both the solvent and solute molecules. In both cases, the governing factor is the greater strength of ion-ion and dipole-dipole interactions as compared with London dispersion forces.

Polar Solvent

<u>polar or ionic solute</u>: polar or ionic solute particles are attracted to the positive and negative poles of the solvent molecules, which results in the solute dissolving.

<u>non-polar solute</u>: non-polar solute particles are not attracted to the solvent molecules. However, the solvent molecules are attracted to each other, and they exclude the solute.

Non-Polar Solvent

<u>polar or ionic solute</u>: polar or ionic solute particles are attracted to each other, but are not attracted to the solvent molecules, so they exclude the solvent and do not dissolve. (They form a precipitate, which means the solute falls (precipitates) to the bottom of the container.)

<u>non-polar solute</u>: neither the solute particles nor solvent molecules are strongly attracted to each other. (Both exhibit only London dispersion forces.) Because neither excludes the other, they spread out and intermingle freely.

A simple one-sentence statement of the above is *"Like dissolves like."*

This statement applies to liquids as well as solids. Polar liquids are miscible with each other; non-polar liquids are miscible with each other; however, non-polar liquids are not miscible with polar liquids. This is why "oil and water do not mix."

Use this space for summary and/or additional notes:

Factors that Affect the Rate of Dissolution

- temperature: solutes dissolve faster at higher temperatures because the molecules have more kinetic energy, which helps to pull the solute molecules apart.

- concentration: as the concentration increases, more and more solvent molecules are occupied with solute particles. The higher the concentration of the solute, the more slowly the solute will dissolve.

- surface area: because the solvent-solute interactions happen at the surface of the solute particles, the greater the surface area, the faster the solvent dissolves.

- pressure: for gases dissolving in liquids, increasing the pressure will force more gas molecules into the liquid, increasing the rate at which the gas dissolves.

- mixing: the faster a solution is mixed/stirred, the faster the solute will dissolve. This is because the mixing pulls dissolved solute particles away from each other, which is similar to lowering the concentration around the individual molecules.

Use this space for summary and/or additional notes:

Big Ideas	Details	Solutions & Dissolution	Page: 377 Unit: Solutions

Homework Problems

For each solute given, indicate whether water (H_2O) or cyclohexane (a nonpolar molecule) would be a better solvent.

1. KNO_3

2. paraffin (long-chain hydrocarbons, such as $C_{20}H_{42}$ or $C_{40}H_{82}$)

3. ethyl alcohol (CH_3–CH_2–OH)

4. acetic acid ($HC_2H_3O_2$)

5. mineral oil

6. ammonia (NH_3)

7. gasoline (short-chain hydrocarbons such as octane, C_8H_{18})

Use this space for summary and/or additional notes:

Solubility

Unit: Solutions

MA Curriculum Frameworks (2016): HS-PS2-7(MA)

MA Curriculum Frameworks (2006): 7.3

Mastery Objective(s): (Students will be able to...)
- Use solubility tables/rules to predict whether a solute will dissolve in water.
- Determine the amount of a solute that can dissolve from a solubility curve.

Success Criteria:
- Predictions about dissolution in water are correct.
- Amounts of solute that can dissolve are determined correctly.

Tier 2 Vocabulary: solution, curve

Language Objectives:
- Explain how solutes dissolve in solvents.

Notes:

In class, you saw a demonstration of the reaction between sodium carbonate (Na_2CO_3) and calcium chloride ($CaCl_2$):

$$Na_2CO_3\,(aq) + CaCl_2\,(aq) \rightarrow NaCl\,(aq) + CaCO_3\,(ppt) \quad (1)$$

When the solutions were mixed, the calcium carbonate that was formed immediately precipitated (formed an insoluble solid). Note that once the calcium carbonate is formed, it doesn't redissolve. *I.e.*, reaction (1) happens, but the reverse reaction (2), doesn't:

$$CaCO_3\,(s) + NaCl\,(aq) \not\rightarrow CaCl_2\,(aq) + Na_2CO_3\,(aq) \quad (2)$$

This is because of the way ionic compounds behave when they are dissolved in water.

If an ionic compound dissolves in water, it <u>dissociates</u> (splits) into its ions. In a chemical equation, we write "(aq)" (meaning "aqueous") after an ionic compound to show that it is dissolved, and is floating around in the solution as separate positive and negative ions.

For example, $CaCl_2$ splits into one Ca^{2+} ion and two Cl^- ions. The Ca^{2+} ions are attracted to the negative part of the H_2O molecule (the oxygen atoms), and Cl^- ions are attracted to the positive parts (the hydrogen atoms).

Use this space for summary and/or additional notes:

The combined attraction between the ions and the water molecules is stronger than the attraction between the Ca^{2+} ion and the Cl^- ion. The stronger attraction wins, which means the $CaCl_2$ dissolves:

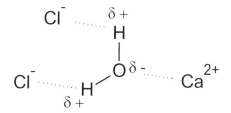

$CaCO_3$, on the other hand, does not dissociate. This must mean that the attraction between the Ca^{2+} ion and the CO_3^{2-} ion is stronger than the combined attraction between the ions and the water molecules. The stronger attraction wins, which means the $CaCO_3$ precipitates.

Note that if you mix the reactants and all of the ions remain in solution, nothing changes. *This means a chemical reaction **did not** occur.*

In other words, *a chemical reaction in an aqueous solution happens **only** if one of the products forms its own distinct phase—either a precipitate, a gas, or a separate liquid phase.*

Use this space for summary and/or additional notes:

Solubility

Solubility Rules

Solubility rules are rules of thumb that describe which compounds are likely to be soluble in water, and which are not.

Recall that the strength of ion-ion intermolecular forces is given by Coulomb's Law:

$$F = \frac{kq_1q_2}{d^2}$$

I.e., the attraction is proportional to the absolute value of the product of the charges ($|q_1q_2|$ — multiply the charges, and then change the sign so that the result is a positive number) and inversely proportional to the square of the distance between the ions.

It is usually (but not always) true that for the solute:

- if $|q_1q_2| \geq 4$, then the **ions'** attraction to each other is usually stronger, and the compound usually precipitates.
- if $|q_1q_2| < 4$, then the **solvent's** attraction to the ions is usually stronger, and the compound usually dissolves.

Note that there are several exceptions to both of these rules. Two examples are:

- hydroxides (OH⁻) and fluorides (F⁻) tend to form precipitates with +2 ions because they are very small ions, so the force of intermolecular attraction (F) is stronger because d^2 is smaller.

- cations (positive ions) of atoms with electronegativities significantly greater than 1 (such as Cu^{+1}, Ag^{+1}, and Pb^{+2}) have a stronger attraction for negative ions, and form precipitates with halogens (Cl⁻, Br⁻, and I⁻).

Use this space for summary and/or additional notes:

Solubility

Big Ideas	Details			

The following is a detailed set of solubility rules:

Ions That Form SOLUBLE Compounds	EXCEPT with	Ions That Form INSOLUBLE Compounds	EXCEPT with
Group 1 ions (Li$^+$, Na$^+$, etc.)		carbonate (CO$_3^{2-}$)	Group 1 ions, NH$_4^+$
ammonium (NH$_4^+$)		chromate (CrO$_4^{2-}$)	
nitrate (NO$_3^-$)		phosphate (PO$_4^{3-}$)	
hydrogen carbonate (HCO$_3^-$)		sulfite (SO$_3^{2-}$)	
chlorate (ClO$_3^-$)		sulfide (S^{2-})	Group 1 ions, NH$_4^+$, Group 2 ions
perchlorate (ClO$_4^-$)			
acetate (C$_2$H$_3$O$_2^-$ or CH$_3$COO$^-$)	Ag$^+$	hydroxide (OH$^-$)	Group I ions, NH$_4^+$, Ba^{2+}, Sr^{2+}, Tl$^+$
halides (Cl$^-$, Br$^-$, I$^-$)	Ag$^+$, Cu$^+$, Pb^{2+}, Hg$_2^{2+}$	oxide (O^{2-})	
sulfates (SO$_4^{2-}$)	Ca^{2+}, Sr^{2+}, Ba^{2+}, Ag$^+$, Pb^{2+}		

Use this space for summary and/or additional notes:

Solubility

Solubility Curves

solubility curve: a graph that shows the solubility of one or more compounds as a function of the temperature. Solubilities are usually expressed either in moles of solute per liter of solution (molarity) or grams of solute per 100 g of solvent.

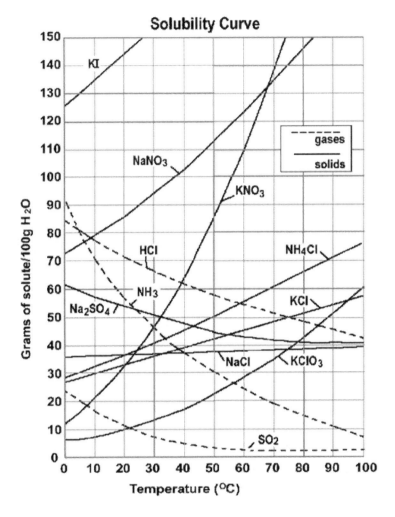

Sample Problem:

Q: How much $NaNO_3$ can dissolve in 50 g of water at 70 °C?

A: From the graph on the following page, the solubility of $NaNO_3$ at 70 °C is 135 g $NaNO_3$/100 g H_2O. Using this number as a conversion factor:

$$50 \text{ g } H_2O \times \frac{135 \text{ g } NaNO_3}{100 \text{ g } H_2O} = 67.5 \text{ g } NaNO_3$$

Use this space for summary and/or additional notes:

Solubility

Big Ideas | **Details**
Page: 383
Unit: Solutions

Homework Problems

For these problems, you will need to use the solubility curves in "Figure I. Solubilities of Selected Compounds" on page 551 of your Chemistry Reference Tables.

1. How much ammonium chloride could you dissolve in 100 g of water at 70 °C?

   Answer: about 61 g NH_4Cl

2. How much HCl could you dissolve in 25 g of water at 45 °C?

   Answer: 15 g HCl

3. If you made a saturated solution of ammonia in 40. g of water at 50. °C, how many grams of ammonia would it contain?

   Answer: 12 g NH_3

4. You want to dissolve 0.75 mol of KCl (F.W. = 74.55 $\frac{g}{mol}$) in 150. mL of water. What is the minimum temperature to which you would have to heat the water to dissolve all of the KCl?

   Answer: 34 °C

Use this space for summary and/or additional notes:

Solubility

Page: 384
Unit: Solutions

5. You have a solution that contains 43 g of an unknown compound dissolved in 100. g of H₂O at a temperature of 55 °C. The unknown compound could be either KCl, Na₂SO₄, KNO₃, or NaNO₃. Describe how you could perform a series of heating or cooling experiments and use a solubility chart to identify the solute in the unknown solution.

6. If you had 95 g of a saturated solution of sodium nitrate at room temperature (25 °C) and you cooled it to 10. °C, how much precipitate would form?
(*Note: the 95 g of solution includes both the NaNO₃ and the water.*)

Answer: 6 g

Use this space for summary and/or additional notes:

Big Ideas	Details

Concentration (Molarity)

Page: 385
Unit: Solutions

Unit: Solutions

MA Curriculum Frameworks (2016): HS-PS2-7(MA)

MA Curriculum Frameworks (2006): 7.3

Mastery Objective(s): (Students will be able to...)
- Calculate the concentration of a solution in $\frac{mol}{L}$.
- Calculate the final concentration of a solution after dilution.

Success Criteria:
- Solutions have the correct quantities substituted for the correct variables.
- Algebra and rounding to appropriate number of significant figures is correct.

Tier 2 Vocabulary: concentration, molar

Language Objectives:
- Explain how concentration is calculated.

Notes:

concentration: how much of something (solute) is dissolved in something else (solvent).

molarity (M): a unit of concentration equal to $\frac{\text{moles of solute}}{\text{L of solution}}$ or $\frac{mol}{L}$.

dilution: the process of decreasing the concentration of a substance by adding more solvent.

dilute: a solution that has a low concentration of solute dissolved in it.

There are three common types of problems involving molarity:

1. Find the molarity of a solution containing ___ grams/moles of solute with a volume of ___ L.

2. "How many moles/grams of a chemical would you need to make ___ L of a ___ M solution?" OR "What volume would you add to ___ moles/grams of a chemical to make a ___ M solution?"

3. What volume of ___ M solution would you add to water to make ___ L of a ___ M solution.

Use this space for summary and/or additional notes:

Concentration (Molarity)

Page: 386
Unit: Solutions

1. **Determining Concentration**

   To calculate the molarity of a solution:

   1. find the moles of solute
   2. find the liters of solution
   3. divide the moles by the liters

   For example: Determine the molarity of a solution made by dissolving 0.25 mol of $CuSO_4$ in enough water to make a total volume of 500 mL (0.5 L) of solution.

   $$\frac{0.25 \text{ mol } CuSO_4}{0.5 \text{ L solution}} = 0.5 \frac{\text{mol } CuSO_4}{L} = 0.5 \text{ M } CuSO_4$$

   (pronounced "0.5 molar copper sulfate").

2. **Determining the mass of solute or the volume of water needed**

   To solve these problems, use the molarity as a conversion factor, rewriting M as $\frac{\text{mol}}{L}$ (for example, 1.75 M would be $\frac{1.75 \text{ mol}}{1 L}$, or 1.75 mol = 1 L).

   For example:

   How many moles of $AgNO_3$ would you need to dissolve in water to make 100. mL (0.100 L) of an 0.50 M solution?

   $$\frac{0.100 \text{ L}}{1} \times \frac{0.50 \text{ mol}}{1 L} = 0.050 \text{ mol}$$

   Note that if the question had asked for <u>grams</u> of $AgNO_3$, you would then need to convert moles of $AgNO_3$ to grams.

Use this space for summary and/or additional notes:

Concentration (Molarity)

3. **Dilution problems**

   The idea is that the moles of solute before dilution must equal the moles of solute after dilution. Because molarity times volume equals moles $\left(\dfrac{mol}{L} \times L = mol\right)$, this means $M_1V_1 = M_2V_2$, where M_1 and M_2 are the molarities before and after dilution, respectively, and V_1 and V_2 are the volumes before and after dilution.

   For example:

   How much 0.50 M HCl would you need to add to water to make 2.0 L of an 0.10 M solution?

   $$M_1V_1 = M_2V_2$$
   $$M_1 = 0.50 \text{ M}$$
   $$V_1 = V_1$$
   $$M_2 = 0.10 \text{ M}$$
   $$V_2 = 2.0 \text{ L}$$
   $$(0.50)V_1 = (0.10)(2.0)$$

   $$V_1 = \dfrac{(0.10)(2.0)}{0.50} = 0.40 \text{ L}$$

Use this space for summary and/or additional notes:

Concentration (Molarity)

Homework Problems

1. What is the molarity of a solution that contains 25.2 g of KNO_3 (F.W. = $101.1 \frac{g}{mol}$) dissolved in enough water to make a total volume of 200. mL of solution?

   Answer: 1.25 M

2. What is the molarity of a solution that contains 22.5 g of NaI (F.W. = $149.98 \frac{g}{mol}$) dissolved in enough water to make a total volume of 500. mL of solution?

   Answer: 0.300 M

3. How many grams of NaOH (F.W. = $40.00 \frac{g}{mol}$) would you dissolve in water to make 1.0 L of a 2.0 M solution?

   Answer: 80. g NaOH

Use this space for summary and/or additional notes:

Concentration (Molarity)

Page: 389
Unit: Solutions

Big Ideas	Details

4. How many grams of KCl (F.W. = 74.55 $\frac{g}{mol}$) would you dissolve in water to make 250. mL of 0.100 M solution?

Answer: 1.86 g KCl

5. How many mL of 12.0 M HCl would you add to water to make 500. mL of a 1.00 M solution?

Answer: 42.0 mL HCl

6. If you put two teaspoons (8.0 g) of sucrose ($C_{12}H_{22}O_{11}$) into 300. mL of coffee, what is the concentration of sugar in the resulting solution?

Answer: 0.078 M

Use this space for summary and/or additional notes:

Big Ideas	Details	Colligative Properties	Page: 390
			Unit: Solutions

Colligative Properties

Unit: Solutions
MA Curriculum Frameworks (2016): HS-PS2-7(MA)
MA Curriculum Frameworks (2006): 7.4
Mastery Objective(s): (Students will be able to...)
- Calculate boiling point elevation, freezing point depression, vapor pressure lowering and osmotic pressure.
- Calculate the molar mass of a solute, based on the grams of a solute added and its effect on the freezing or boiling point of water.

Success Criteria:
- Solutions use the equation appropriate for the information given.
- Solutions have the correct quantities substituted for the correct variables.
- Algebra and rounding to appropriate number of significant figures is correct.

Tier 2 Vocabulary: depression, elevation

Language Objectives:
- Explain why solutes cause changes in freezing and boiling point.

Notes:

<u>colligative properties</u>: properties of a solution that depend on the physical number of particles dissolved, but not on the chemical properties of those particles.

Solutes can affect the physical properties of a solution by "getting in the way" of the solvent molecules.

<u>molality</u> (m): the concentration of a solution measured in grams of solute per 1 000 g of solvent.

Notice that the molality depends only on the masses of the solute and solvent, not on the volume.

<u>van't Hoff factor</u> (i): the number of particles of solute that you get when the solute dissolves. For example, when you dissolve sodium phosphate (Na_3PO_4) in water, it breaks up into three Na^+ ions and one PO_4^{3-} ion, for a total of four particles. This means the van't Hoff factor for Na_3PO_4 is 4.

Note that the van't Hoff factor (i) is a measured quantity. If an ionic compound dissociates completely, the value of i can be approximated from the chemical formula. However, if a compound is a weak electrolyte (dissolves only partially), the value of i must be measured empirically.

Use this space for summary and/or additional notes:

Colligative Properties

Freezing Point Depression

When a solute is added to a solvent, the solvent particles must "push" the solute particles out of the way in order to form a solid, which requires energy. This means that in order to make the solution freeze, the temperature must be *lower*, in order to increase the amount of energy *given off* when the solution forms a solid.

This is why we put salt on ice in winter—the salt particles get in the way of the water freezing, which means the temperature has to be lower in order for the salt water to freeze. As long as the temperature is above this new freezing point, the solution stays liquid (*i.e.*, the ice melts).

$$\Delta T_f = imK_f$$

where:

i = van't Hoff factor

m = molality of the solute $\left(\dfrac{\text{mol solute}}{\text{kg solvent}}\right)$

K_f = freezing point depression constant. For H_2O, $K_f = 1.86\,\dfrac{°C}{m}$ (degrees Celsius per molal)

Sample problem:

What is the freezing point of 25 g Na_2SO_4 dissolved in 500 g of H_2O?

$$\Delta T_f = imK_f$$

$i = 3$ (because $Na_2SO_4 \rightarrow 2\,Na^+ + SO_4^{2-}$, which is a total of 3 ions)

$$m = \dfrac{25\text{ g Na}_2\text{SO}_4}{500\text{ g H}_2\text{O}} \times \dfrac{1\text{ mol Na}_2\text{SO}_4}{142.05\text{ g Na}_2\text{SO}_4} \times \dfrac{1000\text{ g H}_2\text{O}}{1\text{ kg H}_2\text{O}} = 0.352\,m$$

$K_f = 1.86\,\dfrac{°C}{m}$

$$\Delta T_f = imK_f$$

$$\Delta T_f = (3)(0.352\,m)\left(1.86\,\dfrac{°C}{m}\right)$$

$$\Delta T_f = 1.96\,°C$$

The normal boiling point of H_2O is 0 °C, and we just calculated that the freezing point is *lowered* by 1.96 °C. Therefore, the freezing point of the solution is:

$$T_f = -1.96\,°C$$

Use this space for summary and/or additional notes:

Colligative Properties

Big Ideas	Details	Unit: Solutions

Boiling Point Elevation

Solute particles attract solvent molecules as they boil and attempt to escape as a gas. The solution needs extra energy (higher temperature) in order to overcome this extra attraction. This is why solutions—liquids with solutes dissolved in them—boil at higher temperatures.

$$\Delta T_b = i m K_b$$

where:

i = van't Hoff factor (# solute particles from each molecule, sometimes called the "dissociation factor")

m = molality of the solute $\left(\dfrac{\text{mol solute}}{\text{kg solvent}}\right)$

K_b = boiling point elevation constant. For H_2O, $K_b = 0.52 \dfrac{°C}{m}$ (degrees Celsius per molal)

Calculations involving boiling point elevation are done exactly the same way as calculations involving freezing point depression.

Sample problem:

Q: It is often said that salt should be added to boiling water when cooking pasta because the salt will elevate the boiling point of the water, causing the pasta to cook faster. How much would one teaspoon (4 g) of salt raise the boiling point of 4 quarts (about 4 kg) of water?

A: 4 g of NaCl is approximately 0.068 mol.

The molal concentration of salt in the water is therefore
$$\dfrac{0.068 \text{ mol NaCl}}{4.0 \text{ kg } H_2O} = 0.017 \, m.$$

NaCl dissociates into to ions, so $i = 2$. $K_b = 0.52 \dfrac{°C}{m}$. Therefore:

$$\Delta T_b = i m K_b$$
$$\Delta T_b = (2)(0.017)(0.52) = 0.018 \, °C$$

The salt in the water would increase the boiling point from 100 °C to 100.018 °C. We can therefore discount the possibility that boiling point elevation makes any significant contribution to how quickly the pasta cooks.

Use this space for summary and/or additional notes:

Colligative Properties

Raoult's Law (Vapor Pressure Lowering)

Solute particles attract solvent molecules. This attraction is strong enough to prevent some of those solvent molecules from escaping into the vapor phase.

Vapor pressure is the number of molecules of liquid that can escape into the gas phase at a given temperature, expressed as a pressure. Therefore, the presence of solute particles lowers the vapor pressure of the solvent.

Specifically, Raoult's Law states that the partial pressure of *vapor* "i" (P_i) equals the vapor pressure of (pure) "i" ($P^°_{v,i}$) times the mole fraction of *liquid* "i" (χ_i) in the mixture:

$$P_i = P^°_{v,i}\chi_i$$

Sample problem:

A sealed chamber contains a solution of glucose dissolved in water at 22 °C. The vapor pressure of pure water at 22 °C is 2.6 kPa. If the mole fraction of glucose is 0.10, what is the partial pressure of water in the air space above the solution?

Answer:

If the mole fraction of glucose in the solution is 0.10, the mole fraction of water in the solution must be 1 − 0.10 = 0.90. Therefore, the partial pressure of water is:

$$P_{H_2O} = P^°_{v,H_2O}\chi_{H_2O}$$
$$P_{H_2O} = (2.6 \text{ kPa})(0.90) = 2.3 \text{ kPa}$$

Use this space for summary and/or additional notes:

Colligative Properties

Osmotic Pressure

Diffusion is the natural flow of molecules from a region of higher concentration to a region of lower concentration.

Recall from biology that osmosis is a form of diffusion in which solvent molecules are able to travel across a semi-permeable membrane (such as a cell membrane), but solute molecules cannot pass through. Therefore, the higher the concentration of solute molecules on one side of the membrane, the more strongly those solute molecules attract solvent molecules from the other side. The force of this attraction can be measured as a pressure.

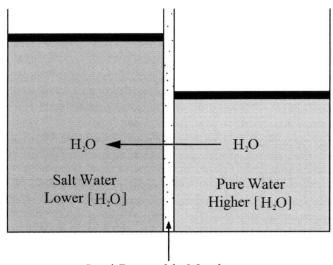

Semi-Permeable Membrane

This is why your skin wrinkles when it gets wet—the solutes inside your skin cells attract the pure water from outside your skin. As the water flows in through the cell membrane, it enlarges your skin cells. As your skin gets larger, the surface area gets larger, which we see as wrinkles. As your skin dries, the water escapes, the cells shrink, and the wrinkles disappear.

Use this space for summary and/or additional notes:

Colligative Properties

<u>osmotic pressure</u> (π): the observed pressure difference across a semi-permeable membrane because of differences in solute concentration. (Yes, it is awkward to think of the Greek letter π as a variable. Chemists are weird.)

Because osmotic pressure is a pressure, and because we are assuming the solute molecules otherwise obey kinetic-molecular theory (*i.e.*, they move freely, more or less like gas molecules), we can apply the ideal gas law. Note that we need to include the van't Hoff factor because *each* of the ions in solution created by dissolving a compound contributes separately to the osmotic pressure. This gives us the following formula:

$$\pi V = inRT$$

where:

π = osmotic pressure (additional pressure due to osmosis)
V = volume of solution
i = van't Hoff factor
n = moles of solute
R = gas constant
T = temperature (Kelvin)

Because molarity equals the moles of solute (n) divided by the volume of solution (V), the above equation can be simplified to:

$$\pi = iMRT$$

where M = molarity of the solute, and everything else is as above.

Use this space for summary and/or additional notes:

Colligative Properties

Homework Problems

1. If 45 grams of sodium chloride were added to 500. grams of water, what would the melting and boiling points be of the resulting solution? $K_b(H_2O) = 0.52 \frac{°C}{m}$ and $K_f(H_2O) = 1.86 \frac{°C}{m}$.

   Answer: M.P. = −5.73 °C   B.P. = 101.6 °C

2. What is the vapor pressure of the solution in problem #1 at 250 °C? The vapor pressure of pure water at 250 °C is 3.17 kPa.

   Answer: 3.08 kPa

3. If the solution in problem #1 (which has a density of $1.056 \frac{g}{mL}$) were placed on one side of a semipermeable membrane, and a 1.00 M solution of NaCl were placed on the other side of the membrane, what would the osmotic pressure be at 27 °C?

   Answer: 12.1 atm

Use this space for summary and/or additional notes:

Colligative Properties

Page: 397
Unit: Solutions

4. Which solution will have a higher boiling point: a solution containing 105 g of sucrose ($C_{12}H_{22}O_{11}$) in 500. g of water, or a solution containing 35 g of NaCl in 500. g of water?

Answer: for the sucrose, T_b = 100.32 °C   for the NaCl, T_b = 101.40 °C

5. 0.546 g of a compound with a van't Hoff factor of 1 was dissolved in 15.0 g of benzene. The freezing point of the solution was found to be 0.240 °C lower than the freezing point of pure benzene. If K_f for benzene is $K_f = 5.12\,\frac{°C}{m}$, what is the molar mass of the compound?

Answer: 776 $\frac{g}{mol}$

Use this space for summary and/or additional notes:

Big Ideas	Details
	# Summary: Solutions

Unit: Solutions

List the main ideas of this chapter in phrase form:

Write an introductory sentence that categorizes these main ideas.

Turn the main ideas into sentences, using your own words. You may combine multiple main ideas into one sentence.

Add transition words to make your writing clearer and rewrite your summary below.

 |

Use this space for summary and/or additional notes:

Introduction: Chemical Reactions

Unit: Chemical Reactions

Topics covered in this chapter:

Chemical Equations 401
Types of Chemical Reactions 405
Predicting the Products of Chemical Reactions 409
Activity (Reactivity) Series 414
Balancing Chemical Equations 417
Net Ionic Equations 425

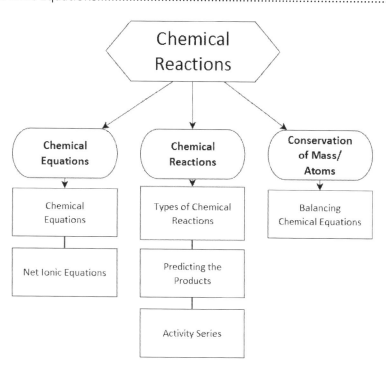

Use this space for summary and/or additional notes:

Introduction: Chemical Reactions

Unit: Chemical Reactions

Standards addressed in this chapter:

Massachusetts Curriculum Frameworks & Science Practices (2016):

HS-PS1-2 Use the periodic table model to predict and design simple reactions that result in two main classes of binary compounds, ionic and molecular. Develop an explanation based on given observational data and the electronegativity model about the relative strengths of ionic or covalent bonds.

HS-PS1-7 Use mathematical representations and provide experimental evidence to support the claim that atoms, and therefore mass, are conserved during a chemical reaction. Use the mole concept and proportional relationships to evaluate the quantities (masses or moles) of specific reactants needed in order to obtain a specific amount of product.

Massachusetts Curriculum Frameworks (2006):

5.1 Balance chemical equations by applying the laws of conservation of mass and constant composition (definite proportions).

5.2 Classify chemical reactions as synthesis (combination), decomposition, single displacement (replacement), double displacement, and combustion.

Use this space for summary and/or additional notes:

Big Ideas	Details
	# Chemical Equations
	Unit: Chemical Reactions
	MA Curriculum Frameworks (2016): HS-PS1-7
	MA Curriculum Frameworks (2006): 5.1
	Mastery Objective(s): (Students will be able to...)
	• Read, write, and interpret chemical equations.
	Success Criteria:
	• Equations have reactants and products on the correct sides of the arrow.
	• Physical states, heat, solvents, catalysts, *etc.* are present when appropriate.
	Tier 2 Vocabulary: equation
	Language Objectives:
	• Define the symbols used in chemical equations.
	Notes:
	<u>chemical equation</u>: a set of symbols that describe a chemical reaction. For example: $$2H_2(g) + O_2(g) \xrightarrow{\Delta} 2H_2O(\ell) + heat$$
	<u>reactants</u>: the starting materials; chemicals (and things like energy) that react. In a chemical equation, the reactants are before the arrow (on the left). In the above equation, the reactants are $H_2(g)$ and $O_2(g)$.
	<u>products</u>: chemicals (and other things like energy) that are produced. In a chemical equation, the products are after the arrow (on the right). In the above equation, the products are $H_2O(\ell)$ and heat.

Use this space for summary and/or additional notes:

Chemical Equations

state of matter: the symbols in parentheses after a compound indicates the physical state of that compound. Some of the common ones are listed in the following table:

States of Matter Used in Chemical Equations

Symbol	Meaning
(s)	solid
(ℓ)	liquid (A script "L" is often used to avoid confusion between the letter "l" and the number "1".)
(g)	gas or vapor
(cd)	condensed phase (*i.e.*, either solid or liquid)
(fl)	fluid phase (*i.e.*, either liquid or gas)
(cr)	crystalline (solid is in the form of crystals)
(lc)	liquid crystal
(vit)	vitreous (glass-like)
(ads)	adsorbed onto a substrate
(sln)	solution
(aq)	aqueous solution (dissolved in water)
(am)	amorphous solid
(ppt)	precipitate (solid) formed by the reaction

reaction conditions: anything that doesn't take part in the reaction, but is needed to make the reaction happen. Reaction condition information is placed above and/or below the arrow. Two common ones are:

- Δ under the arrow means that heat is required in order for the reaction to take place.
- A chemical formula under the arrow usually indicates the solvent that the reaction takes place in.

For example, the equation:

$$2\,H_2\,(g) + O_2\,(g) \xrightarrow{\Delta} 2\,H_2O\,(\ell) + \text{heat}$$

is equivalent to the following statement:

"Two molecules of hydrogen gas and 1 molecule of oxygen gas were heated to produce 2 molecules of liquid water and heat."

Use this space for summary and/or additional notes:

| Big Ideas | Details | Chemical Equations | Page: 403 |
| | | | Unit: Chemical Reactions |

Homework Problems

Write each of the following chemical equations in words.

1. $2\,H_2\,(g) + O_2\,(g) \xrightarrow{\Delta} 2\,H_2O\,(\ell) + 572\text{ kJ}$

2. $CaC_2\,(cr) + 2\,H_2O\,(\ell) \longrightarrow C_2H_2\,(g) + Ca(OH)_2\,(ppt)$

3. $2\,C_2H_2\,(g) + 5\,O_2\,(g) \xrightarrow{\Delta} 4\,CO_2\,(g) + 2\,H_2O\,(\ell) + 2\,600\text{ kJ}$

4. $3\,CaCl_2\,(aq) + 2\,K_3PO_4\,(aq) \longrightarrow Ca_3(PO_4)_2\,(ppt) + 6\,KCl\,(aq)$

Use this space for summary and/or additional notes:

Chemical Equations

Unit: Chemical Reactions

Write each of the following word problems as a chemical equation.

5. One molecule of silicon dioxide gas reacts with four molecules of aqueous hydrofluoric acid to produce a molecule of silicon tetrafluoride gas and two molecules of liquid water.

6. Two moles of aqueous potassium chlorate decompose to produce two moles of aqueous potassium chloride and three moles of oxygen gas.

7. Four moles of solid antimony are heated in the presence of three moles of oxygen gas to form one mole of solid antimony (III) oxide.

8. When 2 moles of liquid octane (C_8H_{18}) are burned in the presence of 25 moles of oxygen gas, 16 moles of carbon dioxide gas and 18 moles of liquid water are formed, and 10 150 kJ of heat is produced.

Use this space for summary and/or additional notes:

Big Ideas	Details	Types of Chemical Reactions	Page: 405
			Unit: Chemical Reactions

Types of Chemical Reactions

Unit: Chemical Reactions

MA Curriculum Frameworks (2016): HS-PS1-7

MA Curriculum Frameworks (2006): 5.1

Mastery Objective(s): (Students will be able to...)
- Recognize & identify the five major classes of chemical reactions.

Success Criteria:
- Reactions are correctly identified.

Tier 2 Vocabulary: reaction

Language Objectives:
- Explain what happens in each of the types of reaction.

Notes:

There are many types of chemical reactions. Five of the most common are:

<u>synthesis</u>: two or more reactants combine to form a single product. For example:

$$Na + Cl_2 \rightarrow NaCl$$

<u>decomposition</u>: one reactant disintegrates (decomposes) to form two or more products:

$$H_2CO_3 \rightarrow H_2O + CO_2$$

<u>single replacement</u> (sometimes called single displacement): atoms of one element replace atoms of another in a compound:

$$\underline{Al} + Cu Cl_2 \rightarrow \underline{Al}Cl_3 + Cu$$

Most of the single replacement reactions you will encounter involve metals reacting with ionic compounds. In this type of single replacement reaction, a positive ion (usually a metal) replaces the other positive ion, or a negative ion (often a non-metal) replaces the other negative ion.

Use this space for summary and/or additional notes:

Types of Chemical Reactions

Big Ideas | Details | Unit: Chemical Reactions

double replacement (sometimes called a double displacement or metathesis reaction): when two positive ions (or two negative ions) switch with each other to form two new compounds. For example:

$$CaCl_2 + Na_2CO_3 \rightarrow CaCO_3 + NaCl$$

Ca starts out paired with Cl, and Na is paired with CO_3. In the reaction, Ca and Na trade places so that Ca is now with CO_3 and Na is now with Cl. (Or you could think of it as Cl and CO_3 trading places—the result is the same.)

combustion: a special kind of reaction in which a hydrocarbon (a compound containing only carbon and hydrogen) reacts with O_2 (burns, or "combusts") to form CO_2 and H_2O. For example:

$$C_3H_8 \;(\ell) + 5\,O_2 \;(g) \rightarrow 3\,CO_2 \;(g) + 4\,H_2O \;(g) + heat$$

All flames are chemical reactions. (The flame itself is the light—photons of energy—produced by the reaction.) Most flames are produced by combustion reactions involving hydrocarbons and oxygen.

The internal combustion engine in your car is a special chemical reactor, in which octane (C_8H_{18}) and other hydrocarbons combust in a chamber (cylinder), producing heat. The heat makes the gases inside the cylinder expand. The expanding gases push the piston, which makes the car go.

Use this space for summary and/or additional notes:

Big Ideas	Details	Types of Chemical Reactions	Page: 407
			Unit: Chemical Reactions

Homework Problems

For each of the following chemical reactions, indicate whether the type of reaction is:

- synthesis
- decomposition
- single replacement
- double replacement
- combustion
- none of the above

1. $H_2 + O_2 \rightarrow H_2O$

2. $S_8 + O_2 \rightarrow SO_2$

3. $HgO \rightarrow Hg + O_2$

4. $Zn + HCl \rightarrow ZnCl_2 + H_2$

5. $Na + H_2O \rightarrow NaOH + H_2$

6. $C_{10}H_{16} + Cl_2 \rightarrow C + HCl$

7. $Si_2H_2 + O_2 \rightarrow SiO_2 + H_2O$

8. $Fe + O_2 \rightarrow Fe_2O_2$

9. $C_7H_6O_2 + O_2 \rightarrow CO_2 + H_2O$

10. $FeS_2 + O_2 \rightarrow Fe_2O_2 + SO_2$

11. $Fe_2O_2 + H_2 \rightarrow Fe + H_2O$

12. $K + Br_2 \rightarrow KBr$

13. $C_2H_2 + O_2 \rightarrow CO_2 + H_2O$

14. $H_2O_2 \rightarrow H_2O + O_2$

15. $C_7H_{16} + O_2 \rightarrow CO_2 + H_2O$

16. $SiO_2 + HF \rightarrow SiF_4 + H_2O$

17. $KClO_2 \rightarrow KCl + O_2$

18. $KClO_2 \rightarrow KClO_4 + KCl$

19. $P_4O_{10} + H_2O \rightarrow H_3PO_4$

20. $Sb + O_2 \rightarrow Sb_4O_6$

21. $C_2H_8 + O_2 \rightarrow CO_2 + H_2O$

22. $Fe_2O_2 + CO \rightarrow Fe + CO_2$

23. $PCl_5 + H_2O \rightarrow HCl + H_3PO_4$

24. $H_2S + Cl_2 \rightarrow S_8 + HCl$

Use this space for summary and/or additional notes:

Types of Chemical Reactions

25. $Fe + H_2O \rightarrow Fe_2O_4 + H_2$

26. $N_2 + H_2 \rightarrow NH_2$

27. $N_2 + O_2 \rightarrow N_2O$

28. $CO_2 + H_2O \rightarrow C_6H_{12}O_6 + O_2$

29. $SiCl_4 + H_2O \rightarrow H_4SiO_4 + HCl$

30. $H_2PO_4 \rightarrow H_4P_2O_7 + H_2O$

31. $CO_2 + NH_2 \rightarrow OC(NH_2)_2 + H_2O$

32. $Al(OH)_2 + H_2SO_4 \rightarrow Al_2(SO_4)_2 + H_2O$

33. $Fe_2(SO_4)_2 + KOH \rightarrow K_2SO_4 + Fe(OH)_2$

34. $H_2SO_4 + HI \rightarrow H_2S + I_2 + H_2O$

35. $Al + FeO \rightarrow Al_2O_2 + Fe$

36. $Na_2CO_2 + HCl \rightarrow NaCl + H_2O + CO_2$

37. $P_4 + O_2 \rightarrow P_2O_5$

38. $K_2O + H_2O \rightarrow KOH$

39. $Al + O_2 \rightarrow Al_2O_3$

40. $Na_2O + H_2O \rightarrow NaOH + O_2$

41. $C + H_2O \rightarrow CO + H_2$

42. $H_3AsO_4 \rightarrow As_2O_5 + H_2O$

43. $Al_2(SO_4)_2 + Ca(OH)_2 \rightarrow Al(OH)_2 + CaSO_4$

44. $FeCl_2 + NH_4OH \rightarrow Fe(OH)_2 + NH_4Cl$

45. $Ca_3(PO_4)_2 + SiO_2 \rightarrow P_4O_{10} + CaSiO_2$

46. $N_2O_5 + H_2O \rightarrow HNO_2$

47. $Al + HCl \rightarrow AlCl_3 + H_2$

48. $H_3BO_2 \rightarrow H_4B_6O_{11} + H_2O$

49. $Mg + N_2 \rightarrow Mg_2N_2$

50. $NaOH + Cl_2 \rightarrow NaCl + NaClO + H_2O$

Use this space for summary and/or additional notes:

Predicting the Products of Chemical Reactions

Unit: Chemical Reactions
MA Curriculum Frameworks (2016): HS-PS1-2
MA Curriculum Frameworks (2006): N/A
Mastery Objective(s): (Students will be able to...)

- Accurately predict the products of single replacement, double replacement, and combustion reactions.

Success Criteria:

- Cation & anion are correct for single and double replacement reactions.
- Products that are ionic compounds have correctly balanced charges.

Tier 2 Vocabulary: product, replacement

Language Objectives:

- Explain how you can tell from the reactants what the reaction is likely to be.

Notes:

Recognizing Reaction Types from the Reactants

If you are familiar with the different types of chemical reactions, you can often tell the reaction type by looking only at the reactants. Once you know the reaction type, it is relatively straightforward to predict what the products should be.

Combustion Reactions

In a combustion reaction, the reactants are always a hydrocarbon (with some unknown number of atoms C, H, and O) and oxygen (O_2). The products are always CO_2 and H_2O.

If you were asked to write a combustion reaction for C_5H_{12}, you would write:

$$C_5H_{12} + O_2 \rightarrow CO_2 + H_2O$$

Use this space for summary and/or additional notes:

Predicting the Products of Chemical Reactions

Single Replacement Reactions

Single and double replacement reactions usually involve ionic compounds (and sometimes water, which we treat as the ionic compound H^+OH^-).

In a single replacement reaction, atoms of an element react with a compound, replacing the atom of the same type. Metals replace metals; non-metals replace non-metals. For example:

$$Na + MgI_2 \rightarrow NaI + Mg$$
(Na replaces Mg.)

$$KBr + Cl_2 \rightarrow KCl + Br_2$$
(Cl replaces Br.)

If an element reacts with a compound, you can predict the products, because the element simply replaces the other element of the same type.

For example, if you were given the problem:

$$Ca + NaCl \rightarrow ?$$

Calcium is a metal, so it will replace sodium. This means calcium will end up with chloride ($CaCl_2$), and sodium will end up by itself (Na). The reaction is therefore:

$$Ca + NaCl \rightarrow CaCl_2 + Na$$

Remember that *we have to balance the charges <u>every time</u> we put two new ions together*. This is the most common mistake beginning students make—forgetting to balance the charges in the new compounds. You can think of breaking apart the reactants as "unbalancing" the charges, which means you need to "re-balance" them when you put them back together in a new arrangement.

In the example above, the Na ion (which has a +1 charge) needed only one Cl^- ion for the charges to balance, but the Ca ion (which has a +2 charge) needs two.

Use this space for summary and/or additional notes:

Predicting the Products of Chemical Reactions

Big Ideas — Details

Unit: Chemical Reactions

Double Replacement Reactions

In a double replacement reaction, the two ions of the same type switch places, as in:

$$KCl + MgO \rightarrow MgCl_2 + K_2O$$

(K and Mg are trading places; in the products, K is now with O and Mg is now with Cl.)

Notice again that we had to balance the charges. We needed only one K^+ ion with Cl^-, but we need 2 K^+ ions with O^{2-}. Similarly, Mg^{2+} needed only one O^{2-} ion, but it needs two Cl^- ions.

If we had the problem:

$$NH_4OH + Ca_3(PO_4)_2 \rightarrow \ ?$$

we would swap NH_4^+ with Ca^{2+}. When we balance the charges, NH_4^+ would go with PO_4^{3-} to form $(NH_4)_3PO_4$, and Ca^{2+} would go with OH^- to form $Ca(OH)_2$. This gives the equation:

$$NH_4OH + Ca_3(PO_4)_2 \rightarrow (NH_4)_3PO_4 + Ca(OH)_2$$

Acid-base reactions are a type of double replacement reaction in which H^+ and OH^- ions combine to form "HOH", which is really H_2O. For example:

$$HCl + Ca(OH)_2 \rightarrow CaCl_2 + H_2O$$

Use this space for summary and/or additional notes:

Big Ideas	Details
	## Predicting the Products of Chemical Reactions
Page: 412
Unit: Chemical Reactions

Homework Problems

Predict the products for each of the following single replacement, double replacement and combustion reactions. (*Don't forget to balance the charges!*)

1. $Ca + AlCl_3 \rightarrow$

2. $BaCl_2 + O_2 \rightarrow$

3. $KCl + Mg(OH)_2 \rightarrow$

4. $Na_3PO_4 + MgSO_4 \rightarrow$

5. $Na + HCl \rightarrow$

6. $Al + CoCl_2 \rightarrow$

7. $Pb(NO_3)_2 + NaI \rightarrow$

8. $SiO_2 + Ca \rightarrow$

9. $Al + FeO \rightarrow$

10. $Zn + HCl \rightarrow$ |

Use this space for summary and/or additional notes:

Predicting the Products of Chemical Reactions

Page: 413
Unit: Chemical Reactions

Big Ideas | Details

11. $Mg + NaNO_3 \rightarrow$

12. $NaOH + HNO_3 \rightarrow$

13. $Mg + CrCl_3 \rightarrow$

14. $Fe(OH)_3 + H_2SO_4 \rightarrow$

15. $H_2S + AuCl_3 \rightarrow$

16. $Cr_2(SO_4)_3 + KOH \rightarrow$

17. $PbCl_2 + Na_2SO_4 \rightarrow$

18. $AgNO_3 + CuCl_2 \rightarrow$

19. $BaCl_2 + Al_2(SO_4)_3 \rightarrow$

20. $TiO_2 + HCl \rightarrow$

21. $Zn + AgNO_3 \rightarrow$

22. $Al + Fe_3O_4 \rightarrow$

Use this space for summary and/or additional notes:

Big Ideas	Details	Activity (Reactivity) Series	Page: 414
			Unit: Chemical Reactions

Activity (Reactivity) Series

Unit: Chemical Reactions
MA Curriculum Frameworks (2016): HS-PS1-2
MA Curriculum Frameworks (2006): N/A
Mastery Objective(s): (Students will be able to...)
- Use the activity series to predict whether or not a single replacement reaction will occur.

Success Criteria:
- Prediction is correct about whether or not a reaction occurs.
- Cation & anion are correct if reaction does occur.
- Products have correctly balanced charges.

Tier 2 Vocabulary: product, replacement, activity

Language Objectives:
- Explain how you can tell using the activity series whether or not a reaction will occur.

Notes:

In the reaction between aluminum metal and copper (II) chloride:

$$Al_{(s)} + CuCl_2{}_{(aq)} \rightarrow AlCl_3{}_{(aq)} + Cu_{(s)} + heat \tag{1}$$

the beaker got hot. This means the reaction gave off heat, which was lost to the surroundings (the water that the chemicals were dissolved in, the beaker, the air, your hand). Once the energy was given off, the chemicals didn't have enough energy to go the other direction. In other words, the reverse reaction does not happen:

$$Cu_{(s)} + AlCl_3{}_{(aq)} \rightarrow no\ reaction \tag{2}$$

Is it possible to predict which direction the reaction will go?

For single replacement reactions, there is a list, called the <u>activity series</u>, (or reactivity series), which lists metals in order from most reactive to least, based on how much energy they give off when they lose electrons to form a positive ion. A metal that's higher on the list can replace anything lower on the list (because more energy is given off), but a metal that's lower on the list doesn't have enough energy to replace one that's higher up.

Use this space for summary and/or additional notes:

Activity (Reactivity) Series

Metal	Ion		Reacts With	Method of Extraction
Cs	Cs^+	↑	cold H_2O dilute acids O_2	electrolysis
Rb	Rb^+			
K	K^+			
Na	Na^+			
Li	Li^+			
Sr	Sr^{2+}			
Ca	Ca^{2+}			
Mg	Mg^{2+}	increasing reactivity	steam dilute acids O_2	metal oxide reduction with carbon or CO_2 smelting with coke
Be	Be^{2+}			
Al	Al^{3+}			
Mn	Mn^{2+}			
Zn	Zn^{2+}			
Cr	Cr^{3+}			
Fe	Fe^{2+}			
Cd	Cd^{2+}			
Co	Co^{2+}		dilute acids O_2	
Ni	Ni^{2+}			
Sn	Sn^{2+}			
Pb	Pb^{2+}			
H_2	H^+		O_2	heat or physical extraction
Cu	Cu^{2+}			
Cu	Cu^+			
Hg	Hg^{2+}			
Ag	Ag^+			
Au	Au^{3+}		some strongly oxidizing acids	
Pt	Pt^{2+}			

To answer the original question, notice that aluminum is higher than copper on the activity series. This means aluminum can replace copper:

$$Al\,(s) + CuCl_2\,(aq) \rightarrow AlCl_3\,(aq) + Cu\,(s) + heat$$

but copper can't replace aluminum:

$$Cu\,(s) + AlCl_3\,(aq) \rightarrow no\ reaction$$

Use this space for summary and/or additional notes:

Activity (Reactivity) Series

Homework Problems

For each of the following single replacement reactions:
a) Check the activity series to see whether the reaction happens.
b) If the reaction happens, predict the products. If the reaction does not happen, write "N.R." ("No Reaction").

1. $K\ (s) + H_2O \rightarrow$

2. $Pb\ (s) + Zn(CH_3COO)_2\ (aq) \rightarrow$

3. $Al\ (s) + Fe_2O_3\ (s) \xrightarrow{\Delta}$

4. $AgNO_3\ (aq) + Ni\ (s) \rightarrow$

5. $Ag\ (s) + H_2SO_4\ (aq) \rightarrow$

6. $NaBr\ (aq) + I_2\ (s) \rightarrow$

7. $Ca\ (s) + MgSO_4\ (aq) \rightarrow$

8. $Ca\ (s) + HCl\ (aq) \rightarrow$

9. $Mg\ (s) + HNO_3\ (aq) \rightarrow$

10. $CuCl_2\ (aq) + Hg\ (\ell) \rightarrow$

11. $Na\ (s) + H_2O\ (\ell) \rightarrow$

Use this space for summary and/or additional notes:

Balancing Chemical Equations

Unit: Chemical Reactions

MA Curriculum Frameworks (2016): HS-PS1-7

MA Curriculum Frameworks (2006): 5.1

Mastery Objective(s): (Students will be able to...)
- Apply the law of definite proportions to balance chemical equations.

Success Criteria:
- Equation is balanced such that there are the same number of atoms (moles) of each element on each side of the equation.

Tier 2 Vocabulary: balance

Language Objectives:
- Explain the law of definite proportions and conservation of mass and relate them to chemical equations.

Notes:

A chemical equation needs to describe the chemical formulas and relative number of molecules involved of each molecule that reacts, and each molecule that is produced.

Remember from Dalton's theory of the atom:

> "Atoms are neither created nor destroyed in any chemical reaction."

Therefore, not only must we have the same kinds of atoms (same elements) on both sides of a chemical reaction, we need to have the *same number* of each kind of atom before and after the reaction..

Use this space for summary and/or additional notes:

For example, consider the chemical equation:

$$S + O_2 \rightarrow SO_3$$

There are 2 oxygen atoms on the left, but 3 on the right. We can't change the formulas of the molecules that take part in the reaction, so we need to specify different numbers of each molecule to "balance" the equation.

The easiest solution would be to split an O_2 molecule in half:

$$S + 1½\,O_2 \rightarrow SO_3$$

But we can't have ½ of a molecule of O_2. Therefore, the smallest set of integers that give us the same number of each atom on both sides would be:

$$2\,S + 3\,O_2 \rightarrow 2\,SO_3$$

This works because there are 2 atoms of S and 6 atoms of O on each side of the equation ("before" and "after").

We balanced this equation by inspection, but for more complicated equations, it helps to have a system.

Use this space for summary and/or additional notes:

Balancing Chemical Equations

Big Ideas | **Details**

To balance an equation, start with one element. Put coëfficients in front of the molecules that contain the element so that you have the same number on each side. Then do the same for every other element. For example, to balance the equation:

$$\_N_2 + \_H_2 \rightarrow \_NH_3$$

we need to figure out the coëfficients that go in the blanks. We can start by balancing any element we want, so let's start with nitrogen (N). The smallest numbers that we can use to balance atoms of N are a "1" in front of N_2 and a 2 in front of NH_3. This gives us:

$$\underline{1}\,N_2 + \_H_2 \rightarrow \underline{2}\,NH_3$$

Now we have 2 atoms of N on each side, so N is balanced. Next, we move on to H. We already have a "2" in front of NH_3, which means we have 6 atoms of H on the right side. To get 6 atoms of H on the left side, we need a "3" in front of H_2. This gives us the equation:

$$\underline{1}\,N_2 + \underline{3}\,H_2 \rightarrow \underline{2}\,NH_3$$

We have coefficients in front of all of the products and reactants, so the equation is balanced. For the final form of the equation, we leave out any coëfficient that is "1". (This is just like algebra—we would write "x" instead of "1x".) This gives us:

$$N_2 + 3\,H_2 \rightarrow 2\,NH_3$$

This equation was equally easy to balance regardless of whether we started with N or H, but for more complicated equations, making good decisions about what order to balance the elements in can make a huge difference.

Use this space for summary and/or additional notes:

Balancing Chemical Equations

Unit: Chemical Reactions
Page: 420

Strategy for Balancing Equations

1. Figure out which elements to balance First, Middle, and Last. We will refer to this method as the "FML" method.* Always start by deciding which elements to save for last:

   - <u>Last</u>: any element that appears by itself (anywhere in the equation)
   - <u>First</u>: elements that appear in only one molecule on each side (if you haven't already saved them for last).
   - <u>Middle</u>: every element that's not already last or first.

2. Start with any element on the "First" list. Add coëfficients to make it balance.

3. Pick another element. (Work your way through the "First," then "Middle," then "Last" lists.) Start with elements that already have at least one coëfficient, but still need at least one.

4. Repeat step #3 until everything is balanced.

Notes:

- Polyatomic ions usually stay together.
- If you end up with a fraction, write it in temporarily, then multiply *all* of your coëfficients by the denominator of the fraction to get back to whole numbers.

---

* Now you can say to yourself, "I have to balance this equation? FML."

Use this space for summary and/or additional notes:

Balancing Chemical Equations

Big Ideas | Details

Page: 421
Unit: Chemical Reactions

Example:

$$H_2SO_4 + HI \rightarrow H_2S + I_2 + H_2O$$

1. Make lists:
   1. <u>Last</u>:   I   I appears by itself as I_2 (on the right)
   2. <u>First</u>:   S, O   S only appears in H_2SO_4 on the left and H_2S on the right; O only appears in H_2SO_4 on the left and H_2O on the right.
   3. <u>Middle</u>:   H appears in two places each on the left and right.

2. Balance "First" elements (S & O; the order doesn't matter):
   a. Let's start with S. Neither H_2SO_4 nor H_2S has a coëfficient, so we choose the smallest pair that works for both: 1 of each:

   $$\underline{1}\,H_2SO_4 + \_HI \rightarrow \underline{1}\,H_2S + \_I_2 + \_H_2O$$

   b. Next balance O. We already have a "1" in front of H_2SO_4, which means we have 4 atoms of O on the left. This means we need a "4" in front of H_2O to have 4 atoms of O on the right.

   $$1\,H_2SO_4 + \_HI \rightarrow 1\,H_2S + \_I_2 + \underline{4}\,H_2O$$

3. Balance "Middle" elements (H). We have a total of 10 H atoms on the right (2 in the $1\,H_2S$ and 8 more in the $4\,H_2O$), and our coëfficients only show 2 H atoms so far on the left. This means we need an "8" in front of HI for the remaining 8 atoms of H.

   $$1\,H_2SO_4 + \underline{8}\,HI \rightarrow 1\,H_2S + \_I_2 + 4\,H_2O$$

4. Balance "Last" elements (I). We have 8 atoms of I on the left, which is 4 molecules of I_2.:

   $$1\,H_2SO_4 + 8\,HI \rightarrow 1\,H_2S + \underline{4}\,I_2 + 4\,H_2O$$

5. For the final answer, leave out any coëfficient of 1:

   $$H_2SO_4 + 8\,HI \rightarrow H_2S + 4\,I_2 + 4\,H_2O$$

Use this space for summary and/or additional notes:

Balancing Chemical Equations

Page: 422
Unit: Chemical Reactions

Homework Problems
Set #1 (Easier)

Balance the following chemical equations.

1. $H_2 + O_2 \rightarrow H_2O$

2. $S_8 + O_2 \rightarrow SO_3$

3. $HgO \rightarrow Hg + O_2$

4. $Zn + HCl \rightarrow ZnCl_2 + H_2$

5. $Na + H_2O \rightarrow NaOH + H_2$

6. $C_{10}H_{16} + Cl_2 \rightarrow C + HCl$

7. $Si_2H_3 + O_2 \rightarrow SiO_2 + H_2O$

8. $Fe + O_2 \rightarrow Fe_2O_3$

9. $C_7H_6O_2 + O_2 \rightarrow CO_2 + H_2O$

Use this space for summary and/or additional notes:

Balancing Chemical Equations

Unit: Chemical Reactions

Big Ideas	Details

10. $FeS_2 + O_2 \rightarrow Fe_2O_3 + SO_2$

11. $Fe_2O_3 + H_2 \rightarrow Fe + H_2O$

12. $K + Br_2 \rightarrow KBr$

13. $C_2H_2 + O_2 \rightarrow CO_2 + H_2O$

14. $H_2O_2 \rightarrow H_2O + O_2$

15. $C_7H_{16} + O_2 \rightarrow CO_2 + H_2O$

16. $SiO_2 + HF \rightarrow SiF_4 + H_2O$

17. $KClO_3 \rightarrow KCl + O_2$

18. $KClO_3 \rightarrow KClO_4 + KCl$

19. $P_4O_{10} + H_2O \rightarrow H_3PO_4$

20. $Sb + O_2 \rightarrow Sb_4O_6$

Use this space for summary and/or additional notes:

Balancing Chemical Equations

Page: 424
Unit: Chemical Reactions

Homework Problems
Set #2 (More Challenging)

Balance the following chemical equations.

1. $Pb(NO_3)_2 \rightarrow PbO + NO_2 + O_2$

2. $Ca_3P_2 + H_2O \rightarrow Ca(OH)_2 + PH_3$

3. $Ca + AlCl_3 \rightarrow CaCl_2 + Al$

4. $H_3PO_3 \rightarrow H_3PO_4 + PH_3$

5. $C_6H_6 + O_2 \rightarrow CO_2 + H_2O$

6. $Al_4C_3 + H_2O \rightarrow CH_4 + Al(OH)_3$

7. $Ag_2S + KCN \rightarrow KAg(CN)_2 + K_2S$

8. $MgNH_4PO_4 \rightarrow Mg_2P_2O_7 + NH_3 + H_2O$

Use this space for summary and/or additional notes:

Big Ideas	Details	Net Ionic Equations	Page: 425
			Unit: Chemical Reactions

Net Ionic Equations

Unit: Chemical Reactions

MA Curriculum Frameworks (2016): HS-PS1-2

MA Curriculum Frameworks (2006): N/A

Mastery Objective(s): (Students will be able to…)

- Write chemical equations as net ionic equations.

Success Criteria:

- Soluble ionic compounds are dissociated.
- Insoluble ionic compounds remain together as solids.
- Spectator ions are identified and omitted from the final net ionic equation.

Tier 2 Vocabulary: net, spectator

Language Objectives:

- Review dissociation. Explain how to dissociate a compound and write its ions separately in an equation.

Notes:

<u>net ionic equation</u>: a chemical equation that shows only ions or pure substances that are changed by the reaction.

<u>spectator ion</u>: an ion that remains in solution and does not participate in a chemical reaction.

If you mixed aqueous solutions of calcium chloride ($CaCl_2$ (aq)) and sodium nitrate ($NaNO_3$ (aq)), you might be tempted to predict that the following (unbalanced) chemical reaction would occur:

$$CaCl_2\,(aq) + NaNO_3\,(aq) \rightarrow Ca(NO_3)_2\,(aq) + NaCl\,(aq)$$

However, recall that aqueous ions dissociate when they dissolve in water:

Symbol	What Actually Happens in H_2O
$CaCl_2$ (aq)	Ca^{2+} (aq) + Cl^- (aq)
$NaNO_3$ (aq)	Na^+ (aq) + NO_3^- (aq)
$Ca(NO_3)_2$ (aq)	Ca^{2+} (aq) + NO_3^- (aq)
$NaCl$ (aq)	Na^+ (aq) + Cl^- (aq)

This means that what we really have in the beaker is:

$$Ca^{2+}\,(aq) + Cl^-\,(aq) + Na^+\,(aq) + NO_3^-\,(aq) \rightarrow Ca^{2+}\,(aq) + NO_3^-\,(aq) + Na^+\,(aq) + Cl^-\,(aq)$$

The above is called a detailed ionic equation.

Use this space for summary and/or additional notes:

Net Ionic Equations

Unit: Chemical Reactions

Big Ideas	Details

In the detailed ionic equation:

$$Ca^{2+}(aq) + Cl^-(aq) + Na^+(aq) + NO_3^-(aq) \rightarrow Ca^{2+}(aq) + NO_3^-(aq) + Na^+(aq) + Cl^-(aq)$$

Notice that the right side and the left side contain exactly the same ions. In other words, *nothing has changed*. If no substances are changed—no chemical bonds are formed or broken—then *no chemical reaction has occurred*!

Now consider the reaction of aqueous silver nitrate with aqueous sodium carbonate:

$$AgNO_3(aq) + Na_2CO_3(aq) \rightarrow Ag_2CO_3(s) + NaNO_3(aq)$$

Notice that one of the products, silver carbonate, forms a solid (precipitate).

The detailed ionic equation would look like this:

$$Ag^+(aq) + NO_3^-(aq) + Na^+(aq) + CO_3^{2-}(aq) \rightarrow Ag_2CO_3(s) + Na^+(aq) + NO_3^-(aq)$$

The spectator ions (ions that remain unchanged by the reaction) are Na^+ and NO_3^-. If we cross those out:

$$Ag^+(aq) + \cancel{NO_3^-(aq)} + \cancel{Na^+(aq)} + CO_3^{2-}(aq) \rightarrow Ag_2CO_3(s) + \cancel{Na^+(aq)} + \cancel{NO_3^-(aq)}$$

we are left with the *unbalanced net ionic equation* for this reaction:

$$Ag^+(aq) + CO_3^{2-}(aq) \rightarrow Ag_2CO_3(s)$$

Of course, we still need to balance the equation! The *balanced net ionic equation* would therefore be:

$$2\,Ag^+(aq) + CO_3^{2-}(aq) \rightarrow Ag_2CO_3(s)$$

Notice that the net ionic reaction is much simpler than the full chemical equation, because the net ionic equation leaves out everything that does not matter in the equation, allowing you to focus only on the details that are important.

If you take AP® Chemistry, you will be expected to write all chemical equations in net ionic form.

Use this space for summary and/or additional notes:

Net Ionic Equations

Unit: Chemical Reactions

Big Ideas	Details
	## Homework Problems
	For each of the following potential double replacement reactions:
	a. Predict the products. (Remember to balance the charges!)
	b. Use your solubility rules to write the phase after each product. If the product is soluble, write (aq) after it. If an ionic compound is not soluble, then it precipitates; write (ppt) after it. If a product is a gas (such as CO_2), then write (g) after it. If a product is a pure liquid (such as H_2O), then write (ℓ) after it.
	c. Rewrite the equation with the aqueous compounds dissociated (split up).
	d. Cancel (cross out) any ions that are the same (unchanged) on both sides.
	e. Write and balance the net ionic equation as your final answer. If it turns out that you have crossed out everything, write "N.R." ("No Reaction") instead.
	1. Na_2CO_3 (aq) + $CaCl_2$ (aq) →
	2. $(NH_4)_3PO_4$ (aq) + NaOH (aq) →
	3. $Ba(C_2H_3O_2)_2$ (aq) + K_3PO_4 (aq) →
	4. $Ca(MnO_4)_2$ (aq) + KOH (aq) →

Use this space for summary and/or additional notes:

Net Ionic Equations

Unit: Chemical Reactions

Big Ideas	Details
	5. $AlCl_3$ (aq) + H_3PO_4 (aq) →
	6. $CaSO_4$ (aq) + $KMnO_4$ (aq) →
	7. NaN_3 (aq) + $Ca(NO_3)_2$ (aq) →
	8. $Sr(NO_3)_2$ (aq) + $K_2Cr_2O_7$ (aq) →
	9. $NaClO_3$ (aq) + $MgSO_4$ (aq) →
	10. Na_3BO_3 (aq) + $ZnSO_4$ (aq) →

Use this space for summary and/or additional notes:

Big Ideas	Details
	Summary: Chemical Reactions — Page: 429 — Unit: Chemical Reactions

Summary: Chemical Reactions

Unit: Chemical Reactions

List the main ideas of this chapter in phrase form:

Write an introductory sentence that categorizes these main ideas.

Turn the main ideas into sentences, using your own words. You may combine multiple main ideas into one sentence.

Add transition words to make your writing clearer and rewrite your summary below.

Use this space for summary and/or additional notes:

Introduction: Oxidation & Reduction

Unit: Oxidation & Reduction

Topics covered in this chapter:

    Oxidation-Reduction (REDOX) Reactions 432

    Electrochemical Cells 439

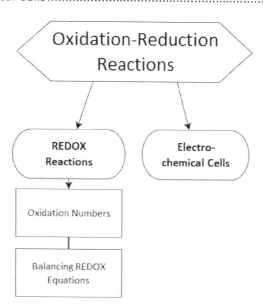

Standards addressed in this chapter:

Massachusetts Curriculum Frameworks & Science Practices (2016):

HS-PS1-10(MA) Use an oxidation-reduction reaction model to predict products of reactions given the reactants, and to communicate the reaction models using a representation that shows electron transfer (redox). Use oxidation numbers to account for how electrons are redistributed in redox processes used in devices that generate electricity or systems that prevent corrosion.

Massachusetts Curriculum Frameworks (2006):

8.4 Describe oxidation and reduction reactions and give some everyday examples, such as fuel burning and corrosion. Assign oxidation numbers in a reaction.

Use this space for summary and/or additional notes:

Oxidation-Reduction (REDOX) Reactions

Unit: Oxidation & Reduction

MA Curriculum Frameworks (2016): HS-PS1-10(MA)

MA Curriculum Frameworks (2006): 8.4

Mastery Objective(s): (Students will be able to...)
- Assign oxidation numbers.
- Write and balance equations for simple REDOX reactions.

Success Criteria:
- Oxidation numbers agree with published/accepted values.
- Oxidation numbers add up to zero for compounds and to the charge for ions.
- Balanced REDOX reactions have the same number of each type of atom and the same number of electrons transferred on each side.

Tier 2 Vocabulary: reduce

Language Objectives:
- Explain electron transfer. Explain the charge that an ion gets when electrons are added or removed.

Notes:

<u>oxidation-reduction reaction</u> (REDOX reaction): a reaction in which one or more electrons are transferred from one atom to another.

In the 1700s, oxidation meant that an atom was combined with oxygen, and was therefore "oxidized". For example:

$$2\,Cu + O_2 \rightarrow 2\,CuO$$

In this reaction, oxygen is more electronegative than copper, so oxygen took electrons away from copper. This means that the copper (which was "oxidized" by oxygen) lost two electrons and ended up with a +2 charge. Oxygen gained two electrons and ended up with a –2 charge. As a result, the term "oxidation" has come to mean "losing electrons".

Also, in the 1700s, scientists found that if they heated the CuO (in which copper has a +2 charge), they ended up with copper metal (with a charge of zero), and the weight was reduced. Thus copper was said to be "reduced". As a result, the term "reduction" has come to mean "gaining electrons".

Use this space for summary and/or additional notes:

Oxidation-Reduction (REDOX) Reactions

Big Ideas	Details
	<u>oxidation:</u> the loss of one or more electrons by an atom in a chemical reaction
	<u>reduction:</u> the gain of one or more electrons by an atom in a chemical reaction.
	Stupid Mnemonics: There are two popular mnemonics for remembering oxidation and reduction, one "Democratic" and one "Republican".
	LEO the lion says 'GER' ("Democratic" mnemonic involving endangered species): LEO stands for "Loss of Electrons is Oxidation" and GER stands for "Gain of Electrons is Reduction"
	OIL RIG ("Republican" mnemonic involving oil companies): OIL stands for "Oxidation Involves Loss (of electrons)", and RIG stands for "Reduction Involves Gain (of electrons)."
	In a redox reaction, at least one element is oxidized, and at least one other element is reduced. *An element cannot be oxidized in a chemical reaction unless some other element is reduced, and vice-versa.* (After all, the electrons have to come from somewhere, and they have to go somewhere.)
	All chemical reactions in which an element becomes part of a compound, or vice-versa, are redox reactions. This includes all single replacement reactions, combustion reactions, and many synthesis and decomposition reactions. However, chemists generally classify a reaction as a redox reaction only when most or all of the energy of the reaction comes from electron transfer.

Use this space for summary and/or additional notes:

Oxidation Numbers

An oxidation number is a measure of how "oxidized" an atom is. An element is neither oxidized nor reduced, so it has an oxidation number of zero.

- An element that has lost electrons (oxidized) gets a positive oxidation number, equal to the number of electrons it has lost.

- An element that has gained electrons (reduced) gets a negative oxidation number, equal to the number of electrons it has gained.

Therefore:

- when an element is oxidized, the oxidation number increases.

- When an element is reduced, the oxidation number is also reduced (decreases).

oxidation number (or "oxidation state"): the charge that an atom would have in a compound if all bonds were completely ionic and every atom in the compound or ion had a charge.

Use this space for summary and/or additional notes:

Oxidation-Reduction (REDOX) Reactions

Assigning Oxidation Numbers

- The oxidation number of a pure element is 0. (Even if it's diatomic.)
- The oxidation numbers in a compound add up to 0
- The oxidation number of an ion is its charge. (Oxidation numbers in a polyatomic ion add up to the charge of the polyatomic ion.)
- In a compound or polyatomic ion:
  - The most electronegative element (the last one in the formula) has a negative oxidation number equal to the number of electrons it would need to fill its valent shell.
  - All other atoms have positive oxidation numbers.
  - Fluorine is always −1.
  - Oxygen is always −2 except in the peroxide ion (O_2^{2-}) and in OF_2.
  - Hydrogen is always +1 except in metal hydrides (such as NaH).
  - Alkali (group 1) metals are always +1.
  - Alkaline Earth (group 2) metals are always +2.
  - Al is always +3, Zn is always +2, and Ag is always +1.
  - Calculate other elements from the above.

Sample Problem:

What are the oxidation numbers of each element in the compound Na_2HPO_4:

- Na_2HPO_4 is an ionic compound made of the ions Na^+ and HPO_4^{2-}.
- The Na^+ ion has a charge of +1, so the oxidation number of Na is +1.
- The HPO_4^{2-} ion has a charge of −2. This means the oxidation numbers of H, P, and O must add up to −2.
  - O = −2. There are 4 O atoms. (4)(−2) = −8
  - H is +1.
  - If the oxidation numbers for the O atoms add up to −8 and H is +1, then the oxidation number for P must be +5 so the total can add up to −2.

Use this space for summary and/or additional notes:

| Big Ideas | Details | Unit: Oxidation & Reduction |

Balancing REDOX Reactions

To fully balance a redox reaction, you must balance:
- Atoms (as you would in a regular equation)
- Electrons lost/gained
- Total charge

Often, redox reactions are shown and balanced as net ionic equations. In this case, balancing them is often a simple matter of making sure that the same number of electrons are produced by the oxidation half-reaction and consumed by the reduction half-reaction.

For example, consider the unbalanced net ionic equation:

$$Al^0(s) + Zn^{2+}(aq) \rightarrow Al^{3+}(aq) + Zn^0(s)$$

In this reaction, Al is oxidized from Al^0 to Al^{+3}, and Zn is reduced from Zn^{+2} to Zn^0. The *atoms* appear balanced, but Zn^{2+} needs only 2 electrons to form Zn^0, but Al^0 produces 3 electrons when oxidized to Al^{3+}.

The two half-reactions are:

$$\text{Oxidation: } Al^0 \rightarrow Al^{3+} + 3\,e^-$$
$$\text{Reduction: } Zn^{2+} + 2\,e^- \rightarrow Zn^0$$

To balance the electrons, we need to multiply the first half-reaction by 2, and the second one by 3, giving:

$$2\,Al^0 \rightarrow 2\,Al^{3+} + 6\,e^-$$
$$3\,Zn^{2+} + 6\,e^- \rightarrow 3\,Zn^0$$

If we combine these and cancel the electrons (because we have the same number on both sides), we get the balanced net ionic equation:

$$2\,Al^0(s) + 3\,Zn^{2+}(aq) \rightarrow 2\,Al^{3+}(aq) + 3\,Zn^0(s)$$

REDOX reactions can get a lot more complicated, especially when acid-base reactions are also taking place and the water that the ions are dissolved in participates in the reaction. Balancing complex REDOX reactions is beyond the scope of this course, but is covered in AP® Chemistry.

Use this space for summary and/or additional notes:

Homework Problems

For each of the following compounds or ions, write the oxidation number of each element.

1. FeO

2. N_2

3. $KMnO_4$

4. NH_4OH

5. $(NH_4)_3PO_4$

6. $Cr_2O_7^{2-}$

Use this space for summary and/or additional notes:

Oxidation-Reduction (REDOX) Reactions

Unit: Oxidation & Reduction

Balance the following (redox) equations. A superscript "0" indicates a pure element.

7. $Zn^0 + Tl^+ \rightarrow Tl^0 + Zn^{2+}$

8. $Li^0 + Cr^{3+} \rightarrow Cr^0 + Li^+$

9. $K^0 + Mg^{2+} \rightarrow K^+ + Mg^0$

Use this space for summary and/or additional notes:

Electrochemical Cells

Unit: Oxidation & Reduction

MA Curriculum Frameworks (2016): HS-PS1-10(MA)

MA Curriculum Frameworks (2006): N/A

Mastery Objective(s): (Students will be able to...)
- Explain how an electrochemical cell (such as a battery) works.

Success Criteria:
- Explanations account for each of the parts of the electrochemical cell.

Tier 2 Vocabulary: battery, bridge

Language Objectives:
- Explain how a battery works.

Notes:

<u>electrochemistry</u>: using chemical (redox) reactions to produce electricity or *vice-versa*. In an electrochemical reaction, oxidation and reduction reactions occur in separate containers, and the electrons that travel from one reaction to the other pass through an electric circuit.

Use this space for summary and/or additional notes:

Electrochemical Cells

galvanic cell: (also called a voltaic cell) a chemical apparatus that uses an electrochemical reaction to produce electricity. (A battery is a type of galvanic cell.)

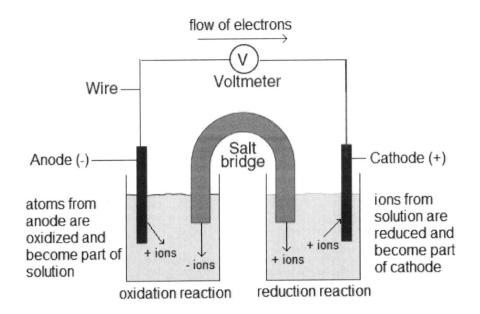

half cell: either of the two halves of a galvanic cell.

electrolytic cell: a cell similar to a galvanic cell, except that the reaction is not spontaneous, and electricity is used to add the energy needed to make the reaction occur. (Electrolysis of water is an example.)

electrode: a solid metal strip where either oxidation or reduction occurs. The metal strips also conduct the electrons into or out of the electric circuit.

anode: the negatively (−) charged electrode. At the anode:

- Oxidation happens. (Atoms from the anode are oxidized to positive ions.)
- These metal ions become part of the solution. (*I.e.*, the anode loses mass.)
- The electrons produced by oxidation move up the wire into the electric circuit.

cathode: the positively (+) charged electrode. At the cathode:

- Reduction happens. (Ions from the solution are reduced to neutral metal atoms.)
- These metal ions become part of the cathode. (*I.e.*, the cathode gains mass.)
- The electrons needed for reduction move from the electric circuit through the wire and into the cathode.

Use this space for summary and/or additional notes:

Electrochemical Cells

Unit: Oxidation & Reduction

Note that in physics, electric "current" is defined to be the direction that a <u>positive</u> particle would move. This means that the "current" flows in the <u>opposite</u> direction from the electrons.

<u>salt bridge</u>: a salt solution that is connected to both half cells. The salt bridge provides ions for the two half-cells to keep the charges balanced. (If the charges are not allowed to balance, opposite charges would build up in both cells and the reaction would stop.) The salt solution must be made of ions that do not take part in the reactions at the cathode or anode. (KNO_3 is commonly used.)

<u>electroplating</u>: using an electrolytic cell to add a layer of metal to something. The cathode is attached to the object to be electroplated. An electric current reduces metal ions from the solution, which are deposited onto the object.

Use this space for summary and/or additional notes:

Summary: Oxidation & Reduction Reactions

Unit: Oxidation & Reduction Reactions

List the main ideas of this chapter in phrase form:

Write an introductory sentence that categorizes these main ideas.

Turn the main ideas into sentences, using your own words. You may combine multiple main ideas into one sentence.

Add transition words to make your writing clearer and rewrite your summary below.

Use this space for summary and/or additional notes:

Introduction: Stoichiometry

Unit: Stoichiometry

Topics covered in this chapter:

Stoichiometry 445
Stoichiometry: Mass-Mass Problems 450
Limiting Reactant 454
Percent Yield 462
Marathon Problems 465

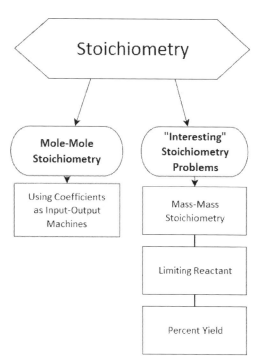

Use this space for summary and/or additional notes:

Introduction: Stoichiometry

Unit: Stoichiometry

Standards addressed in this chapter:

Massachusetts Curriculum Frameworks & Science Practices (2016):

HS-PS1-7 Use mathematical representations and provide experimental evidence to support the claim that atoms, and therefore mass, are conserved during a chemical reaction. Use the mole concept and proportional relationships to evaluate the quantities (masses or moles) of specific reactants needed in order to obtain a specific amount of product.

Massachusetts Curriculum Frameworks (2006):

5.5 Calculate the mass-to-mass stoichiometry for a chemical reaction.

5.6 Calculate percent yield in a chemical reaction.

Use this space for summary and/or additional notes:

Stoichiometry

Unit: Stoichiometry
MA Curriculum Frameworks (2016): HS-PS1-7
MA Curriculum Frameworks (2006): 5.5
Mastery Objective(s): (Students will be able to...)
- Solve mole-mole stoichiometry problems.

Success Criteria:
- For each compound in the chemical equation, the ratio of the coëfficients is the same as the ratio of the moles.
- Solutions have the correct quantities substituted for the correct variables.
- Algebra and rounding to appropriate number of significant figures is correct.

Tier 2 Vocabulary: mole, coëfficient

Language Objectives:
- Explain how the coefficients in a chemical equation are like the numbers in a pre-algebra "input-output machine."

Notes:

stoichiometry: measurement of how much of each reactant is used and how much of each product is produced in a chemical reaction.

stoichiometry problem: a chemistry problem in which you are given a balanced chemical equation and the quantity of one compound, and you are asked to find the quantity of another compound produced or consumed in the same equation.

For example, in the chemical reaction:

$$3\ CaCl_2 + 2\ Na_3PO_4 \rightarrow 6\ NaCl + Ca_3(PO_4)_2$$

3 molecules of $CaCl_2$ would produce 1 molecule of $Ca_3(PO_4)_2$. Because a mole is always the same number of molecules, this means 3 moles of $CaCl_2$ produces 1 mole of $Ca_3(PO_4)_2$.

Stoichiometry is simply the process of using the coëfficients in a balanced chemical equation to convert from moles of one compound to moles of another.

Use this space for summary and/or additional notes:

Sample problem:

Q: Suppose 1.75 mol of $CaCl_2$ reacts according to the following equation:

$$3\ CaCl_2 + 2\ Na_3PO_4 \rightarrow 6\ NaCl + Ca_3(PO_4)_2$$

How many moles of $Ca_3(PO_4)_2$ would be produced?

A: What makes this a stoichiometry problem is that you are "coming into the equation" with information about the amount of one compound (4.5 mol of $CaCl_2$), and you are being asked to find the amount of a different compound (how many moles of $Ca_3(PO_4)_2$).

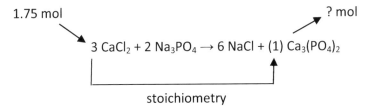

stoichiometry

The coefficients (3 $CaCl_2$ and 1 $Ca_3(PO_4)_2$) are in a 3:1 ratio. This means the moles of $CaCl_2$ reacted : $Ca_3(PO_4)_2$ produced must always be in a 3:1 ratio.

In the equation above, we can use *any pair* of coëfficients to make a conversion factor. There are eight possible conversion factors you could get from the equation $3\ CaCl_2 + 2\ Na_3PO_4 \rightarrow 6\ NaCl + Ca_3(PO_4)_2$.

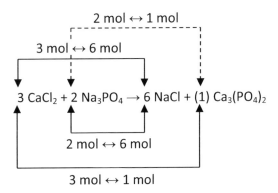

Use this space for summary and/or additional notes:

Stoichiometry

Big Ideas | **Details**

In elementary school, you may have been taught pre-algebra problems using "input/output machines". For example, a "times 2 machine" would multiply anything that goes through it by 2:

$$3.4 \text{ mol} \to \boxed{\times 2} \to 6.8 \text{ mol}$$

Getting from one coefficient to another is just like using one of those multiplication or division "machines":

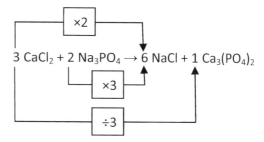

For example, suppose we had 1.75 mol of $CaCl_2$ and we wanted to know how much $Ca_3(PO_4)_2$ would be made.

1.75 mol $CaCl_2$ ÷ 3 = 0.583 mol $Ca_3(PO_4)_2$

Sometimes it's easier to think of the stoichiometry "machine" as a sequence of division and multiplication "machines".

In the following example, we can get from moles of O_2 to moles of CO_2 either in one step (the top pathway) by multiplying by $^4/_7$, or in two steps (the bottom pathway) by dividing by 7 (to get rid of the old coëfficient) and then multiplying by 4 (to get the new one).

$$\boxed{\times {}^4/_7}$$
$$2\,C_2H_6 + 7\,O_2 \to 4\,CO_2 + 6\,H_2O$$
$$\boxed{\div 7 \;\; \times 4}$$

For example, if we started with 15.4 mol O_2, the "machines" tell us that:

15.4 mol $O_2 \to \boxed{\times {}^4/_7} \to$ 8.8 mol CO_2 or   15.4 mol $O_2 \to \boxed{\div 7} \to \boxed{\times 4} \to$ 8.8 mol CO_2

Use this space for summary and/or additional notes:

Stoichiometry

Homework Problems

1. Determine how much of each product would be made when 4.0 mol of $Pb(NO_3)_2$ decomposes in the reaction: $2\ Pb(NO_3)_2 \rightarrow 2\ PbO + 4\ NO_2 + O_2$

   Answer: 4.0 mol PbO; 8.0 mol NO_2; 2.0 mol O_2

2. Determine how much of each product would be made when 1.33 mol of Ca_3P_2 reacts with excess water in the reaction:
   $Ca_3P_2 + 6\ H_2O \rightarrow 3\ Ca(OH)_2 + 2\ PH_3$

   Answer: 3.99 mol $Ca(OH)_2$; 2.66 mol PH_3

3. Determine how much $AlCl_3$ would you need to completely react with 1.5 mol Ca in the reaction: $3\ Ca + 2\ AlCl_3 \rightarrow 3\ CaCl_2 + 2\ Al$

   Answer: 1.0 mol $AlCl_3$

4. Determine how much of each product would be made when 1.50 mol H_3PO_3 decomposes in the reaction: $4\ H_3PO_3 \rightarrow 3\ H_3PO_4 + PH_3$

   Answer: 1.13 mol H_3PO_4; 0.375 mol PH_3

Use this space for summary and/or additional notes:

Stoichiometry

Unit: Stoichiometry

5. Determine how many moles of KCl would be produced from 0.175 mol of K and excess Cl_2 in the reaction: $2\,K + Cl_2 \rightarrow 2\,KCl$

   Answer: 0.175 mol KCl

6. Determine how many moles of Na_2O would be required to produce 0.275 mol of NaOH in the reaction: $Na_2O + H_2O \rightarrow 2\,NaOH$

   Answer: 0.138 mol Na_2O

7. Determine how many moles of O_2 will be produced by 8.75 mol of $NaClO_3$ in the reaction: $2\,NaClO_3 \rightarrow 2\,NaCl + 3\,O_2$

   Answer: 26.3 mol O_2

8. Determine how many moles of NaCl are produced in the following reaction when 45.4 L of O_2 are produced at S.T.P. in the reaction:
   $2\,NaClO_3 \rightarrow 2\,NaCl + 3\,O_2$
   (*Hint: you will need to convert 45.4 L of gas at S.T.P. into moles first.*)

   Answer: 1.33 mol NaCl

Use this space for summary and/or additional notes:

Stoichiometry: Mass-Mass Problems

Unit: Stoichiometry

MA Curriculum Frameworks (2016): HS-PS1-7

MA Curriculum Frameworks (2006): 5.5

Mastery Objective(s): (Students will be able to...)

- Solve stoichiometry problems that require mole conversions.

Success Criteria:

- Conversions between moles and other quantities are set up and executed correctly.
- For each compound in the chemical equation, the ratio of the coëfficients is the same as the ratio of the moles.
- Algebra and rounding to appropriate number of significant figures is correct.

Tier 2 Vocabulary: mole, coëfficient

Language Objectives:

- Explain the order of operations: convert to moles, do stoichiometry, convert from moles to desired units.

Notes:

stoichiometry: measurement of how much of each reactant is used and how much of each product is produced in a chemical reaction.

Remember that stoichiometry has to be done in moles.

- If you are given amounts in any other unit, you need to convert to moles before doing stoichiometry.
- If your answer needs to be in another unit, you need to convert after doing stoichiometry.

mass-mass problem: a stoichiometry problem that requires mole conversions from mass of a reactant to moles, and then moles of a product back to mass.

Note that there are many other similar problems that would work the same way—e.g., from volume of a gas (using the ideal gas law) to moles, from volume of a liquid that has a certain concentration (in $\frac{mol}{L}$) to moles, etc.

Use this space for summary and/or additional notes:

Stoichiometry: Mass-Mass Problems

Sample Problem:

How many grams of copper metal would be produced from 13.5 g of aluminum and excess copper chloride solution in the chemical reaction:

$$2 \text{ Al (s)} + 3 \text{ CuCl}_2 \text{ (aq)} \rightarrow 2 \text{ AlCl}_3 \text{ (aq)} + 3 \text{ Cu (s)}$$

Strategy:

1. Convert grams of Al to moles.
2. Use stoichiometry to convert moles of Al to moles of Cu.
3. Convert moles of Cu to grams.

Setup:

$$\frac{13.5 \text{ g Al}}{1} \times \frac{1 \text{ mol Al}}{27.0 \text{ g Al}} \times \frac{3 \text{ mol Cu}}{2 \text{ mol Al}} \times \frac{63.5 \text{ g Cu}}{1 \text{ mol Cu}}$$

Answer:

$$\frac{(13.5)(3)(63.5)}{(27.0)(2)} = \frac{2572}{54.0} = 47.6 \text{ g Cu}$$

<u>theoretical yield</u>: the amount of a product you could make based on stoichiometry calculations, assuming that at least one of the reactants is completely used up.

<u>excess</u>: having more of a reactant than is needed. This means simply that there is "enough that you don't have to worry about using it all up." We will see problems in which this is not the case in the next section ("Limiting Reactant," starting on page 454.)

By this point in this course, you have undoubtedly figured out that most of the challenging problems you will encounter are created by stringing together a sequence of easy problems until it becomes hard to keep track of what you're doing. Stoichiometry is easy (once you get the hang of it). Mole conversions are easy (assuming you've got the hang of them). Combining the two looks hard, but it's just a sequence of easy problems.

Use this space for summary and/or additional notes:

Stoichiometry: Mass-Mass Problems

Homework Problems

1. In the chemical reaction:

    $$2\ K + Cl_2 \rightarrow 2\ KCl$$

    a. How many *moles* of KCl (F.W. 74.55 $\frac{g}{mol}$) would be produced from 2.50 g of K and excess Cl_2?

    Answer: 0.0639 mol KCl

    b. How many *grams* of KCl would be produced?

    Answer: 4.76 g KCl

2. In the chemical reaction:

    $$Na_2O + H_2O \rightarrow 2\ NaOH$$

    a. If 124 g of Na_2O (F.W. 61.98 $\frac{g}{mol}$) is reacted with excess H_2O, how many grams of NaOH (F.W. 40.00 $\frac{g}{mol}$) will be made?

    Answer: 160. g NaOH

    b. If, instead, you wanted to make 100. g of NaOH, how many grams of Na_2O would you need?

    Answer: 77.5 g NaOH

Use this space for summary and/or additional notes:

Stoichiometry: Mass-Mass Problems

3. In the decomposition reaction:

   $$2\ NaClO_3 \rightarrow 2\ NaCl + 3\ O_2$$

   If you reacted 26.6 g of $NaClO_3$ (F.W. 106.44 $\frac{g}{mol}$), what volume of O_2 would you make at a pressure of 1.03 atm and a temperature of 30 °C?
   (*Hint: Use PV = nRT.*)

   Answer: 9.06 L O_2 (g)

4. Given the precipitation reaction:

   $$3\ CaCl_2\ (aq) + 2\ Na_3PO_4\ (s) \rightarrow Ca_3(PO_4)_2\ (ppt) + 6\ NaCl\ (aq)$$

   If you added an excess of powdered Na_3PO_4 to 100. mL of an 0.200 $\frac{mol}{L}$ solution of $CaCl_2$, how many grams of precipitate would form?
   (Assume that all of the $Ca_3(PO_4)_2$ precipitates, and that all of the Na_3PO_4 dissolves.)

   Answer: 2.07 g $Ca_3(PO_4)_2$ (ppt)

5. How many grams of precipitate would form if 94.6 g of $FeCl_3 \cdot 6\ H_2O$ crystals were added to an aqueous solution containing an excess of Na_2SiO_3?
   (*Hint: you will need to predict the products and balance the equation in order to do the stoichiometry.*)

   Answer: 50.5 g $Fe_2(SiO_3)_3$ (ppt)

Use this space for summary and/or additional notes:

Limiting Reactant

Unit: Stoichiometry
MA Curriculum Frameworks (2016): HS-PS1-7
MA Curriculum Frameworks (2006): 5.5
Mastery Objective(s): (Students will be able to...)
- Identify the limiting reactant in a stoichiometry problem.
- Perform stoichiometry calculations in a problem that involves a limiting reactant.
- Determine the amount(s) of the non-limiting reactant(s) left over.

Success Criteria:
- Limiting reactant correctly identified
- Stoichiometry calculations performed correctly (correct amount of desired compound in the desired units).
- Algebra and rounding to appropriate number of significant figures is correct.

Tier 2 Vocabulary: limiting

Language Objectives:
- Explain why a chemical reaction runs out of something.

Notes:

Q: *What happens when a chemical reaction runs out of something?*

A: *The reaction stops.*

A reaction in which you run out of something is called a limiting reactant problem.

The reactant that you run out of is called the limiting reactant (or limiting reagent) because running out of it is what limits how much product you can make.

Use this space for summary and/or additional notes:

Limiting Reactant

Page: 455
Unit: Stoichiometry

Consider the following reaction:

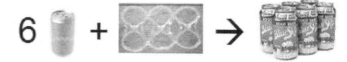

Suppose you have 51 cans and four six-pack rings. There are two possibilities:

1. We use up all of the cans. (Situation A)
2. We use up all of the six-pack rings. (Situation B)

Situation	Cans	Six-Pack Rings	Six-Packs	
A	Have 51	need at least 8.5	could make 8.5	X
B	need at least 24	Have 4	could make 4	☺

As you can see, we have only enough of *both* reactants to make 4 six-packs. Once we have made 4 six-packs, we have used up all of the six-pack rings, and we cannot make any more.

This means six-pack rings are the limiting reactant, and we use all of them.

We used up 24 cans (the non-limiting reactant), and we had 27 cans left over.

On the other hand, suppose we have only 15 cans and 4 six-pack rings. Again, there are two possibilities:

1. We use up all of the cans. (Situation C)
2. We use up all of the six-pack rings. (Situation D)

Situation	Cans	Six-Pack Rings	Six-Packs	
C	have 15	need at least 2.5	could make 2.5	☺
D	need at least 24	have 4	could make 6	X

This time, we can make 2.5 six-packs and then we run out of cans, so now the limiting reactant is cans. If we make 2.5 six-packs, we would use 2.5 six-pack rings, which means we would have 1.5 six-pack rings left over.

Use this space for summary and/or additional notes:

Limiting Reactant

The secret to solving limiting reactant problems is to do a stoichiometry problem on each reactant (using the factor-label method) to see how much of one of the products you make if you used it all up. The limiting reactant is the one that can make the least amount of product (gets used up first).

Steps for Solving Limiting Reactant Problems

1. Does the problem give you amounts for more than one reactant? (If not, it's not a limiting reactant problem.)

2. Convert the amount of each reactant to moles.

3. For each of the reactants, use stoichiometry to figure out how much of one of the products you could make if you used the reactant up.

4. The reactant that can make the *least* amount of product is the one that gets used up first—it is the limiting reactant.

5. Perform *all* of your stoichiometry calculations using the number of moles of the limiting reactant that you identified in step **Error! Reference source not found.**.

6. If the problem asks how much of one of the non-limiting reactants is left over, use the moles of the limiting reactant to find out how many moles of the other reactant got used up. Subtract this number from the moles you started with to find out how much is left over.

7. If the problem is asking for a quantity other than moles (such as grams), convert from moles to the desired unit.

Use this space for summary and/or additional notes:

Limiting Reactant

Big Ideas | **Details**

Page: 457
Unit: Stoichiometry

Sample Problem:

Q: Given the following reaction:

$$16 \text{ Cu} + S_8 \rightarrow 8 \text{ Cu}_2S$$

If we had 27.5 moles of copper and 1.4 moles of S_8, how much Cu_2S would we make?

A: Determine how many moles of Cu_2S we could make from each reactant:

$$\frac{27.5 \text{ mol Cu}}{1} \times \frac{8 \text{ mol Cu}_2S}{16 \text{ mol Cu}} = 13.75 \text{ mol Cu}_2S \quad \leftarrow \text{more}$$

$$\frac{1.4 \text{ mol } S_8}{1} \times \frac{8 \text{ mol Cu}_2S}{1 \text{ mol } S_8} = 11.2 \text{ mol Cu}_2S \quad \leftarrow \text{less}$$

We can make $\boxed{11.2 \text{ mol of Cu}_2S}$ and then we run out of S_8. This means S_8 is the limiting reactant.

Q: How much of the non-limiting reactant would be left over?

A: S_8 was limiting, and we had 1.4 moles of it. We need to find out how much Cu got used up.

$$\frac{1.4 \text{ mol } S_8}{1} \times \frac{16 \text{ mol Cu}}{1 \text{ mol } S_8} = 22.4 \text{ mol Cu}_2S$$

Now we subtract to find how much was left:

$$27.5 \text{ mol} - 22.4 \text{ mol} = \boxed{5.1 \text{ mol Cu left over}}$$

Use this space for summary and/or additional notes:

Limiting Reactant

Page: 458
Unit: Stoichiometry

Homework Problems

Set #1: Scaffolded

1. Consider the reaction: $2\ Si_2H_2 + 5\ O_2 \rightarrow 4\ SiO_2 + 2\ H_2O$

    a. If you had 8 mol Si_2H_2, how many moles of O_2 would you need for the above reaction?

    Answer: 20 moles O_2

    b. If you had 15 mol O_2, how many moles of Si_2H_2 would you need for the above reaction?

    Answer: 6 mol Si_2H_2

    c. If you had 8 mol Si_2H_2 and 15 mol O_2, which reactant would be limiting?

    Answer: O_2

    d. How many moles of the non-limiting reactant would be left over?

    Answer: 2 mol Si_2H_2 left over

    e. What is the theoretical yield of SiO_2, in moles? (*I.e.,* how many moles of SiO_2 would you make?)

    Answer: 12 mol SiO_2

Use this space for summary and/or additional notes:

Limiting Reactant

Page: 459
Unit: Stoichiometry

2. Consider the reaction: 3 Ca (s) + 2 AlCl$_3$ (aq) → 3 CaCl$_2$ (aq) + 2 Al (s)

   a. If you had 6 mol Ca, how many moles of AlCl$_3$ would you need for the above reaction?

   Answer: 4 mol AlCl$_3$

   b. If you had 8 mol AlCl$_3$, how many moles of Ca would you need for the above reaction?

   Answer: 12 mol Ca

   c. If you had 6 mol Ca and 8 mol AlCl3, which reactant would be limiting?

   Answer: Ca

   d. How many moles of the non-limiting reactant would be left over?

   Answer: 4 mol AlCl$_3$ left over

   e. What is the theoretical yield of CaCl$_2$, in moles? (*I.e.,* how many moles of CaCl$_2$ would you make?)

   Answer: 6 mol CaCl$_2$

Use this space for summary and/or additional notes:

Big Ideas	Details

Limiting Reactant

Page: 460
Unit: Stoichiometry

Homework Problems

Set #2: Unscaffolded

1. How many moles of H_2O would be produced if 3.5 mol H_2 react with 1.5 mol O_2 in the reaction:

$$2 H_2 + O_2 \rightarrow 2 H_2O$$

   (*Note: because amounts were given for both reactants, this is a limiting reactant problem.*)

   Answer: 3.0 mol H_2O

2. If 12.0 mol S_8 reacted with 100. mol O_2 in the *unbalanced* equation:

$$S_8 + O_2 \rightarrow SO_3$$

   Which reactant is limiting, and how much of the other reactant would be left over?

   Answer: O_2 is limiting; there will be 3.7 mol S_8 left over.

3. 325 g of H_2O is poured onto a 450. g block of sodium metal. The equation for this reaction is:

$$2 Na + 2 H_2O \rightarrow 2 NaOH + H_2$$

   a. What is the limiting reactant?

      Answer: H_2O

   b. If the reaction temperature is 227 °C (500. K) at a pressure of 1 atm, how many liters of H_2 gas are produced?

      (*Hint: find the moles of H_2 produced and use the ideal gas law to calculate the volume.*)

      Answer: 369 L H_2

Use this space for summary and/or additional notes:

Chemistry 1

Mr. Bigler

Limiting Reactant

Unit: Stoichiometry

4. 5.00 g Zn are reacted with 100. mL of 1.00 M HCl in the reaction:

   $Zn\ (s) + 2\ HCl\ (aq) \rightarrow ZnCl_2\ (aq) + H_2\ (g)$

   a. Determine which reactant is limiting.

   Answer: HCl

   b. Determine the number of *grams* of $ZnCl_2$ that will be produced.

   Answer: 6.82 g $ZnCl_2$

   c. If the reaction conditions are 177 °C (*remember to convert to Kelvin!*) and 1 atm pressure, determine the number of liters of H_2 gas that will be produced.

   Answer: 1.85 L H_2

   d. Determine the mass in grams of the non-limiting reactant that will be left over.

   Answer: 1.73 g Zn

Use this space for summary and/or additional notes:

Percent Yield

Unit: Stoichiometry

MA Curriculum Frameworks (2016): HS-PS1-7

MA Curriculum Frameworks (2006): 5.6

Mastery Objective(s): (Students will be able to...)
- Calculate the percent yield of a reaction.

Success Criteria:
- Theoretical yield calculated correctly using stoichiometry calculations.
- Algebra and rounding to appropriate number of significant figures is correct.

Tier 2 Vocabulary: yield

Language Objectives:
- Explain how to turn fractions into percentages.

Notes:

<u>theoretical yield</u>: the amount of a product predicted, based only on stoichiometry calculations.

<u>actual yield</u>: the actual amount of product recovered in the laboratory.

<u>percent yield</u>: the amount of product recovered, expressed as a percentage of the theoretical yield.

When you do a stoichiometry calculation, the answer to the question "how much product should be produced" is the theoretical yield.

Actual yield depends on several factors. Many reactions do not go to completion, but instead reach an equilibrium condition where the amount of reactants and products is constant. Sometimes it is not possible to recover all of the product because of challenges associated with separating it from the other reactants and products. *Etc.*

Because of these factors, the actual yield is determined by performing the reaction in a laboratory and measuring the amount of product you got.

Once you have the actual and theoretical yield numbers, the percent yield is:

$$\frac{\text{actual yield}}{\text{theoretical yield}} \times 100 = \text{percent yield}$$

Use this space for summary and/or additional notes:

Percent Yield

Big Ideas | Details
Page: 463
Unit: Stoichiometry

Sample Problem:

Q: Suppose you perform the reaction:

$$3 \text{ Al (s)} + 3 \text{ CuCl}_2 \text{ (aq)} \rightarrow 2 \text{ AlCl}_3 \text{ (aq)} + 3 \text{ Cu (s)}$$

If you start with 9.0 g of Al and you recover 28 g of Cu, what was your percent yield?

A: First, calculate the theoretical (predicted) yield of Cu using stoichiometry:

$$\frac{9.0 \text{ g Al}}{1} \times \frac{1 \text{ mol Al}}{27.0 \text{ g Al}} \times \frac{3 \text{ mol Cu}}{2 \text{ mol Al}} \times \frac{63.6 \text{ g Cu}}{1 \text{ mol Cu}} = 31.8 \text{ g Cu}$$

Then calculate: $\dfrac{\text{actual yield}}{\text{theoretical yield}} \times 100 = \text{percent yield}$:

$$\frac{28 \text{ g Cu recovered}}{31.8 \text{ g Cu predicted}} = 0.88 \times 100 = 88\%$$

Note that the percent yield cannot exceed 100 %. Conservation of mass tells us that if we had enough aluminum and copper chloride in the above reaction to make 31.8 g of copper metal, there's no way we could actually make more than that. If you calculate a percent yield greater than 100 %, you should:

- Double-check your calculations to make sure you didn't make a mistake.
- Look for other compounds that might have gotten into the product that you measured. For example:
    - If you collect a precipitate on a piece of filter paper, it will be with with everything else that was in the beaker. Even if you evaporate the water, it will leave behind other compounds that had been dissolved. Also, if you let the product dry in the air on a humid day, the compound could be hygroscopic and/or form a hydrate.
    - If you collect a gas, the most common way is to use a eudiometer (gas collection tube) that starts filled with water, and to have the gas displace the water. However, there will also be water vapor in the tube, so you need to account for the water based on the vapor pressure of water at the temperature the gas was collected.

Use this space for summary and/or additional notes:

Percent Yield

Homework Problems

In order to isolate percent yield problems, these questions refer to the mass-mass stoichiometry homework problems starting on page 452.

1. In problem #1, part b on page 452, suppose that 3.85 g KCl was recovered in the lab. What was the percent yield?

   Answer:

2. In problem #2, part a on page 452, suppose that 125 g of NaOH was recovered. What was the percent yield?

   Answer:

3. In problem #3 on page 453 suppose that 10 L of O_2 was recovered.

   a. What was the percent yield?

      Answer:

   b. What might have happened in the lab that could account for the fact that you got a percent yield higher than 100 %?

4. In problem #4 on page 453 suppose you started with 3.00 g of precipitate, which was still wet from the solution in the reaction. You let it dry in the lab until the mass stopped changing, and you recorded it to be 2.50 g. Is this a good answer? If not, what else could you do?

Use this space for summary and/or additional notes:

Marathon Problems

Unit: Stoichiometry

MA Curriculum Frameworks (2016): HS-PS1-2, HS-PS1-7, HS-PS2-8(MA)

MA Curriculum Frameworks (2006): 4.1, 4.6, 5.1, 5.3, 5.4, 5.5, 5.6, 6.2, 7.2

Mastery Objective(s): (Students will be able to...)

- Solve challenging problems that combine several aspects of chemistry.

Success Criteria:

- Problems correctly utilize strategies from various topics throughout the year.
- Solutions use the equation appropriate for the information given.
- Solutions have the correct quantities substituted for the correct variables.
- Algebra and rounding to appropriate number of significant figures is correct.

Language Objectives:

- Explain what each part of each problem is asking and which topic it relates to.

Notes:

These are intentionally challenging problems that relate topics we studied throughout the year, including gas laws, solutions & concentration, solubility, naming compounds and writing formulas, predicting products, activity series, balancing equations, stoichiometry, limiting reactant, and percent yield.

A couple of the answers are provided so you can check your work at key points in the process, but you're on your own for the rest!

Use this space for summary and/or additional notes:

Marathon Problems

Page: 466
Unit: Stoichiometry

Homework Problems

1. 0.75 L of 2.5 M sodium phosphate is mixed with 1.25 L of 2.0 M calcium chloride.

    a. Predict the products, write and balance the chemical equation.

    b. Use your solubility rules to determine whether a chemical reaction happens based on whether a precipitate forms.
    (*If no reaction happens, you may skip the rest of this question. Hint: a reaction does happen.* ☺)

    c. Calculate the number of moles of each reactant and determine which one is limiting.

    d. How many *grams* of the precipitate are produced?

    Answer: 260 g

    e. If 150. g of precipitate was recovered on the filter paper, what is the percent yield?

    f. If one of the products remains in solution (*Hint: it does.*), what is its concentration in $\frac{mol}{L}$?
    (*Hint: you will need to add the volumes of the two solutions that you started with to find the total volume.*)

Use this space for summary and/or additional notes:

Marathon Problems

Page: 467
Unit: Stoichiometry

2. In a laboratory experiment 115 g of sulfur (S_8) was reacted with 89.6 L of oxygen gas (O_2) at S.T.P., in a synthesis reaction, producing only compound X. This compound contains 50 % sulfur by mass, and its empirical formula is the same as its molecular formula.

   a. Use percent composition data to determine the chemical formula of compound X.

   b. Write a balanced chemical equation for the reaction.

   c. Which of the reactants was limiting?

   d. What is the theoretical yield of compound X, in grams?

      Answer: 229 g

   e. If 189 g of compound X was actually recovered, what was the percent yield of X?

      Answer: 82.4 %

Use this space for summary and/or additional notes:

Summary: Stoichiometry

Unit: Stoichiometry

List the main ideas of this chapter in phrase form:

Write an introductory sentence that categorizes these main ideas.

Turn the main ideas into sentences, using your own words. You may combine multiple main ideas into one sentence.

Add transition words to make your writing clearer and rewrite your summary below.

Use this space for summary and/or additional notes:

Introduction: Thermochemistry (Heat)

Unit: Thermochemistry (Heat)

Topics covered in this chapter:

Heat & Temperature	471
Specific Heat Capacity & Calorimetry	474
Phase Changes & Heating Curves	481
Thermodynamics	489
Enthalpy of Formation	496
Heat of Reaction	500
Bond Energies	505

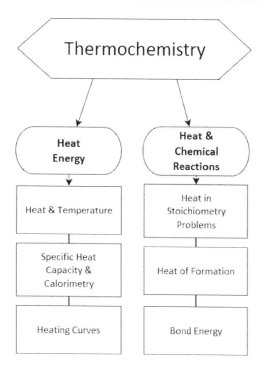

Use this space for summary and/or additional notes:

Introduction: Thermochemistry (Heat)

Unit: Thermochemistry (Heat)

Standards addressed in this chapter:

Massachusetts Curriculum Frameworks & Science Practices (2016):

HS-PS1-3 Cite evidence to relate physical properties of substances at the bulk scale to spatial arrangements, movement, and strength of electrostatic forces among ions, small molecules, or regions of large molecules in the substances. Make arguments to account for how compositional and structural differences in molecules result in different types of intermolecular or intramolecular interactions.

HS-PS1-4 Develop a model to illustrate the energy transferred during an exothermic or endothermic chemical reaction based on the bond energy difference between bonds broken (absorption of energy) and bonds formed (release of energy).

HS-PS3-4b Provide evidence from informational text or available data to illustrate that the transfer of energy during a chemical reaction in a closed system involves changes in energy dispersal (~~enthalpy~~ entropy[*] change) and heat content (~~entropy~~ enthalpy change) while assuming the overall energy in the system is conserved.

Massachusetts Curriculum Frameworks (2006):

6.4 Describe the law of conservation of energy. Explain the difference between an endothermic process and an exothermic process.

6.5 Recognize that there is a natural tendency for systems to move in a direction of disorder or randomness (entropy).

---

[*] The MA 2016 Curriculum Frameworks reversed the parenthetical references to entropy and enthalpy. I have corrected them in these notes.

Use this space for summary and/or additional notes:

Heat & Temperature

Unit: Thermochemistry (Heat)

MA Curriculum Frameworks (2016): HS-PS1-4

MA Curriculum Frameworks (2006): 6.4

Mastery Objective(s): (Students will be able to...)

- Explain the difference between heat and temperature.
- Describe what is happening at the molecular level when a system is in thermal equilibrium.

Success Criteria:

- Explanation accounts for total energy as well as direction of energy flow ("driving force").
- Description accounts for and relates macroscopic observations to microscopic phenomena.

Tier 2 Vocabulary: heat, temperature

Language Objectives:

- Explain the difference between heat and temperature.

Notes:

heat: energy that can be transferred when moving atoms or molecules collide with each other.

temperature: a measure of the average kinetic energy of the particles (atoms or molecules) of a system.

thermometer: a device that measures temperature, most often via thermal expansion/contraction of a liquid or solid.

Use this space for summary and/or additional notes:

Heat & Temperature

Unit: Thermochemistry (Heat)

Note that heat is the energy itself, whereas temperature is a measure of the "quality of the heat"—the average of the kinetic energies of the individual molecules:

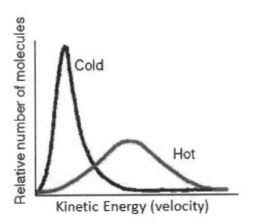

When objects are placed in contact, heat is transferred when the molecules collide. Molecules that have a lot of energy tend to transfer more energy than they receive. Molecules that have little energy tend to receive more energy than they transfer. This means two things:

1. On a macroscopic scale, heat always transfers from objects with a higher temperature (more kinetic energy) to objects with a lower temperature (less kinetic energy).

2. If you wait long enough, all of the molecules will have the same temperature (*i.e.,* the same average kinetic energy).

In other words, the temperature of one object relative to another determines which direction the heat will flow.

Use this space for summary and/or additional notes:

Heat & Temperature

Page: 473
Unit: Thermochemistry (Heat)

Big Ideas | Details

As an analogy, heat transfer is a lot like flowing water.

- Water flows from a higher elevation to a lower one, just like heat flows from a higher temperature to a lower one.

- The total energy of the water going over the waterfall depends on both the height of the waterfall (the average gravitational potential energy of the water molecules) and the total mass of water going over it. Similarly, the total heat (energy) contained in an object depends on both the mass of the object and its temperature.

Use this space for summary and/or additional notes:

| Big Ideas | Details |

Specific Heat Capacity & Calorimetry

Unit: Thermochemistry (Heat)
MA Curriculum Frameworks (2016): HS-PS2-6, HS-PS3-1
MA Curriculum Frameworks (2006): N/A
Mastery Objective(s): (Students will be able to...)

- Calculate the heat transferred when an object with a known specific heat capacity is heated.
- Perform calculations related to calorimetry.
- Describe what is happening at the molecular level when a system is in thermal equilibrium.

Success Criteria:

- Variables are correctly identified and substituted correctly into the correct equations.
- Algebra is correct and rounding to appropriate number of significant figures is reasonable.

Tier 2 Vocabulary: heat, specific heat capacity, coffee cup calorimeter

Language Objectives:

- Explain what the specific heat capacity of a substance measures.
- Explain how heat is transferred between one substance and another.

Labs, Activities & Demonstrations:

- Calorimetry lab.

Notes:

Different objects have different abilities to hold heat. For example, if you enjoy pizza, you may have noticed that the sauce holds much more heat (and burns your mouth much more readily) than the cheese or the crust.

The amount of heat that a given mass of a substance can hold is based on its specific heat capacity.

Use this space for summary and/or additional notes:

Specific Heat Capacity & Calorimetry

Big Ideas | Details | Unit: Thermochemistry (Heat)

specific heat capacity (C): a measure of the amount of heat required per gram of a substance to produce a specific temperature change in the substance.

C_p: specific heat capacity, measured at constant pressure. For gases, this means the measurement was taken allowing the gas to expand as it was heated.

C_v: specific heat capacity, measured at constant volume. For gases, this means the measurement was made in a sealed container, allowing the pressure to rise as the gas was heated.

For solids and liquids, $C_p \approx C_v$ because the pressure and volume change very little as they are heated. For gases, $C_p > C_v$ (always). For ideal gases, $C_p - C_v = R$, where R is a constant known as "the gas constant."

When there is a choice, C_p is more commonly used than C_v because it is easier to measure. When dealing with solids and liquids, most physicists just use C for specific heat capacity and don't worry about the distinction.

Calculating Heat from a Temperature Change

The amount of heat gained or lost when an object changes temperature is given by the equation:

$$Q = mC\Delta T$$

where:

Q = heat (J or kJ)

m = mass (g or kg)

C = specific heat capacity ($\frac{J}{g \cdot °C}$)

ΔT = temperature change (K or °C)

Note that $1\frac{J}{g \cdot °C} \equiv 1\frac{kJ}{kg \cdot °C} \equiv 1\frac{J}{g \cdot °C}$.

You need to be careful with the units. If the mass is given in kilograms (kg), your specific heat capacity will have units of $\frac{kJ}{kg \cdot °C}$ and the heat energy will come out in kilojoules (kJ). If mass is given in grams, you will use units of $\frac{J}{g \cdot °C}$ and the heat energy will come out in joules (J).

Use this space for summary and/or additional notes:

Specific Heat Capacity & Calorimetry

Unit: Thermochemistry (Heat)

Specific Heat Capacities of Some Substances

Substance	Specific Heat Capacity $(\frac{J}{g \cdot °C})$	Substance	Specific Heat Capacity $(\frac{J}{g \cdot °C})$
water at 20 °C	4.181	aluminum	0.897
ethylene glycol (anti-freeze)	2.460	glass	0.84
		iron	0.450
ice at −10 °C	2.080	copper	0.385
steam at 100 °C	2.11	brass	0.380
steam at 130 °C	1.99	silver	0.233
vegetable oil	2.00	lead	0.160
air	1.012	gold	0.129

Calorimetry

calorimetry: the measurement of heat flow

In a calorimetry experiment, heat flow is calculated by measuring the mass and temperature change of an object and applying the specific heat capacity equation.

calorimeter: an insulated container for performing calorimetry experiments.

coffee cup calorimeter: a calorimeter that is only an insulated container—it does not include a thermal mass (such as a mass of water). It is usually made of styrofoam, and is often nothing more than a styrofoam coffee cup.

bomb calorimeter: a calorimeter for measuring the heat produced by a chemical reaction. A bomb calorimeter is a double-wall metal container with water between the layers of metal. The heat from the chemical reaction makes the temperature of the water increase. Because the mass and specific heat of the calorimeter (water and metal) are known, the heat produced by the reaction can be calculated from the increase in temperature of the water.

It has a great name, but a bomb calorimeter doesn't involve actually blowing anything up. ☺

Use this space for summary and/or additional notes:

Specific Heat Capacity & Calorimetry

Unit: Thermochemistry (Heat)

Solving Coffee Cup Calorimetry Problems

Most coffee cup calorimetry problems involve placing a hot object in contact with a colder one. Many of them involve placing a hot piece of metal into cold water.

To solve the problems, assume that both objects end up at the same temperature.

If we decide that heat gained (going into a substance) by each object that is getting hotter is positive, and heat lost (coming out of a substance) by every substance that is getting colder is negative, then the basic equation is:

$$\text{Heat Lost} + \text{Heat Gained} = \text{Change in Thermal Energy}$$
$$\sum Q_{lost} + \sum Q_{gained} = \Delta Q$$

If the calorimeter is insulated, then no heat is gained or lost by the entire system (which means $\Delta Q = 0$).

If we have two substances (#1 and #2), one of which is getting hotter and the other of which is getting colder, then our equation becomes:

$$\text{Heat Lost} + \text{Heat Gained} = \text{Change in Thermal Energy}$$
$$\sum Q_{lost} + \sum Q_{gained} = \Delta Q = 0$$
$$m_1 C_1 \Delta T_1 + m_2 C_2 \Delta T_2 = 0$$

In this example, ΔT_1 would be negative and ΔT_2 would be positive.

To solve a calorimetry problem, there are six quantities that you need: the two masses, the two specific heat capacities, and the two temperature changes. (You might be given initial and final temperatures for either or both, in which case you'll need to subtract. Remember that if the temperature increases, ΔT is positive, and if the temperature decreases, ΔT is negative.) The problem will usually give you all but one of these and you will need to find the missing one.

If you need to find the final temperature, use $\Delta T = T_f - T_i$ on each side. You will have both T_i numbers, so the only variable left will be T_f. (The algebra is straightforward, but ugly.)

Use this space for summary and/or additional notes:

Specific Heat Capacity & Calorimetry

Unit: Thermochemistry (Heat)

Sample Problems:

Q: An 0.050 kg block of aluminum is heated and placed in a calorimeter containing 0.100 kg of water at 20.°C. If the final temperature of the water was 30.°C, to what temperature was the aluminum heated?

A: To solve the problem, we need to look up the specific heat capacities for aluminum and water in the table on page 476. The specific heat capacity of aluminum is $0.898 \frac{J}{g \cdot °C}$, and the specific heat capacity for water is $4.181 \frac{J}{g \cdot °C}$.

We also need to realize that we are looking for the initial temperature of the aluminum. ΔT is always **final – initial**, which means $\Delta T_{Al} = 30 - T_{i,Al}$. (Because the aluminum starts out at a higher temperature, this will give us a negative number, which is what we want.)

$$m_{Al}C_{Al}\Delta T_{Al} + m_w C_w \Delta T_w = 0$$
$$(0.050)(0.897)(30 - T_i) + (0.100)(4.181)(30 - 20) = 0$$
$$0.0449(30 - T_i) + 4.181 = 0$$
$$1.3455 - 0.0449 T_i + 4.181 = 0$$
$$5.5265 = 0.0449 T_i$$
$$T_i = \frac{5.5265}{0.0449} = 123.2 \, °C$$

Q: An 0.025 kg block of copper at 95°C is dropped into a calorimeter containing 0.075 kg of water at 25°C. What is the final temperature?

A: We solve this problem the same way. The specific heat capacity for copper is $0.385 \frac{J}{g \cdot °C}$, and $\Delta T_{Cu} = T_f - 95$ and $\Delta T_w = T_f - 25$. This means T_f will appear in two places. The algebra will be even uglier, but it's still a straightforward Algebra 1 problem:

$$m_{Cu}C_{Cu}\Delta T_{Cu} + m_w C_w \Delta T_w = 0$$
$$(0.025)(0.385)(T_f - 95_i) + (0.075)(4.181)(T_f - 25) = 0$$
$$0.009625(T_f - 95) + 0.3138(T_f - 25) = 0$$
$$0.009625 T_f - (0.009625)(95) + 0.3136 T_f - (0.3138)(25) = 0$$
$$0.009625 T_f - 0.9144 + 0.3138 T_f - 7.845 = 0$$
$$0.3234 T_f = 8.759$$
$$T_f = \frac{8.759}{0.3234} = 27 \, °C$$

Use this space for summary and/or additional notes:

Specific Heat Capacity & Calorimetry

Page: 479
Unit: Thermochemistry (Heat)

Homework Problems

You will need to look up specific heat capacities in Table Z. Selected Properties of the Elements, starting on page 556.

1. 375 kJ of heat is added to a 25.0 kg granite rock. If the temperature increases by 19.0 °C, what is the specific heat capacity of granite?

   Answer: $0.790 \frac{J}{g \cdot °C}$

2. A 0.040 kg block of copper at 95 °C is placed in 0.105 kg of water at an unknown temperature. After equilibrium is reached, the final temperature is 24 °C. What was the initial temperature of the water?

   Answer: 21.5 °C

3. A sample of metal with a specific heat capacity of $0.50 \frac{J}{g \cdot °C}$ is heated to 98 °C and then placed in an 0.055 kg sample of water at 22 °C. When equilibrium is reached, the final temperature is 35 °C. What was the mass of the metal?

   Answer: 0.0948 kg

Use this space for summary and/or additional notes:

Specific Heat Capacity & Calorimetry

Unit: Thermochemistry (Heat)

4. A 0.280 kg sample of a metal with a specific heat capacity of $0.430 \frac{J}{g \cdot °C}$ is heated to 97.5 °C then placed in an 0.0452 kg sample of water at 31.2 °C. What is the final temperature of the metal and the water?

Answer: 57 °C

Use this space for summary and/or additional notes:

Phase Changes & Heating Curves

Unit: Thermochemistry (Heat)

MA Curriculum Frameworks (2016): HS-PS1-2, HS-PS1-3, HS-PS3-4b

MA Curriculum Frameworks (2006): 6.3

Mastery Objective(s): (Students will be able to…)

- Determine the amount of heat required for all of the phase changes that occur over a given temperature range.

Success Criteria:

- Variables are correctly identified and substituted correctly into the correct equations.
- Algebra is correct and rounding to appropriate number of significant figures is reasonable.

Tier 2 Vocabulary: specific heat capacity, heating curve

Language Objectives:

- Explain what the heat is used for in each step of a heating curve.

Labs, Activities & Demonstrations:

- Evaporation from washcloth.
- Fire & ice (latent heat of paraffin).

Notes:

phase: a term that relates to how rigidly the atoms or molecules in a substance are connected.

solid: molecules are rigidly connected. A solid has a definite shape and volume.

liquid: molecules are loosely connected—bonds are continuously forming and breaking. A liquid has a definite volume, but not a definite shape.

gas: molecules are not connected. A gas has neither a definite shape nor a definite volume. Gases will expand to fill whatever space they occupy.

plasma: the system has enough heat to remove electrons from atoms, which means the system is comprised of particles with rapidly changing charges.

phase change: when an object or substance changes from one phase to another through gaining or losing heat.

Use this space for summary and/or additional notes:

Phase Changes & Heating Curves

Breaking bonds requires energy. Forming bonds releases energy. This is true for the bonds that hold a solid or liquid together as well as for chemical bonds (regardless of what previous teachers may have told you!)

I.e., you need to add energy to turn a solid to a liquid (melt it), or to turn a liquid to a gas (boil it). Energy is released when a gas condenses or a liquid freezes. (*E.g.,* ice in your ice tray needs to give off heat in order to freeze. Your freezer needs to remove that heat in order to make this happen.)

The reason evaporation causes cooling is because the system (the water) needs to absorb heat from its surroundings (*e.g.,* your body) in order to make the change from a liquid to a gas (vapor). When the water absorbs heat from you and evaporates, you have less heat, which means you have cooled off.

Calculating the Heat of Phase Changes

<u>heat of fusion</u> (ΔH_{fus}) (sometimes called "latent heat" or "latent heat of fusion"): the amount of heat required to melt one kilogram of a substance. This is also the heat released when one kilogram of a liquid substance freezes. For example, the heat of fusion of water is $334 \frac{J}{g}$. The heat required to melt a sample of water is therefore:

$$Q = m\Delta H_{fus} = m(334 \tfrac{J}{g})$$

<u>heat of vaporization</u> (ΔH_{vap}): the amount of heat required to vaporize (boil) one kilogram of a substance. This is also the heat released when one kilogram of a gas condenses. For example, the heat of vaporization of water is $2260 \frac{J}{g}$. The heat required to boil a sample of water is therefore:

$$Q = m\Delta H_{vap} = m(2260 \tfrac{J}{g})$$

Use this space for summary and/or additional notes:

Phase Changes & Heating Curves

heating curve: a graph of temperature *vs.* heat added. The following is a heating curve for water:

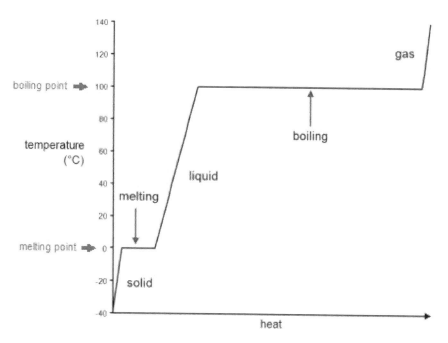

In the "solid" portion of the curve, the sample is solid water (ice). As heat is added, the temperature increases. The specific heat capacity of ice is $2.11 \frac{J}{g \cdot °C}$, so the heat required is:

$$Q_{solid} = mC\Delta T = m(2.11 \tfrac{J}{g \cdot °C})\Delta T$$

In the "melting" portion of the curve, the sample is a mixture of ice and water. As heat is added, the ice melts, but the temperature remains at 0°C until all of the ice is melted. The heat of fusion of ice is $334 \frac{J}{g}$, so the heat required is:

$$Q_{melt} = m\Delta H_{fus} = m(334 \tfrac{J}{g})$$

In the "liquid" portion of the curve, the sample is liquid water. As heat is added, the temperature increases. The specific heat capacity of liquid water is $4.181 \frac{J}{g \cdot °C}$, so the heat required is:

$$Q_{liquid} = mC\Delta T = m(4.181 \tfrac{J}{g \cdot °C})\Delta T$$

Use this space for summary and/or additional notes:

Phase Changes & Heating Curves

Unit: Thermochemistry (Heat)

In the "boiling" portion of the curve, the sample is a mixture of water and water vapor (steam). As heat is added, the water boils, but the temperature remains at 100°C until all of the water has boiled. The heat of vaporization of water is $2260 \frac{J}{g}$, so the heat required is:

$$Q_{melt} = m\Delta H_{vap} = m(2260 \tfrac{J}{g})$$

In the "gas" portion of the curve, the sample is water vapor (steam). As heat is added, the temperature increases. The specific heat capacity of steam is approximately $2.08 \frac{J}{g \cdot °C}$. (This is at 100 °C; the specific heat capacity of steam decreases as the temperature increases.) The heat required is:

$$Q_{gas} = mC\Delta T = m(2.08 \tfrac{J}{g \cdot °C})\Delta T$$

Steps for Solving Heating Curve Problems

A heating curve problem is a problem in which a substance is heated across a temperature range that passes through the melting and/or boiling point of the substance, which means the problem includes heating or cooling steps and melting/freezing or boiling/condensing steps.

1. Sketch the heating curve for the substance over the temperature range in question. Be sure to include the melting and boiling steps as well as the heating steps.

2. From your sketch, determine whether the temperature range in the problem passes through the melting and/or boiling point of the substance.

3. Split the problem into:
    a. Heating (or cooling) steps within each temperature range.
    b. Melting or boiling (or freezing or condensing) steps.

4. Find the heat required for each step.
    a. For the heating/cooling steps, use the equation $Q = mC\Delta T$.
    b. For melting/freezing steps, use the equation $Q = m\Delta H_{fus}$.
    c. For boiling/condensing steps, use the equation $Q = m\Delta H_{vap}$.

5. Add the values of Q from each step to find the total.

Use this space for summary and/or additional notes:

Phase Changes & Heating Curves

Unit: Thermochemistry (Heat)

Sample Problem

Q: How much heat would it take to raise the temperature of 15.0 g of H_2O from −25.0 °C to +130.0 °C?

A: The H_2O starts out as ice. We need to:

1. Heat the ice from −25.0 °C to its melting point (0 °C).
2. Melt the ice.
3. Heat the water up to its boiling point (from 0 °C to 100 °C).
4. Boil the water.
5. Heat the steam from 100 °C to 130 °C.
6. Add up the heat for each step to find the total.

heat solid: $Q_1 = mC\Delta T = (15)(2.11)(25) = 791.25 \text{ J}$

melt the ice: $Q_2 = m\Delta H_{fus} = (15)(334) = 5010 \text{ J}$

heat liquid: $Q_3 = mC\Delta T = (15)(4.181)(100) = 6270 \text{ J}$

boil: $Q_4 = m\Delta H_{vap} = (15)(2260) = 33\,900 \text{ J}$

heat gas: $Q_5 = mC\Delta T = (15)(2.08)(30) = 936 \text{ J}$

$$Q = Q_1 + Q_2 + Q_3 + Q_4 + Q_5$$
$$Q = 791 + 5010 + 6270 + 33\,900 + 936 = 46\,910 \text{ J}$$

Use this space for summary and/or additional notes:

Phase Changes & Heating Curves

Unit: Thermochemistry (Heat)

Homework Problems

For the following problems, use data from the following table:

	C (sol.) ($\frac{kJ}{kg \cdot °C}$)	M.P. (°C)	ΔH_{fus} ($\frac{kJ}{kg}$)	C (liq) ($\frac{kJ}{kg \cdot °C}$)	B.P. (°C)	ΔH_{vap} ($\frac{kJ}{kg}$)	C_p (gas) ($\frac{kJ}{kg \cdot °C}$)
water	2.11	0	334	4.18	100	2260	2.08*
potassium	0.560	62	61.4	1.070	760	2025	0.671
mercury	0.142	−39	11.3	0.140	357	293	0.104
silver	0.217	962	111	0.318	2212	2360	—

*Note that because of the volume change from heating, the specific heat capacity of gases, C_p, increases with increasing temperature.

1. A 0.0250 kg sample of water is heated from −40.0°C to 150. °C.

   a. Sketch the heating curve for the above process. Label the starting temperature, melting point, boiling point, and final temperature on the y-axis.

   b. Calculate the heat required for each step of the heating curve, and the total heat required.

   Answer: 80.01 kJ

Use this space for summary and/or additional notes:

Phase Changes & Heating Curves

Page: 487
Unit: Thermochemistry (Heat)

2. A 0.085 kg sample of mercury is heated from 25°C to 500. °C.

   a. Sketch the heating curve for the above process. Label the starting temperature, melting point, boiling point, and final temperature on the *y*-axis.

   b. Calculate the heat required for each step of the heating curve, and the total heat required.

   Answer: 30.12 kJ

Use this space for summary and/or additional notes:

Phase Changes & Heating Curves

Unit: Thermochemistry (Heat)

3. A 0.045 kg block of silver at a temperature of 22°C is heated with 20.0 kJ of energy.

   a. Calculate the total heat required by calculating the heat for each step until the entire 20.0 kJ is accounted for.

   b. What is the final temperature and what is the physical state (solid, liquid, gas) of the silver at that temperature?

   Answer: liquid, 1 369°C

Use this space for summary and/or additional notes:

Thermodynamics

Unit: Thermochemistry (Heat)
MA Curriculum Frameworks (2016): HS-PS3-4b
MA Curriculum Frameworks (2006): N/A
Mastery Objective(s): (Students will be able to...)
- Explain the laws of thermodynamics.
- Apply the laws of thermodynamics conceptually to hypothetical situations.

Success Criteria:
- Explanations account for enthalpy and entropy differences.
- Explanations account for the conservation of energy.

Tier 2 Vocabulary: system, free

Language Objectives:
- Explain the laws of thermodynamics.

Notes:

In the universe, energy is what "makes things happen". In chemistry, we use the following terms to distinguish between the molecules we are talking about vs. the ones we aren't:

<u>system</u>: the collection of molecules under consideration for a given situation.

<u>surroundings</u>: everything that is not part of the system

E.g., for a bunch of chemicals in a beaker, the chemicals would be the <u>system</u>, and the beaker, the air in the room, and everything else would be the <u>surroundings</u>.

Use this space for summary and/or additional notes:

Thermodynamics

Heat Flow

We generally use the variable Q to represent heat.

Heat flow is always represented in relation to the <u>system</u>.

Heat Flow	Sign of Q	System	Surroundings
from surroundings to system	+ (positive)	gains heat (gets warmer)	lose heat (get colder)
from system to surroundings	− (negative)	loses heat (gets colder)	gain heat (get hotter)

A positive value of Q means heat is flowing *into* the system. Because the heat is transferred from the molecules outside the system to the molecules in the system, the temperature of the system increases, and the temperature of the surroundings decreases.

A negative value of Q means heat is flowing *out of* the system. Because the heat is transferred from the molecules in the system to the molecules outside the system, the temperature of the system decreases, and the temperature of the surroundings increases.

endothermic reaction: a chemical reaction in which heat energy in the system is used to make the reaction proceed. This causes the system to get colder, which then causes heat to flow into the system from the surroundings. (Positive value of Q.)

exothermic reaction: a chemical reaction in which heat energy is released as the reaction proceeds. This causes the system to get hotter, which then causes heat to flow out of the system into the surroundings. (Negative value of Q.)

This is confusing for most people.

If you add sodium hydrogen carbonate (baking soda) to a strong acid, the solution will get colder, and the beaker will feel cold to the touch. The reaction took heat away from the solution. What makes this an endothermic reaction is that the system (the solution) is now colder than the surroundings, which means heat flows from the surroundings into the system to warm it back up.

If you add sodium hydroxide (lye) to a strong acid, the solution will get hotter, and the beaker will feel hot to the touch. The reaction released heat into the solution. What makes this an exothermic reaction is that the system (the solution) is now hotter than the surroundings, which means heat flows from the system into the surroundings to heat up the surroundings (and therefore cool off the system).

Use this space for summary and/or additional notes:

Thermodynamics

Unit: Thermochemistry (Heat)

A quick and simple way to think of this is by thinking of a glass of ice water. The ice water is the system and your hand is part of the surroundings. When you pick up the glass, your hand gets colder because heat is flowing from your hand (the surroundings) into the system (the ice water). This means the system is gaining heat, and the surroundings are losing heat. The value of Q would be positive in this example.

<u>thermal equilibrium</u>: when all of the particles in a system have the same average kinetic energy (temperature).

When a system is at thermal equilibrium, no net heat is transferred. (*I.e.*, collisions between particles still transfer energy back and forth, but the average temperature of the particles in the system—the macroscopic quantity that we can measure with a thermometer—is not changing.)

Use this space for summary and/or additional notes:

Usable vs. Non-Usable Energy

In any system, some of the energy

<u>enthalpy</u>: the total usable thermal (heat) energy of a system.

<u>entropy</u>: thermal energy that exists in a system, but cannot be transferred to other molecules or objects.

Energy can only be transferred in a reasonable amount of time if there is enough of an energy difference between the particles that are supplying the energy and the particles receiving it. For example, if your body temperature is 37 °C (98.6 °F) and you jump into water that is 10 °C (50 °F), you will lose heat very quickly. However, if you climb into a 36.9 °C spa ("hot tub"), you can stay in for an hour and your body temperature will be unaffected.

Entropy is therefore the energy that has been dispersed ("lost") into the surroundings, because there is so little difference between its heat content and the heat content of the surroundings that the energy cannot be transferred to other particles.

However, in many cases it is possible to reduce the entropy of a system by adding energy.

For example, if you open a bottle of ethyl acetate (the solvent in nail polish), the molecules will gradually diffuse into the room, and the entire room will smell like nail polish. The entropy of the gases in the room has increased, because the molecules of ethyl acetate (and the energy they contain) are more spread than they were before.

You could build a machine to pump all of the air in the room through a −40 °C condenser. This would condense 99 % of the ethyl acetate to a liquid, which you could then pour back into the bottle. This would indeed reduce entropy, but you would have to use much more energy to condense the ethyl acetate than the amount of energy you would recover in the form of reduced entropy.

Use this space for summary and/or additional notes:

Thermodynamics

In chemistry, we categorize energy in the following ways:

<u>kinetic energy</u> (*K*): the energy contained in the particles due to motion of the particles or sub-particles (atoms, molecules, atomic nuclei, electrons, *etc.*) The equation for kinetic energy is $K = \tfrac{1}{2}mv^2$, so kinetic energy depends on the masses and velocities (speeds) of the particles.

<u>temperature</u> (*T*): the average kinetic energy of the particles in a system. You can think of it as "heat per molecule". Increasing the temperature makes the molecules move faster; decreasing the temperature makes the molecules move more slowly.

<u>internal energy</u> (*U*): the total kinetic energy[*] of the molecules in a system.

Internal energy depends on the average kinetic energy of the molecules (temperature, *T*) and the number of molecules (or number of moles of molecules, *n*). For the units to work out so the energy comes out in units of joules, we multiply by the gas constant, giving the equation:

$$U = \tfrac{3}{2}nRT$$

<u>work</u> (*W*): the energy that a gas can transfer by its pressure causing a change in volume. Work can be done on the surroundings (the gas decreases its energy), or work can be done by the surroundings on the gas (the gas increases its energy).

$$W = P\Delta V$$

<u>enthalpy</u> (*H*): the total usable energy of a system, which includes its internal energy plus energy in the form of work that could be transferred between the system and surroundings.

$$H = U + PV$$

---

[*] Actually, internal energy also includes microscopic potential energy, which the energy of chemical and nuclear particle bonds, and the other physical force fields within the system (internal induced electric or magnetic dipole moment, stress-strain energy due to deformation of solids, *etc.*). Microscopic potential energy is beyond the scope of this course.

Use this space for summary and/or additional notes:

Thermodynamics

Unit: Thermochemistry (Heat)

<u>entropy</u> (S): energy that exists in a system, but cannot be transferred to other molecules or objects.

<u>free energy</u>: the energy of a system that is "free" (available) to do work. (Sometimes called the usable energy of the system.)

<u>Helmholtz free energy</u> (A): the usable energy of a system with a constant volume. Named after the German physicist Hermann von Helmholtz.
(Useful in physics; rarely used in chemistry.)

$$A = U - TS$$

<u>Gibbs free energy</u> (G): the usable energy of a system with constant pressure. Named after the American physicist J. Willard Gibbs.
(Useful in chemistry; rarely used in physics.)

$$G = H - TS$$

In chemistry, Gibbs free energy is the most useful in predicting whether (and to what extent) a chemical reaction will happen.

A chemical reaction is <u>spontaneous</u> if the change in Gibbs free energy is negative ($\Delta G < 0$), i.e., if the reaction *releases* energy and the total energy of the products is less than the total energy of the reactants.

A chemical reaction is <u>non-spontaneous</u> if the change in Gibbs free energy is positive ($\Delta G > 0$), i.e., if the reaction *consumes* energy and the total energy of the products is more than the total energy of the reactants.

A rule of thumb is that if a chemical reaction releases more than approximately 20 kJ of free energy (meaning $\Delta G < -20$ kJ), then the reaction "goes to completion," which means there will essentially be no detectable amount of the reactants left after the reaction takes place.

Similarly, if a chemical reaction would require more than 20 kJ of free energy (meaning $\Delta G \geq +20$ kJ), then "no reaction occurs," meaning there will essentially be no detectable amount of products made.

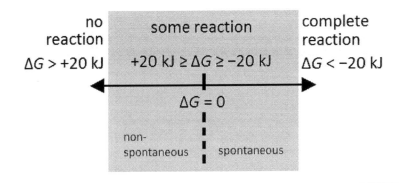

Use this space for summary and/or additional notes:

Thermodynamics

Unit: Thermochemistry (Heat)

Laws of Thermodynamics

Thermodynamics allows us to predict much of what happens in the universe based on energy changes. These predictions can be summed up in the four[*] laws of thermodynamics:

0. If you allow objects/systems in contact with each other to exchange heat for an infinite amount of time, they will have the same temperature.
   ("You have to play the game.")

1. Chemical reactions proceed in a way that releases energy, *i.e.*, usable heat energy (enthalpy) moves from particles with higher temperature to particles with lower temperature. Heat energy cannot flow from a lower temperature to a higher temperature.
   ("You can't win.")

2. Some heat energy is always lost to the surroundings (entropy) and cannot be recovered. Therefore, the entropy of the universe is always increasing.
   ("You can't break even.")

3. In a completely closed system, the total energy (enthalpy plus entropy) is constant. A change that would cause an increase in enthalpy (positive ΔH) would normally not happen spontaneously, but it can occur spontaneously if there is a sufficient increase in entropy (positive ΔS).
   ("You can't get out of the game.")

What You Need to Know

This is a lot of information, and yet only the tip of the iceberg; there are entire college courses on thermodynamics. Most of the information in this section is included to explain the bigger picture; in this course, you do not need to use any of these equations. The important points you need to understand in a first-year chemistry class are:

- Some energy, called enthalpy (H), exists in the form of usable heat that can be taken up or given off in chemical reactions, and/or used to do work.
- Some energy, called entropy (S), exists in the form of heat that is dispersed ("lost") to the surroundings and cannot be recovered.
- The combination of enthalpy and entropy, called Gibbs free energy (G), determines whether and to what extent a chemical reaction occurs.

---

[*] Originally there were three laws numbered 1–3. The "zero" law was added later, and was numbered zero in order to preserve the numbering of the first, second, and third laws.

Use this space for summary and/or additional notes:

Big Ideas	Details	Unit: Thermochemistry (Heat)

Enthalpy of Formation

Unit: Thermochemistry (Heat)

MA Curriculum Frameworks (2016): HS-PS1-4

MA Curriculum Frameworks (2006): N/A

Mastery Objective(s): (Students will be able to...)
- Determine the enthalpy of formation for selected compounds by looking up data in a table.
- Identify the formation of compounds as spontaneous or non-spontaneous based on the sign of the enthalpy of formation.

Success Criteria:
- Enthalpy of formation has the correct sign and the correct units.
- Compounds are correctly identified as forming spontaneously and not forming spontaneously based on the sign of their enthalpies of formation.

Tier 2 Vocabulary: formation, spontaneous

Language Objectives:
- Explain how to find the enthalpy of formation of a compound.
- Explain how to determine whether formation of a compound is spontaneous or non-spontaneous.

Notes:

Although free energy is the best predictor of whether a reaction happens, focusing on just the changes in enthalpy can be more practical for several reasons.

1. Entropy cannot be measured, but must be calculated based on other measurements and calculations, such as enthalpy and equilibrium. For this reason, entropy numbers are difficult to obtain and are often not available for reactions of interest.

2. Free energy depends on the temperature. (Recall that $G = H - TS$) If a reaction is producing or consuming heat, the temperature will be changing during the reaction, which means free energy calculations will require careful measurements and complex, calculus-based equations. Enthalpy does not depend on temperature, which means enthalpy numbers can be used directly regardless of the temperature at which they were measured.

3. Enthalpy measurements can be used directly to calculate the thermal energy (heat) produced or consumed by a reaction.

Use this space for summary and/or additional notes:

Enthalpy of Formation

Unit: Thermochemistry (Heat)

The concept of enthalpy of formation is that if we define the enthalpy content of a pure element to be zero[*], then:

- The enthalpy content of a compound can be measured by the reactions that produce the compound.

- The enthalpy content of a compound equals the heat energy released by forming its chemical bonds, and therefore also equals the heat energy needed to break those bonds.

standard enthalpy of formation: (ΔH_f°) the amount of enthalpy (recoverable/usable heat) that is released when a compound is formed.

In the "Intermolecular Forces" section starting on page 340, we saw that it takes energy to break intramolecular bonds (bonds between one molecule and another), and forming those bonds releases energy. The same is true for intermolecular bonds (bonds within a molecule).

If a chemical compound forms spontaneously, it forms because the compound has less energy than the elements that it is formed from. When the compound is formed, the excess energy is given off as heat. Once that energy is released, there is no longer enough energy for the compound to spontaneously disintegrate, unless heat is added. These compounds have a negative enthalpy of formation.

In some cases, energy is required to form a chemical compound. In these cases, the compound is unstable, and when a small amount of energy is added to break the bonds, a larger amount of energy is released. These compounds have a positive enthalpy of formation.

---

[*] Recall that enthalpy includes internal energy, which includes the energy of chemical and nuclear bonds, internal induced electric or magnetic dipole moment, stress-strain energy due to deformation of solids, *etc.*, as well as energy due to motion of the molecules, atoms, nuclei, electrons, *etc.* Saying that the enthalpy is zero does not mean we are ignoring these factors—remember that enthalpy can be negative. We are simply choosing the "zero point" to be the total of those energies in a pure element.

Use this space for summary and/or additional notes:

Enthalpy of Formation

Unit: Thermochemistry (Heat)

Determining Enthalpy of Formation

The enthalpy of formation is defined as the energy that would need to be *added* in order to form a compound or ion directly from its elements.

This means that pure elements in their natural state by definition have an enthalpy of formation of zero, because it takes no energy to form an element from its elements. This is true even for elements that are polyatomic in their natural state, such as N_2, O_2, Cl_2, F_2, Br_2, I_2, P_4 and S_8; even though these elements contain one or more chemical bonds, their enthalpy of formation is still defined to be zero.

If a compound or ion forms spontaneously, the process must *release* energy. This means that compounds or ions that form spontaneously have *negative* enthalpies of formation. (Adding negative energy is mathematically the same as releasing energy.)

Enthalpies of formation (ΔH_f^o) (and also entropies of formation) of selected compounds and elements are listed in "Table BB. Thermodynamic Data" in your Chemistry Reference Tables, on page 559.

Examples

The standard enthalpy of formation of $CaCl_2$ is $-795.8 \frac{kJ}{mol}$. This means we would need to add $-798.2\,kJ$ to make one mole of $CaCl_2$ from elemental Ca and Cl_2. Adding a negative amount of heat means the reaction actually *releases* $795.8 \frac{kJ}{mol}$ of heat. (Note that 795 kJ, or 795 800 J, is a lot of energy.)

The standard enthalpy of formation of C_2H_2 (acetylene) is $+226.7 \frac{kJ}{mol}$. This means that acetylene does not form spontaneously, and we need to add 226.7 kJ of heat energy to produce one mole of acetylene gas. That energy will be released as heat when acetylene combusts to form CO_2 and H_2O.

Use this space for summary and/or additional notes:

Enthalpy of Formation

Unit: Thermochemistry (Heat)

Homework Problem

Based on enthalpy of formation (ΔH_f^o) data in "Table BB. Thermodynamic Data" on page 559 of your Chemistry Reference Tables, rank the following ten compounds in order, from least stable to most stable.

$AgCl$, Al_2O_3, C_2H_2, H_2O, CO_2, N_2O, $CuSO_4$, H_2, $MgSO_4$, Si

1. _____

2. _____

3. _____

4. _____

5. _____

6. _____

7. _____

8. _____

9. _____

10. _____

Use this space for summary and/or additional notes:

Heat of Reaction

Unit: Thermochemistry (Heat)
MA Curriculum Frameworks (2016): HS-PS1-4
MA Curriculum Frameworks (2006): N/A
Mastery Objective(s): (Students will be able to...)
- Identify chemical reactions as exothermic or endothermic.
- Calculate the heat produced or consumed by a reaction based on heats of formation.
- Use the heat of reaction correctly in stoichiometry calculations.

Success Criteria:
- Heat of reaction accounts for heats of formation of all reactants and products.
- Calculation uses the correct signs for heats of formation of reactants and products, based on whether the compounds are being consumed or formed.
- Heat of formation is treated like a coëfficient (with its correct sign and units) in stoichiometry calculations.

Tier 2 Vocabulary: reaction, product

Language Objectives:
- Explain how heat of reaction is calculated.
- Explain how heat of reaction is used in stoichiometry calculations.

Notes:

Because a chemical reaction is the conversion of reactants into products, we can think of a chemical reaction as "taking apart" the reactants and then using the atoms to "build" the products.

"Taking apart" the reactants would involve breaking the bonds in compounds to form the elements they were made from. Because heats of formation give the energies to form the compounds, the energy needed to take apart the compounds would have the opposite sign. For example, if the heat of formation of H_2O is $-285.8 \frac{kJ}{mol}$, this indicates that 285.8 kJ of energy is released when one mole of H_2O is formed, which means we have to supply 285.8 kJ of energy to turn one mole of H_2O into hydrogen and oxygen.

Yes, this is confusing. Heat of formation is the amount of energy we have to *put in* to form the compound. A negative number means that the energy *comes out* when we form the compound instead of putting it in. But then, if we want to break apart the compound and turn it back into its elements, we have to *put in that same amount* of energy.

Use this space for summary and/or additional notes:

Heat of Reaction

Unit: Thermochemistry (Heat)

Sample Problem:

Q: Calculate the energy produced by the reaction:

$$Ca + FeO \rightarrow CaO + Fe$$

A: First we look up our heats of formation. We need to remember to change the sign of any compounds that we are breaking apart (reactants), and keep the sign the same for compounds that we are creating (products):

$$\begin{array}{ccccccc}
Ca & + & FeO & \rightarrow & CaO & + & Fe \\
0\,\tfrac{kJ}{mol} & & +272\,\tfrac{kJ}{mol} & & -634.92\,\tfrac{kJ}{mol} & & 0\,\tfrac{kJ}{mol}
\end{array}$$

sign changed because we're "taking apart" FeO

To find the heat of reaction, we simply add up all of the heat of formation numbers for each compound in the reaction:

$$(+272\,\tfrac{kJ}{mol}) + (-634.92\,\tfrac{kJ}{mol}) = -361\,\tfrac{kJ}{mol}$$

Because the coëfficients are all one, that means we can cancel moles to give:

$$(+272\,kJ) + (-634.92\,kJ) = -361\,kJ$$

The last step is to change the sign of the answer. Our calculation tells us that the enthalpy change for the reaction is −361 kJ. That means the products have 361 kJ *less* enthalpy than the reactants did. That energy was *released* as heat, which means the heat was *produced* by the reaction.

We always represent heat as a positive quantity in chemical reactions. If heat is produced, it is a product so it goes on the right. This gives us:

$$Ca + FeO \rightarrow CaO + Fe + 361\,kJ$$

Use this space for summary and/or additional notes:

Heat of Reaction

If we had gotten a positive number for the heat of reaction, that would mean that the products have more enthalpy content than the reactants, which means we would have had to add heat in order to make the reaction happen. In this case, heat would be a reactant and would go on the left.

For example, in the reaction:

$$N_2 + O_2 \rightarrow 2\ NO$$

We see that N_2 and O_2 are elements with an enthalpy of formation of zero, and NO has an enthalpy of formation of $+90.2\ \frac{kJ}{mol}$. Because the coëfficient for NO is 2, this means the enthalpy of reaction is +180.4 kJ. (Using heat of reaction with stoichiometry will be explained on the following pages.) The products have more enthalpy than the reactants, which means we have to add energy to make the reaction happen. This means heat is a reactant, so it goes on the left:

$$N_2 + O_2 + 180.4\ kJ \rightarrow 2\ NO$$

Use this space for summary and/or additional notes:

Heat of Reaction

Page: 503
Unit: Thermochemistry (Heat)

Heat of Reaction and Stoichiometry

Heat of reaction works just like the coëfficients in a stoichiometry problem. Because heats of reaction are expressed in $\frac{kJ}{mol}$, we need to multiply each substance's heat of formation by its coëfficient in order to calculate its contribution to the heat of reaction.

Sample Problem:

Q: Calculate the heat produced by the reaction:

$$3\ Ca + 2\ AlCl_3 \rightarrow 3\ CaCl_2 + 2\ Al$$

A: Again we look up our heats of formation. However, now, we need to *both* remember to use positive *vs.* negative numbers correctly *and* multiply each heat of formation by its coëfficient in the chemical equation:

3 Ca	+	2 AlCl$_3$	→	3 CaCl$_2$	+	2 Al
$3(0\ \frac{kJ}{mol})$		$2(+704\ \frac{kJ}{mol})$		$3(-795.8\ \frac{kJ}{mol})$		$2(0\ \frac{kJ}{mol})$
0 kJ		+1 408 kJ		−2 387.4 kJ		0 kJ

As before, to find the heat of reaction, we add the heats of formation:

$$(+1\,408\ kJ) + (-2\,387.4\ kJ) = -979\ kJ$$

The result tells us that the reaction produces 979 kJ of heat when the number of moles of each reactant and product are the same as the coëfficients. (This is sometimes described as "979 kJ per mole of reaction".) We can write the equation as follows:

$$3\ Ca + 2\ AlCl_3 \rightarrow 3\ CaCl_2 + 2\ Al + 979\ kJ$$

Again, notice that we changed the sign of the heat when we wrote it in the reaction because heat is produced, which means (positive) heat is a product of the reaction.

In this reaction, suppose we had used 4.5 mol of Ca (and excess AlCl$_3$). To find the heat produced, we would calculate it as if the 979 kJ were the coëfficient for heat in the chemical equation.

Using our stoichiometry "input/output machine", 3 mol of Ca produces 979 kJ of heat. This means that to get from 3 to 979, we have to divide by 3 and multiply by 979. This gives:

$$\frac{4.5\ \cancel{mol\ Ca}}{1} \cdot \frac{979\ kJ}{3\ \cancel{mol\ Ca}} = 1\,469\ kJ \text{ of heat produced}$$

Use this space for summary and/or additional notes:

Big Ideas	Details	Heat of Reaction	Page: 504

Unit: Thermochemistry (Heat)

Homework Problems

Using enthalpy of formation (ΔH_f^o) data in "Table BB. Thermodynamic Data" in your Chemistry Reference Tables, on page 559, calculate the heat of reaction for each of the following chemical reactions and write out the reaction with the heat in the correct place.

1. Pb + FeSO$_4$ → PbSO$_4$ + Fe

   Answer: $\Delta H^o = +9\,kJ$ Pb + FeSO$_4$ + 9 kJ → PbSO$_4$ + Fe

2. 2 C$_2$H$_2$ + 5 O$_2$ → 4 CO$_2$ + 2 H$_2$O

   Answer: $\Delta H^o = -2\,511\,kJ$ 2 C$_2$H$_2$ + 5 O$_2$ → 4 CO$_2$ + 2 H$_2$O + 2 511 kJ

3. C$_6$H$_{12}$O$_6$ + 6 O$_2$ → 6 CO$_2$ + 6 H$_2$O

   Answer: $\Delta H^o = -2\,537\,kJ$ C$_6$H$_{12}$O$_6$ + 6 O$_2$ → 6 CO$_2$ + 6 H$_2$O + 2 537 kJ

Use this space for summary and/or additional notes:

Big Ideas	Details	Bond Energies	Page: 505
			Unit: Thermochemistry (Heat)

Bond Energies

Unit: Thermochemistry (Heat)

MA Curriculum Frameworks (2016): HS-PS1-4

MA Curriculum Frameworks (2006): N/A

Mastery Objective(s): (Students will be able to...)

- Identify the bonds in molecular compounds and look up their bond energies.
- Calculate the heat produced or consumed by a reaction based on bond energy calculations.

Success Criteria:

- Bond energies have the correct sign and the correct units based on where they appear in the chemical equation.
- Heat of reaction calculated from bond energies agrees with heat of reaction calculated from heat of formation data.

Tier 2 Vocabulary: bond

Language Objectives:

- Explain how heat of reaction is calculated using bond energies.

Notes:

Ionic compounds form crystals that have straightforward ratios of cations to anions, which means it is practical to publish a table of the heats of formation of common ionic compounds and to calculate the heat of reaction from heat of formation data. Molecular compounds, however, can form long chains, and there are thousands upon thousands of common molecular compounds.

For most molecular compounds, instead of attempting to publish an exhaustive list of heat of formation data, it is much simpler to publish a table of the energy it takes to dissociate the individual bonds in molecular compounds. This way, it is possible to estimate the heat of reaction by simply adding up the energies for breaking or forming the individual bonds in a compound.

These bond energies are shown in "Table AA. Bond Dissociation Energies & Bond Lengths" in your Chemistry Reference Tables on page 558. Note that the table is "bond *dissociation* energies." This means that the numbers represent the amount of energy needed to *dissociate (break)* the bonds listed. If it takes energy (a positive number) to break a bond, that means energy is released (a negative number) when the bond is formed. This means we can use bond energies to calculate the enthalpy of formation of a compound, but we need to remember that the bond energies are negative when bonds are formed.

Use this space for summary and/or additional notes:

Bond Energies

Unit: Thermochemistry (Heat)

Here are bond dissociation energies for some bonds commonly found in organic compounds. A more extensive list can be found in "Table AA. Bond Dissociation Energies & Bond Lengths" on page 558 of your Chemistry Reference Tables.

Bond	C–C	C=C	C≡C	C–H	C–O	C=O	O=O	O–H
Bond Dissociation Energy* ($\frac{kJ}{mol}$)	346	602	835	411	358	799	494	459

Sample problem:

Use bond dissociation energy data to predict the energy of reaction for the combustion of methane:

$$CH_4 + 2\ O_2 \rightarrow CO_2 + 2\ H_2O$$

This actually means:

break 4 C–H bonds	+	break 2 1 O=O bond	→	form 2 C=O bonds	+	form 2 O–H bonds
4(411 kJ)	+	2(494 kJ)	→	2(−799 kJ)	+	4(−459 kJ)
1 644 kJ	+	988 kJ	+	−1 598 kJ	+	−1 836 kJ

Note that numbers on the left are positive, because we are *dissociating (breaking)* those bonds, and numbers on the right are negative because we are *forming* those bonds.

As with heat of reaction calculations, we add the energies for each of the individual molecules:

$$1\ 644\ kJ + 988\ kJ + (-1\ 598\ kJ) + (-1\ 836\ kJ) = -802\ kJ$$

---

* These are bond dissociation energies are for homolytic dissociation (*i.e.*, the electrons are equally split between the two atoms). Heterolytic bond dissociation energies (where the electrons split unequally) are always higher.

Use this space for summary and/or additional notes:

Bond Energies

We can now write the reaction as:

$$CH_4 + 2\,O_2 \rightarrow CO_2 + 2\,H_2O + 802\text{ kJ}$$

This agrees with the heat of reaction that we would calculate from enthalpy of formation data:

Compound	$\Delta H_f \left(\frac{kJ}{mol}\right)$
CH_4 (g)	−74.8
O_2 (g)	0
CO_2 (g)	−393.5
H_2O (g)	−241.8

$$\begin{array}{ccccccc}
CH_4 & + & 2\,O_2 & \rightarrow & CO_2 & + & 2\,H_2O \\
+74.8\,\tfrac{kJ}{mol} & & 2(0\,\tfrac{kJ}{mol}) & & -393.5\,\tfrac{kJ}{mol} & & 2(-241.8\,\tfrac{kJ}{mol}) \\
+74.8\text{ kJ} & & 0\text{ kJ} & & -393.5\text{ kJ} & & -483.6\text{ kJ}
\end{array}$$

$$74.8\text{ kJ} + 0 + (-393.5\text{ kJ}) + (-483.6\text{ kJ}) = -802.3\text{ kJ}$$

Note that adding the bond dissociation energies for a compound does not give the compound's enthalpy of formation. As an example, recall that the heat of formation of O_2 is zero, but it takes $494\,\tfrac{kJ}{mol}$ of heat energy to break an O=O double bond.

Use this space for summary and/or additional notes:

Bond Energies

Limitations of Using Bond Energy to Calculate Heat of Reaction

The above example with methane is one of the few instances where bond energies and heats of formation give the same values. This is often not the case. For example, if you calculate the heat of reaction for the combustion of acetylene:

$$2\ H-C\equiv C-H + 5\ O_2 \rightarrow 4\ CO_2 + 2\ H_2O + \text{heat}$$

you would get −2 511 kJ using heat of formation data, but −4 914 kJ using bond energies.

Enthalpies of formation are well-defined and precise, meaning that different people could measure them and expect to get the same answer. Average bond enthalpies, however, can vary widely depending on which molecules were used to gather the data. (In fact, the numbers may vary significantly from one published table to another.)

For example, there are many different molecules with C–H bonds. The bond energy of a C–H bond in CH_4 is very different from the bond energy of a C–H bond in acetylene.

The bond energies found in the bond energy tables for the bonds involved in the combustion of methane are based on:

Bond	C–H	C=O	O=O	O–H
Energy $\left(\frac{kJ}{mol}\right)$	411	799	494	459
Based on	CH_4	CO_2	O_2	H_2O

Needless to say, the basis for each of these bond energies is exactly the same as the bonds broken or formed in the combustion of methane. It is therefore not surprising that using bond energies gives the same result. However, it takes much more energy to break the C–H bond in acetylene, which is part of the reason that using bond energies does not give the correct answer for acetylene.

Because average bond enthalpies are much less accurate than enthalpies of formation, calculating heat of reaction from bond energies should only be used for a quick estimation, or as a last resort when heat of formation data are not available.

Use this space for summary and/or additional notes:

Bond Energies

Unit: Thermochemistry (Heat)

Homework Problem

1. Propane (C_3H_8) has the following structure:

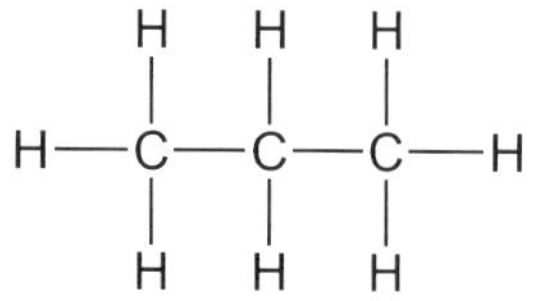

```
      H   H   H
      |   |   |
  H — C — C — C — H
      |   |   |
      H   H   H
```

Using the bond energy data from this chapter or on page 558 of your Chemistry reference tables, compute the heat of reaction for the complete combustion of propane:

$$C_3H_8 + 5\,O_2 \rightarrow 3\,CO_2 + 4\,H_2O + \text{heat}$$

Answer: −2 016 kJ (Note: calculating from ΔH_f° values gives 2 044 kJ)

Use this space for summary and/or additional notes:

Summary: Thermochemistry (Heat)

Unit: Thermochemistry (Heat)

List the main ideas of this chapter in phrase form:

Write an introductory sentence that categorizes these main ideas.

Turn the main ideas into sentences, using your own words. You may combine multiple main ideas into one sentence.

Add transition words to make your writing clearer and rewrite your summary below.

Use this space for summary and/or additional notes:

Introduction: Kinetics & Equilibrium

Unit: Kinetics & Equilibrium

Topics covered in this chapter:

Collision Theory 513
Rate of Reaction (Kinetics) 516
Equilibrium 520
Le Châtelier's Principle 527

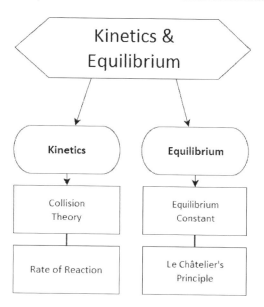

Use this space for summary and/or additional notes:

Introduction: Kinetics & Equilibrium

Standards addressed in this chapter:

Massachusetts Curriculum Frameworks & Science Practices (2016):

HS-PS1-4 Develop a model to illustrate the energy transferred during an exothermic or endothermic chemical reaction based on the bond energy difference between bonds broken (absorption of energy) and bonds formed (release of energy).

HS-PS1-5 Construct an explanation based on kinetic molecular theory for why varying conditions influence the rate of a chemical reaction or a dissolving process. Design and test ways to slow down or accelerate rates of processes (chemical reactions or dissolving) by altering various conditions.

HS-PS1-6 Design ways to control the extent of a reaction at equilibrium (relative amount of products to reactants) by altering various conditions using Le Châtelier's principle. Make arguments based on kinetic molecular theory to account for how altering conditions would affect the forward and reverse rates of the reaction until a new equilibrium is established.

Massachusetts Curriculum Frameworks (2006):

7.5 Identify the factors that affect the rate of a chemical reaction (temperature, mixing, concentration, particle size, surface area, catalyst).

7.6 Predict the shift in equilibrium when a system is subjected to a stress (Le Châtelier's principle) and identify the factors that can cause a shift in equilibrium (concentration, pressure, volume, temperature).

Use this space for summary and/or additional notes:

Collision Theory

Unit: Kinetics & Equilibrium

MA Curriculum Frameworks (2016): HS-PS1-4, HS-PS1-5

MA Curriculum Frameworks (2006): 7.5

Mastery Objective(s): (Students will be able to…)

- Use collision theory to explain activation energy and when a reaction does or does not occur.

Success Criteria:

- Explanations refer to Kinetic Molecular Theory (KMT).
- Explanations account for the role of activation energy.

Tier 2 Vocabulary: collision

Language Objectives:

- Explain the process of reactants colliding with enough energy to react and form products.

Notes:

Recall that Kinetic Molecular Theory (KMT) states:

- Gases are made of very large numbers of molecules.
- Molecules are constantly moving (obeying Newton's laws of motion), and their speeds are constant.
- Molecules are very far apart compared with their diameter.
- Molecules collide with each other and walls of container in elastic collisions.
- Molecules behaving according to KMT are not reacting or exerting any other forces (attractive or repulsive) on each other.

Use this space for summary and/or additional notes:

Collision Theory

Big Ideas | **Details**

Page: 514
Unit: Kinetics & Equilibrium

When we studied KMT earlier this year, we mentioned that the last bullet point did not mean that reactions did not occur, but that they were covered by collision theory, not KMT.

Collision theory states that:

- Molecules can be modeled as rigid spheres.
- Molecules move according to KMT.
- When molecules collide, kinetic energy (proportional to temperature) is transferred.
- If the molecules collide with enough energy to break existing chemical bonds, and if they are oriented in a way that allows the new bonds to form, a reaction occurs.
- If a reaction does not occur, the reactants "bounce" off each other as described by KMT.

For example, consider the reaction: $NO_3 + CO \rightarrow NO_2 + CO_2$

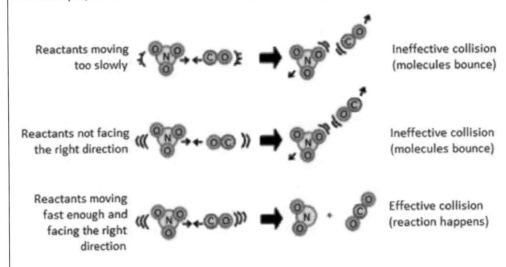

effective collision: a collision with enough energy that it leads to formation of products

ineffective collision: a collision that does not have enough energy to lead to formation of products.

mechanism: the details of which molecules collide and in what order and orientation in order for the reaction to take place.

Use this space for summary and/or additional notes:

Collision Theory

Unit: Kinetics & Equilibrium

The progress of the reaction vs. energy can be graphed in a "reaction coördinate" diagram:

<u>reaction coördinate diagram</u>: a graph that shows energy *vs.* progress of a chemical reaction.

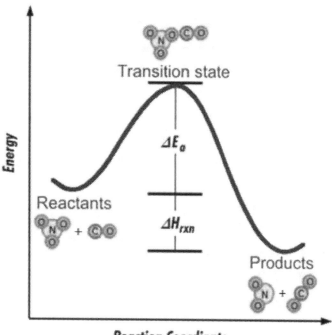

<u>enthalpy of reaction</u> (ΔH_{rxn}): the difference between the enthalpy of the products and the enthalpy of the reactants. The enthalpy change is the overall heat produced or consumed by the chemical reaction.

<u>activation energy</u> (ΔE_a): minimum kinetic energy needed for a collision to produce the transition state (and therefore proceed to form products).

<u>transition state</u> (or <u>activated complex</u>): the configuration of all molecules at the instant an effective collision happens. In the transition state, all of the energy needed by the reaction has been converted to enthalpy.

Use this space for summary and/or additional notes:

Rate of Reaction (Kinetics)

Unit: Kinetics & Equilibrium

MA Curriculum Frameworks (2016): HS-PS1-5

MA Curriculum Frameworks (2006): 7.5

Mastery Objective(s): (Students will be able to...)
- List and explain factors that affect the rate of a chemical reaction.

Success Criteria:
- Descriptions convey how each factor affects the rate of reaction.

Tier 2 Vocabulary: intermediate

Language Objectives:
- Explain what it means for a reaction to happen faster *vs.* slower, and how each factor affects the reaction rate.

Labs, Activities & Demonstrations:
- Drop of food coloring in hot *vs.* cold water.

Notes:

<u>reactants</u>: the compounds consumed in the chemical reaction; compounds that *react*.

<u>products</u>: the compounds created by the chemical reaction; compounds that are *produced*.

<u>intermediates</u>: compounds that are produced in one step of a multi-step reaction and consumed by a later step.

Use this space for summary and/or additional notes:

Rate of Reaction (Kinetics)

Unit: Kinetics & Equilibrium

<u>reaction rate</u> (k): the rate at which products are formed in a chemical reaction, usually expressed in units of: $\frac{mol}{L \cdot s}$ or $\frac{M}{s}$ (where M = molarity = $\frac{mol}{L}$)

The reaction rate is related to the activation energy. A reaction with higher activation energy will happen more slowly, because fewer of the collisions will have enough energy to enable the molecules to react. Conversely, a reaction with lower activation energy will happen more quickly.

The equation for rate of reaction is: $\ln(k) = -\frac{RT}{E_a}$ or $k = e^{-RT/E_a}$

Quantitative rate calculations are studied in AP® Chemistry. In this course, you need to understand how the equation shows that a higher temperature will speed up the reaction (larger value of k), and a higher activation energy will slow down the reaction (smaller value of k).

<u>rate-limiting step</u> (or <u>rate-determining step</u>): the step that determines the overall rate of the reaction. In a multi-step reaction, the rate-limiting step is the slowest step.

For example, in the multi-step reaction:

$$A \xrightarrow{\text{fast}} B \xrightarrow{\text{slow}} C \xrightarrow{\text{fast}} D$$

- A → B will happen faster than B can get used up, so B will accumulate and the first reaction will not affect the overall rate.
- C → D will happen fast, which means as soon as C is produced, it will react to produce D.

Therefore, the rate of B → C, which happens slowly, is what determines the overall rate of the reaction A → D.

<u>catalyst</u>: a substance that speeds up a reaction by lowering the activation energy of (and therefore speeding up) the rate-limiting (slowest) step.

Use this space for summary and/or additional notes:

Rate of Reaction (Kinetics)

Factors that Affect Reaction Rates

- <u>concentration of reactants:</u> higher concentration means more frequent collisions = faster rate. (Only applies to molecules involved in the rate-determining step.) For gases, higher pressure = higher concentration.

- <u>surface area of reactants:</u> more surface area means higher probability of a collision = faster rate.

- <u>temperature:</u> higher temperature = faster because faster-moving molecules collide more often, and because faster-moving molecules have more kinetic energy to overcome the activation energy.

- <u>nature of the reactants:</u> weak bonds are easier to break than strong bonds. Reactions involving dissolved ions are very fast, because bonds are already broken.

- <u>catalysts:</u> catalysts *speed up reactions* in any of several ways:
  - bring molecules into the correct orientation for an effective collision (equivalent to increasing the concentration and/or surface area)
  - assist in breaking of bonds in the reactant(s) and/or formation of bonds in the products (equivalent to changing the nature of the reactants and/or lowering the activation energy)

Catalysts are not reactants; they are <u>not</u> consumed by the reaction.

Use this space for summary and/or additional notes:

Rate of Reaction (Kinetics)

Homework Problems

Consider the following decomposition reaction:

$$2\,N_2O_5 \rightarrow 2\,N_2 + 5\,O_2$$

This reaction happens in three steps:

1. $2\,N_2O_5 + 2\,H_2O \rightarrow 4\,HNO_3$ fast
2. $2\,HNO_3 \rightarrow N_2 + 3\,O_2 + H_2$ slow
3. $2\,H_2 + O_2 \rightarrow 2\,H_2O$ very fast

Answer the following questions:

1. Which compounds are intermediates in this reaction?

2. If you wanted to speed up the overall reaction, which of the three steps would you try to speed up? Explain why, and give an example of how you might do this.

Use this space for summary and/or additional notes:

Equilibrium

Unit: Kinetics & Equilibrium
MA Curriculum Frameworks (2016): HS-PS1-6
MA Curriculum Frameworks (2006): 7.6
Mastery Objective(s): (Students will be able to...)
- List and explain factors that affect the equilibrium of a chemical reaction.
- Write equilibrium expressions from chemical equations and chemical equations from equilibrium expressions.

Success Criteria:
- Equilibrium expressions have products on top and reactants on the bottom.
- Equilibrium expressions have coëfficients shown as exponents.

Tier 2 Vocabulary: forward, backward, expression

Language Objectives:
- Explain how different factors affect the equilibrium of a chemical reaction.

Notes:

<u>reversible reaction</u>: a reaction that proceeds in both forward and backward directions. Usually written with double half-arrows:

$$N_2 \, (g) + 3\,H_2 \, (g) \rightleftharpoons 2\,NH_3 \, (g)$$

<u>dynamic chemical equilibrium</u>: reaction is happening in both directions, but the changes balance each other, so the concentrations of reactants & products remain constant. At equilibrium:

$$\text{rate}_{\text{forward reaction}} = \text{rate}_{\text{reverse reaction}} \quad \text{or} \quad k_f = k_r$$

<u>concentration</u>: the amount of a compound dissolved in a given amount of solution, usually expressed in $\frac{mol}{L}$ (moles of the compound per liter of solution).

A shorthand way to write the concentration of a compound is to place the chemical formula in square brackets. For example, [NH_3] means "the concentration of NH_3 in $\frac{mol}{L}$".

Use this space for summary and/or additional notes:

Equilibrium

Unit: Kinetics & Equilibrium

Arrows Used in Chemical Equations

Arrow	Meaning
$A + B \rightarrow C + D$	A + B react to produce C + D. Either there is little or no reverse reaction, or no information is given about equilibrium.
$A + B \rightleftharpoons C + D$	A + B are in equilibrium with C + D. No information is given about whether products or reactants are favored.
$A + B \rightleftharpoons C + D$	A + B are in equilibrium with C + D. Products are favored. (I.e., the concentrations of products are higher than the concentrations of reactants.)
$A + B \rightleftharpoons C + D$	A + B are in equilibrium with C + D. Reactants are favored. (I.e., the concentrations of reactants are higher than the concentrations of products.)
$A \leftrightarrow B$	A and B are different resonance structures of the same compound. This is different from a chemical reaction that is at equilibrium.

Use this space for summary and/or additional notes:

Equilibrium

Unit: Kinetics & Equilibrium

equilibrium expression: a mathematical expression relating the concentrations of the products and reactants at equilibrium. For the reaction:

$$N_2 + 3H_2 \rightleftharpoons 2NH_3$$

the equilibrium expression is:

$$K_{eq} = \frac{[NH_3]^2}{[N_2][H_2]^3}$$

where [NH$_3$] means the molarity of NH$_3$ (the concentration in $\frac{mol}{L}$).

The equilibrium expression comes from collision theory. In order for N$_2$ and H$_2$ to react, one N$_2$ and 3 H$_2$ molecules need to find each other. The probability of this 4-way collision is related to the product of the concentrations of each of the molecules involved, *i.e.*, [N$_2$] [H$_2$] [H$_2$] [H$_2$] or [N$_2$] [H$_2$]^3. Notice that the coëfficients in the equation became exponents in the equilibrium expression.

The reverse reaction requires 2 NH$_3$ molecules to collide; the probability of this collision is the product of the concentrations, [NH$_3$] [NH$_3$] or [NH$_3$]^2. Again, the coëfficient "2" became the exponent.

For the equilibrium expression, we write the concentrations of products on the top and reactants on the bottom. (The reason for this comes from the rate laws for the forward and reverse reactions. Rate laws are beyond the scope of this course, but are studied in AP Chemistry.) However, for our purposes, it also means that an equilibrium constant greater than 1 means we have more products, and an equilibrium constant less than 1 means we have more reactants.

equilibrium constant (K_{eq}): when we plug the concentrations (in $\frac{mol}{L}$) for each of the products and reactions into the equilibrium expression, we get a value for K_{eq}.

If $K_{eq} > 1$ then there are more products than reactants, and we say that the "equilibrium lies to the right".

If $K_{eq} < 1$ then there are more reactants than products, and we say that "equilibrium lies to the left".

If $K_{eq} = 1$ then there are equal amounts of reactants and products, and we say that the reactants and products are equally favored.

K_{eq} depends on temperature, but is constant for a given reaction at a given temperature.

Use this space for summary and/or additional notes:

Equilibrium Constant *vs.* Free Energy

In the section on Thermodynamics, starting on page 489, we saw that the extent of the reaction (how much reaction occurs) depends on the change in Gibbs free energy of the reaction.

The equilibrium constant is related to the Gibbs free energy by the equation:

$$\Delta G = -RT\ln(K_{eq}) \quad \text{or} \quad K_{eq} = e^{-\Delta G/RT}$$

While the calculations are beyond the scope of this course, you should understand that a higher value of ΔG corresponds with a higher value of the equilibrium constant K_{eq}:

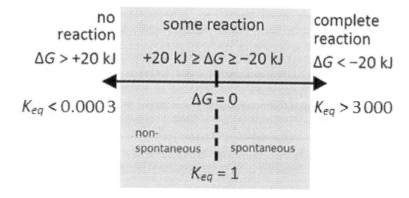

Sample Problems:

Q: Write the equilibrium expression for the chemical reaction $2H_2 + Cl_2 \rightleftharpoons 2HCl$

A: Products go on the top and reactants go on the bottom. Coëfficients become exponents.

$$K_{eq} = \frac{[HCl]^2}{[H_2]^2[Cl_2]}$$

Q: Write the chemical equation for the equilibrium expression: $K_{eq} = \dfrac{[CO_2]^4[H_2O]^2}{[C_2H_2]^2[O_2]^5}$

A: The denominator becomes the reactants (on the left), and the numerator becomes the products (on the right). Exponents become coëfficients:

$$2C_2H_2 + 5O_2 \rightleftharpoons 4CO_2 + 2H_2O$$

Use this space for summary and/or additional notes:

Equilibrium

Unit: Kinetics & Equilibrium

Q: Calculate the value of the equilibrium constant for the reaction $2\,NOBr \rightleftharpoons 2\,NO + Br_2$, if the concentration of NOBr is 3.00 M, the concentration of NO is 0.750 M, and the concentration of Br_2 is 0.200 M

A: The equilibrium expression is:

$$K_{eq} = \frac{[NO]^2[Br_2]}{[NOBr]^2}$$

Plugging in the concentrations, we get:

$$K_{eq} = \frac{(0.750)^2(0.200)}{3.00^2} = 0.0125$$

Note that the units for equilibrium constants are tricky, because each concentration is in $\frac{mol}{L}$. This means that if a concentration is squared, the units become $\left(\frac{mol}{L}\right)^2 = \frac{mol^2}{L^2}$. This means that every equilibrium constant has its own units, depending on the number of molecules that take part in the forward and reverse reactions.

From the equilibrium expression, the units of the equilibrium constant for this expression happens to be:

$$\frac{\left(\frac{mol}{L}\right)^2 \left(\frac{mol}{L}\right)}{\left(\frac{mol}{L}\right)^2} = \frac{mol}{L}$$

Working with the units for the equilibrium constant is beyond the scope of this course.

Use this space for summary and/or additional notes:

Equilibrium

Homework Problems

Write the expression for the equilibrium constants for each of the following reactions.

1. $Xe + 3F_2 \rightleftharpoons 2 XeF_6$

2. $CH_4 + 2H_2S \rightleftharpoons CS_2 + 4H_2$

3. $3CO_2 + 4H_2O \rightleftharpoons C_3H_8 + 5O_2$

4. Write the chemical equation for the equilibrium system given by the expression:

$$K_{eq} = \frac{[H_2O]^2[O_2]}{[H_2O_2]^2}$$

5. Write the chemical equation for the equilibrium system given by the expression:

$$K_{eq} = \frac{[NH_3]^2}{[N_2][H_2]^3}$$

6. Write the chemical equation for the equilibrium system given by the expression:

$$K_{eq} = \frac{[HCl]^4[O_2]}{[H_2O]^2[Cl_2]^2}$$

Use this space for summary and/or additional notes:

Equilibrium

Page: 526
Unit: Kinetics & Equilibrium

7. A reaction vessel contains 0.150 M CH_4, 0.233 M H_2O, 0.259 M H_2, and 0.513 M CO. If the equilibrium reaction is $CH_4 + H_2O \rightleftharpoons CO + 3H_2$, write the equilibrium expression and calculate the value of K_{eq}.

Answer: $K_{eq} = 0.255$

8. A 10 L flask contains 0.128 mol of CO, 0.155 mol of H_2 and 0.0244 mol of CH_3OH. If the equilibrium reaction is $CH_3OH \rightleftharpoons CO + 2H_2$, write the equilibrium expression and calculate the value of K_{eq}.

   (Note: you will need to divide each number of moles by 10 L to get the concentrations in $\frac{mol}{L}$.)

Answer: $K_{eq} = 0.00126$

Use this space for summary and/or additional notes:

Le Châtelier's Principle

Unit: Kinetics & Equilibrium

MA Curriculum Frameworks (2016): HS-PS1-6

MA Curriculum Frameworks (2006): 7.6

Mastery Objective(s): (Students will be able to...)

- Use Le Châtelier's Principle to predict a shift in equilibrium in response to a change.

Success Criteria:

- Prediction correctly describes the shift in equilibrium when the concentration of one chemical species is changed.

Tier 2 Vocabulary: stress

Language Objectives:

- Explain how a change provokes a response.

Notes:

If a reaction is at equilibrium, the reaction will resist any change with a corresponding change that shifts the reaction back to its equilibrium. Because K_{eq} is a constant, after the equilibrium shifts, the value of K_{eq} will be the same as it was before the change.

In plain English, if you change something, the equilibrium will shift to partly undo the change. This principle is called Le Châtelier's Principle, named after the French chemist Henry Louis Le Châtelier who first proposed the idea.

For example, consider the reaction:

$$N_2 \text{ (g)} + 3\,H_2 \text{ (g)} \leftrightharpoons 2\,NH_3 \text{ (g)} + 92.1 \text{ kJ}$$

For this reaction, $K_{eq} = \dfrac{[NH_3]^2}{[N_2][H_2]^3} = 835$ at 25 °C.

Suppose we started with $[N_2]$ = 0.05 M, $[H_2]$ = 0.3 M, and $[NH_3]$ = 1.06 M.

If we add more $[H_2]$, the reaction would use more H_2, and make more NH_3. If we kept adding H_2 until $[H_2]$ = 0.4 M, we would have $[N_2]$ = 0.026 M, and $[NH_3]$ = 1.18 M. As you can see, adding more H_2 caused the reaction to use up more N_2 and make more NH_3.

Use this space for summary and/or additional notes:

Le Châtelier's Principle

Le Châtelier's Principle tells us that we don't have to perform the equilibrium calculation to qualitatively predict what will happen. We can just look at the equation:

$$N_2 \text{ (g)} + 3\, H_2 \text{ (g)} \rightleftharpoons 2\, NH_3 \text{ (g)} + 92.1 \text{ kJ}$$

if we add more H_2, the equilibrium will shift to use more of it up. This means the equilibrium will shift to the right, also using up more N_2 and making more NH_3.

On the other hand, if we added NH_3, the equilibrium would instead shift to the left to use up some of the NH_3, and make more N_2 and H_2.

Action	Equilibrium shift
Add N_2 or H_2	to the right
Remove N_2 or H_2	to the left
Add NH_3	to the left
Remove NH_3	to the right
Increase the temperature (add heat)	to the left

Note that the value of K_{eq} is different at different temperatures. Adding reactants or products doesn't change the value of K_{eq}, but changing the temperature does. Le Châtelier tells us that adding heat must shift the equilibrium to the *left*. The equilibrium shift occurs because increasing the temperature results in a lower value of K_{eq} for this equation.

Quantitative equilibrium calculations and the relationship between the equilibrium constant and thermodynamics are studied in more depth in AP® Chemistry.

Use this space for summary and/or additional notes:

Le Châtelier's Principle

Homework Problems

Consider the chemical equation:

$$6H_2\text{ (g)} + P_4\text{ (g)} \rightleftharpoons 4PH_3\text{ (g)} + 53.5\text{ kJ}$$

1. Indicate which direction the equilibrium would shift as a result of each of the following:

   a. Adding P_4

   b. Removing PH_3

   c. Removing H_2

   d. Decreasing the temperature

2. Write the equilibrium expression for the above reaction.

3. The value of K_{eq} for this reaction is 4.44 at 25 °C. If the reaction is at equilibrium at 25 °C, the concentration of H_2 is 1.00 M and the concentration of P_4 is 0.025 M, what is the concentration of PH_3?

   Answer: $[PH_3] = 0.58$ M

4. If the reaction is cooled to 4 °C, the value of the equilibrium constant increases to 4.77. Is this consistent with the prediction made by Le Châtelier's Principle in question #1d above? Explain.

Use this space for summary and/or additional notes:

Summary: Kinetics & Equilibrium

Unit: Kinetics & Equilibrium

List the main ideas of this chapter in phrase form:

Write an introductory sentence that categorizes these main ideas.

Turn the main ideas into sentences, using your own words. You may combine multiple main ideas into one sentence.

Add transition words to make your writing clearer and rewrite your summary below.

Use this space for summary and/or additional notes:

Introduction: Acids & Bases

Unit: Acids & Bases

Topics covered in this chapter:

    Acids & Bases 532

    pH & Indicators 537

    pKa & Buffers 544

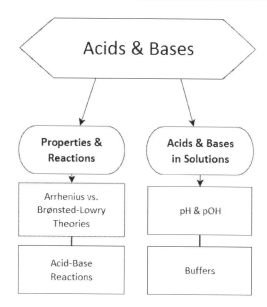

Standards addressed in this chapter:

Massachusetts Curriculum Frameworks & Science Practices (2016):

HS-PS1-9(MA) Relate the strength of an aqueous acidic or basic solution to the extent of an acid or base reacting with water as measured by the hydronium ion concentration (pH) of the solution. Make arguments about the relative strengths of two acids or bases with similar structure and composition.

Massachusetts Curriculum Frameworks (2006):

8.1 Define the Arrhenius theory of acids and bases in terms of the presence of hydronium and hydroxide ions in water and the Brønsted-Lowry theory of acids and bases in terms of proton donors and acceptors.

8.2 Relate hydrogen ion concentrations to the pH scale and to acidic, basic, and neutral solutions. Compare and contrast the strengths of various common acids and bases (*e.g.,* vinegar, baking soda, soap, citrus juice).

8.3 Explain how a buffer works.

Use this space for summary and/or additional notes:

Acids & Bases

Unit: Acids & Bases
MA Curriculum Frameworks (2016): HS-PS1-9(MA)
MA Curriculum Frameworks (2006): 8.1
Mastery Objective(s): (Students will be able to...)
- Define acids and bases based on both the Arrhenius and Brønsted-Lowry theories and give examples.
- Classify acids and bases as strong or weak.
- Identify conjugate acid and base pairs.

Success Criteria:
- Prediction correctly describes the shift in equilibrium when the concentration of one chemical species is changed.

Tier 2 Vocabulary: stress

Language Objectives:
- Explain how a change provokes a response.

Notes:

Acids are one of the first substances that come to mind when we think of chemistry. Acids are the dangerous chemicals that mad scientists in movies throw at people, and the chemicals that impressively dissolve metals and other substances right before your eyes.

Acids have held this sort of fascination for centuries. The American chemist Ira Remsen wrote the following wonderful anecdote of his first encounter with them:

> While reading a text book of chemistry, I came upon the statement, "nitric acid acts upon copper." I was getting tired of reading such absurd stuff and I determined to see what this meant. Copper was more or less familiar to me, for copper cents were then in use. I had seen a bottle marked "nitric acid" on a table in the doctor's office where I was then "doing time!" I did not know its peculiarities, but I was getting on and likely to learn. The spirit of adventure was upon me. Having nitric acid and copper, I had only to learn what the words "act upon" meant. Then the statement "nitric acid acts upon copper," would be something more than mere words.

Use this space for summary and/or additional notes:

Acids & Bases

Big Ideas	Details
	All was still. In the interest of knowledge I was even willing to sacrifice one of the few copper cents then in my possession. I put one of them on the table; opened the bottle marked "nitric acid;" poured some of the liquid on the copper; and prepared to make an observation. But what was this wonderful thing which I beheld? The cent was already changed, and it was no small change either. A greenish blue liquid foamed and fumed over the cent and over the table. The air in the neighborhood of the performance became colored dark red. A great cloud arose: This was disagreeable and suffocating—how should I stop this?
	I tried to get rid of the objectionable mess by picking it up and throwing it out the window, which I had meanwhile opened. I learned another fact—nitric acid not only acts upon copper but it acts upon fingers. The pain led to another unpremeditated experiment. I drew my fingers across my trousers and another fact was discovered. Nitric acid acts upon trousers.
	Taking everything into consideration, that was the most impressive experiment, and, relatively, probably the most costly experiment I have ever performed. I tell of it even now with interest. It was a revelation to me. It resulted in a desire on my part to learn more about that remarkable kind of action. Plainly the only way to learn about it was to see its results, to experiment, to work in a laboratory.
	However, not all acids are this dangerous, especially when they are more dilute. We eat and drink vinegar (dilute acetic acid), orange juice (which contains citric acid), and Coca-Cola (which contains phosphoric acid).
	acid: a substance that can produce H_3O^+ ions in water, release H^+ ions[*] in solution, and/or accept electrons from another substance.
	base: the "opposite" of an acid; a substance that can produce OH^- ions in water, accept H^+ ions in solution and/or donate electrons to another substance.
	_____
	[*] Note that an H^+ ion is a proton. Chemists often use the term "proton" in place of "H^+ ion" for convenience. Thus an acid is a compound that releases protons in water. These protons are just the H^+ ions—there's no nuclear weirdness going on!

Use this space for summary and/or additional notes:

Acids & Bases

Some Properties of Acids & Bases

Acids

- taste sour
- react with some metals (*i.e.*, the ones above hydrogen on the activity series)
- dissolve plants
- produce H_3O^+ ions in water (Arrhenius definition)
- release H^+ ions (Brønsted-Lowry definition)
- accept electrons (Lewis definition)

Bases

- taste bitter
- feel "slippery" (like soap)
- dissolve people (skin)
- produce OH^- ions in water (Arrhenius definition)
- accept H^+ ions (Brønsted-Lowry definition)
- give electrons (Lewis definition)

<u>dissociation</u>: to dissolve by splitting into positive and negative ions

Acids & bases dissociate in water.

Strong acids & bases dissociate completely; weak acids & bases only dissociate partially.

For example:

$$HCl \rightarrow H^+ + Cl^-$$
$$NaOH \rightarrow Na^+ + OH^-$$

<u>neutralization</u>: a reaction in which an acid and a base react to produce a salt (a type of ionic compound) plus water. For example:

$$HNO_3 + KOH \rightarrow KNO_3 + H_2O$$
$$(acid) + (base) \rightarrow (salt) + (water)$$

Use this space for summary and/or additional notes:

Acids & Bases

Big Ideas	Details
	strong acid: an acid that dissociates completely in water and produces H⁺ ions, which then convert H_2O molecules to H_3O^+ ions. Strong acids include HCl, HBr, HI, H_2SO_4, HNO_3, and $HClO_4$.
	weak acid: an acid that only partially dissociates in water. HF is an example of a weak acid: $$HF \rightleftharpoons H^+ + F^-$$
	strong base: a base that dissociates completely in water and produces OH⁻ ions. Strong bases include all of the group 1 hydroxides (LiOH, NaOH, KOH, *etc.*), plus the group 2 hydroxides $Ca(OH)_2$, $Sr(OH)_2$, and $Ba(OH)_2$.
	weak base: a base that only partially dissociates in water. NH_3 is an example of a weak base: $$NH_3 + H_2O \rightleftharpoons NH_4^+ + OH^-$$
	conjugates: the acid & base forms of a compound. The acid form has an extra H⁺ that can dissociate. The base form is the same compound without the H⁺.
	conjugate base: the base formed by removing H⁺ from an acid. For example, the conjugate base of HCl is Cl⁻.
	conjugate acid: the acid formed by adding H⁺ to a base. For example, the conjugate acid of NH_3 is NH_4^+.
	polyprotic: an acid that can lose more than one H⁺. For example, H_2SO_4 can lose one H⁺ to dissociate into H⁺ and HSO_4^-. HSO_4^- can then lose a second H⁺ to dissociate into H⁺ and SO_4^{2-}.
	Remember that an H⁺ ion is just a proton. A polyprotic acid is just an acid with more than one proton that it can lose by dissociation.
	amphoteric: a substance that "can go either way"—*i.e.*, it has both a conjugate acid and a conjugate base. For example, the HSO_4^- ion is amphoteric: $$H_2SO_4 \rightleftharpoons HSO_4^- \rightleftharpoons SO_4^{2-}$$

Use this space for summary and/or additional notes:

Acids & Bases

Homework Problems

Give the conjugate base for each of the following acids:

1. HCl
2. H_2S
3. HCO_3^-
4. $H_2PO_4^-$
5. HSO_4^-
6. H_2SO_3
7. NH_3
8. HS^-

Give the conjugate acid for each of the following bases:

9. HSO_4^-
10. SO_3^{2-}
11. ClO_4^-
12. $H_2PO_4^-$
13. SO_4^{2-}
14. PO_4^{3-}
15. CH_3NH_2
16. F^-

Use this space for summary and/or additional notes:

pH & Indicators

Unit: Acids & Bases

MA Curriculum Frameworks (2016): HS-PS1-9(MA)

MA Curriculum Frameworks (2006): 8.2

Mastery Objective(s): (Students will be able to…)

- Calculate pH from [H⁺] and pOH from [OH⁻].
- Identify acids and bases from their pK_a values.
- Select an appropriate indicator for a desired pH range.

Success Criteria:

- pH and pOH are calculated correctly.
- Acids and bases are correctly identified from their pK_a values.
- Indicator changes color in a pH range that includes the pH of the given acid or base.

Tier 2 Vocabulary: acid, base, indicator

Language Objectives:

- Explain why higher [H⁺] results in a lower pH.

Notes:

In water, a very small amount of H_2O dissociates into H^+ and OH^- ions:

$$H_2O \rightleftharpoons H^+ + OH^-$$

The amount of dissociation of any compound in a solvent is a constant that is determined by the attractions of the ions for each other *vs.* the attraction between the ions and the solvent.

In water at 25 °C, the product of the concentrations of H^+ and OH^- ions (in $\frac{mol}{L}$) is 1.0×10^{-14}. This number is called the "water dissociation constant" K_w[*]. In other words, in water at 25 °C:

$$K_w = [H^+][OH^-] = 1.0 \times 10^{-14}$$

---

[*] K_w is actually the equilibrium constant for the dissociation reaction. $K_{eq} = \dfrac{[H^+][OH^-]}{[H_2O]}$.

However, because H_2O is a pure liquid, the concentration of H_2O in pure H_2O is constant—it's just the density divided by the molar mass, which works out to 55.6 M. Therefore, we leave [H_2O] out of the equilibrium expression.

Use this space for summary and/or additional notes:

pH & Indicators

Unit: Acids & Bases

Recall that acids create H^+ (or H_3O^+) in water, and bases create OH^- in water. In the dissociation equation:

$$H_2O \rightleftharpoons H^+ + OH^-$$

Le Châtelier's principle predicts that if we add acid, $[H^+]$ increases. This shifts the equilibrium to the left, which means $[OH^-]$ decreases, and it is still true that $[H^+][OH^-] = 1.0 \times 10^{-14} = K_w$.

Similarly, if we add base, $[OH^-]$ increases and $[H^+]$ decreases and $[H^+][OH^-] = 1.0 \times 10^{-14} = K_w$.

If we have exactly the same amount of acid and base, then $[H^+] = [OH^-]$ and both are equal to the square root of 1×10^{-14}, which is 1×10^{-7} M. A solution with the same amount of acid and base is said to be _neutral_.

Working with concentrations in scientific notation that vary over 14 powers of ten is unwieldy, so we define a function "p" which means "take the logarithm of the quantity and multiply the result by −1." (See the "Logarithms" topic starting on page 102 for a brief description of the logarithm mathematical function.)

Therefore, the quantity "pH" would be $-\log[H^+]$.

<u>pH</u>: a measure of the strength of an acidic or basic solution. Equal to $-\log[H^+]$.

Examples:

if $[H^+] = 0.001$ M, then pH $= -\log(0.001) = 3$
if $[H^+] = 0.000\,000\,01$ M ($= 1 \times 10^{-8}$ M) then pH $= -\log(1 \times 10^{-8}) = 8$

<u>pOH</u>: another measure of the strength of an acidic or basic solution. Equal to $-\log[OH^-]$. Much less commonly used that pH.

Examples:

if $[OH^-] = 0.001$ M, then pOH $= -\log(0.001) = 3$
if $[OH^-] = 0.000\,000\,01$ M ($= 1 \times 10^{-8}$ M) then pOH $= -\log(1 \times 10^{-8}) = 8$

pH & pOH Equations

$$pH = -\log[H^+] \qquad\qquad pOH = -\log[OH^-]$$
$$[H^+] = 10^{-pH} \qquad\qquad [OH^-] = 10^{-pOH}$$

Because the (multiplication) product of $[H^+][OH^-] = 1 \times 10^{-14}$, this means that:

$$pH + pOH = 14$$

Use this space for summary and/or additional notes:

pH & Indicators

Details

Note that the higher the concentration of H⁺ ions (higher value of $[H^+]$), the lower the pH.

Low pH = acidic = more H^+ = less OH^-
High pH = basic = less H^+ = more OH^-

$[H^+]$	$[OH^-]$	pH	pOH	Acidic/Basic?
1 M (= 1×10^0 M)	1×10^{-14} M	0	14	very acidic
0.1 M (= 1×10^{-1} M)	1×10^{-13} M	1	13	
0.01 M (= 1×10^{-2} M)	1×10^{-12} M	2	12	
1×10^{-3} M	1×10^{-11} M	3	11	
1×10^{-4} M	1×10^{-10} M	4	10	
1×10^{-5} M	1×10^{-9} M	5	9	slightly acidic
1×10^{-6} M	1×10^{-8} M	6	8	
1×10^{-7} M	1×10^{-7} M	7	7	neutral
1×10^{-8} M	1×10^{-6} M	8	6	
1×10^{-9} M	1×10^{-5} M	9	5	slightly basic
1×10^{-10} M	1×10^{-4} M	10	4	
1×10^{-11} M	1×10^{-3} M	11	3	
1×10^{-12} M	0.01 M (= 1×10^{-2} M)	12	2	
1×10^{-13} M	0.1 M (= 1×10^{-1} M)	13	1	
1×10^{-14} M	1 M (= 1×10^0 M)	14	0	very basic

Sample Problems:

Q: What is the pH of a solution with $[H^+] = 2.5 \times 10^{-4}$ M?

A: $-\log(2.5 \times 10^{-4}) = 3.60$

Q: What is the concentration of H^+ ions in a solution with a pH of 11.4?

A: $10^{-11.4} = 3.98 \times 10^{-12}$ M

Use this space for summary and/or additional notes:

pH & Indicators

Unit: Acids & Bases

An aqueous solution is neutral when the concentration of H⁺ and OH⁻ are equal. This occurs in water at pH 7.00 at a temperature of 25 °C. However, remember that temperature affects equilibrium; as the temperature increases, more H⁺ and OH⁻ dissociate. This means [H⁺] and [OH⁻] *both* increase with higher temperatures, which means K_w increases. When that happens, [H⁺] and [OH⁻] are still equal in a neutral solution, but both are larger, and because [H⁺] and [OH⁻] are larger, the pH and pOH are both lower.

Temp. (°C)	K_w	pH of a neutral solution
0	0.114×10^{-14}	7.47
10	0.293×10^{-14}	7.27
20	0.681×10^{-14}	7.08
25	1.008×10^{-14}	7.00
30	1.471×10^{-14}	6.92
40	2.916×10^{-14}	6.77
50	5.476×10^{-14}	6.63
100	51.3×10^{-14}	6.14

In other words, despite what your previous teachers may have taught you, a pH of 7 is only neutral at 25 °C. In fact, in warm-blooded animals with body temperatures around 37 °C, a neutral pH would be approximately 6.8.

This also means that pH + pOH = 14 is only correct at 25 °C.

Use this space for summary and/or additional notes:

pH & Indicators

Indicators

<u>indicator</u>: a substance that changes color in a specific range of pH values. Indicators are used as a visual way to measure pH.

The following table lists some common indicators.

Name of Indicator	color in acid	color in base	pH range where color change occurs
bromophenol blue	yellow	purple	3.0–4.6
methyl red	red	yellow	4.4–6.2
litmus	red	blue	5.5–8.2
bromothymol blue	yellow	blue	6.0–7.6
phenol red	yellow	red	6.8–8.4
phenolphthalein	clear	pink	8.2–10.0

There are many others, and multiple indicators can be used in order to have different color changes over a broader pH range.

In fact, some clever chemists have developed a "universal indicator," which is typically composed of water, propanol, phenolphthalein, sodium hydroxide, methyl red, bromothymol blue, and thymol blue. This mixture indicates pH over a range from 3 to 11, in ROYGBIV (rainbow) order:

pH range	Description	Color
< 3	Strong acid	red
3–6	Weak acid	orange or yellow
7	Neutral	green
8–11	Weak base	blue
> 11	Strong base	indigo (dark blue) or violet (purple)

Use this space for summary and/or additional notes:

pH & Indicators

Page: 542
Unit: Acids & Bases

Homework Problems

For each of the following solutions, calculate the information indicated. Choose pH indicators from the "Common Acid-Base Indicators" table in your reference packets.

1. $[H+] = 2.5 \times 10^{-4}$ M

   a. pH =

   b. Is the solution acidic, basic, or neutral?

   c. Which pH indicator would be best for this solution?

2. $[H+] = 4.59 \times 10^{-7}$ M

   a. pH =

   b. Is the solution acidic, basic, or neutral?

   c. Which pH indicator would be best for this solution?

3. pH = 9.1

   a. [H+] =

   b. Is the solution acidic, basic, or neutral?

   c. Which pH indicator would be best for this solution?

Use this space for summary and/or additional notes:

pH & Indicators

Unit: Acids & Bases

4. pH = 5.5

    a. [H+] =

    b. Is the solution acidic, basic, or neutral?

    c. Which pH indicator would be best for this solution?

5. [OH−] = 7.9 × 10^{-7} M

    a. [H+] =

    b. pH =

    c. Is the solution acidic, basic, or neutral?

    d. Which pH indicator would be best for this solution?

Use this space for summary and/or additional notes:

pK_a & Buffers

Unit: Acids & Bases

MA Curriculum Frameworks (2016): N/A

MA Curriculum Frameworks (2006): 8.3

Mastery Objective(s): (Students will be able to...)

- Calculate pH from $[H^+]$ and pOH from $[OH^-]$.
- Identify acids and bases from their pK_a values.
- Select an appropriate indicator for a desired pH range.

Success Criteria:

- pH and pOH are calculated correctly.
- Acids and bases are correctly identified from their pK_a values.
- Indicator changes color in a pH range that includes the pH of the given acid or base.

Tier 2 Vocabulary: acid, base, indicator

Language Objectives:

- Explain why higher $[H^+]$ results in a lower pH.

Notes:

Acid-base chemistry is largely equilibrium chemistry in which the solvent, usually H_2O, plays a significant role.

As stated earlier, water dissociates into H^+ and OH^- ions. Acids and bases change the concentrations of H^+ and OH^- ions in solution, which can have significant effects on the behavior of the solution.

<u>acid dissociation constant</u> (K_a): is the equilibrium constant for the dissociation of an acid. For the "generic" acid HA:

$$K_a = \frac{[H^+][A^-]}{[HA]}$$

The greater the K_a value, the stronger the acid. (Remember your negative exponents! E.g., 10^{-5} is <u>greater</u> than 10^{-7}.)

Use this space for summary and/or additional notes:

pKa & Buffers

Unit: Acids & Bases

pK_a = −log K_a (analogous to pH). The lower (or more negative) the pK_a, the stronger the acid.

When exactly 50 % of the acid HA is neutralized, [HA] = [A⁻], and the above formula reduces to K_a = [H⁺]. This means that pH = pK_a when the acid is half-neutralized.

base dissociation constant (K_b): is the equilibrium constant for the dissociation of a base. For the "generic" base B:

$$K_b = \frac{[HB^+][OH^-]}{[B]}$$

We can use the concept of pK_a to add to our definitions of strong acids and bases:

strong acid: an acid with a pK_a lower than that of H_3O^+ (1.0). Strong acids include HCl, HBr, HI, H_2SO_4 and HNO_3.

Strong acids dissociate completely into H⁺ and the corresponding anion. The dissociated H⁺ converts H_2O molecules to H_3O^+ ions.

weak acid: an acid with a pK_a higher than that of H_3O^+ (1.0), but less than 7.0 (the pH of a neutral solution at 25 °C).

strong base: a base whose conjugate acid is weaker than H_2O (i.e., whose conjugate acid has a pK_a higher than 14). Hydroxides are strong bases because they release OH⁻. However, note that aqueous $Mg(OH)_2$ *acts* more like a weak base because the limited solubility of $Mg(OH)_2$ results in a concentration of OH⁻ that is similar to that produced by a weak base.

Strong bases either release OH⁻ ions directly into solution, or form OH⁻ ions by pulling H⁺ off of H_2O molecules.

weak base: a base whose conjugate acid has a pK_a higher than 7.0 but less than 14.

Use this space for summary and/or additional notes:

pKa & Buffers

Unit: Acids & Bases

Buffers

<u>buffer</u>: a weak acid or base that prevents the pH of a solution from changing drastically until it neutralizes the buffer.

For example, if you have a fish tank, you want to keep the pH from getting too low, you could add $NaHCO_3$. The reaction:

$$H^+ + HCO_3^- \rightleftharpoons H_2CO_3$$

occurs around pH 6.4. As acid accumulates in your fish tank, it will react with the HCO_3^- ions, and the pH will remain above 6.4 until all of the HCO_3^- ions have been converted to H_2CO_3.

Buffers can work in either direction—to absorb acid or base. If you use a combination of two buffers (one above and one below your desired pH), you can keep the pH within a narrow range.

In fact, water acts as a buffer, but over a very wide pH range. The pH of an aqueous solution is limited, because stronger acids just convert more H_2O to H_3O^+, and stronger bases just convert more H_2O to OH^-. The presence of water effectively keeps the pH between 1 and 14. In fact, the reason your biology teacher taught you that the pH range goes from 1–14 is because acid-base reactions in biology all happen in aqueous environments.

Use this space for summary and/or additional notes:

pKa & Buffers

Page: 547
Unit: Acids & Bases

Big Ideas | Details

Homework Problems

1. Rank the following acids from strongest to weakest, based on their pK_a values. Refer to of your Chemistry Reference Tables.

Based on pKa values in "Table P. *pKa* Values for Common Acids" on page 553 of your Chemistry Reference Tables, rank the following ten compounds in order, from the strongest acid to the strongest base.

HF, HCN, HCl, HPO_4^{2-}, HNO_3, H_2O, CH_3COOH, NH_4^+, H_2SO_4, H_2CO_3

1. _____   5. _____   8. _____

2. _____   6. _____   9. _____

3. _____   7. _____   10. _____

4. _____

11. The wastes from fish in a fish tank produce acids, which cause the pH of the water in the tank to decrease over time. Which acid-base pair from the table of *pKa* Values for Common Acids would be most effective at keeping the pH from dropping below 7.0. Explain.

Use this space for summary and/or additional notes:

Chemistry 1   Mr. Bigler

Summary: Acids & Bases

Unit: Acids & Bases

List the main ideas of this chapter in phrase form:

Write an introductory sentence that categorizes these main ideas.

Turn the main ideas into sentences, using your own words. You may combine multiple main ideas into one sentence.

Add transition words to make your writing clearer and rewrite your summary below.

Use this space for summary and/or additional notes:

Appendix: Chemistry Reference Tables

Table A. Standard Temperature and Pressure 549
Table B. Selected Units 549
Table C. Selected Prefixes 549
Table D. Physical Constants for Water 550
Table E. Vapor Pressure and Density of Water 550
Figure F. Phase Diagram for Water 550
Table G. Solubility Guidelines 551
Table H. K_{sp} Values for Some Insoluble Salts at 25 °C 551
Figure I. Solubilities of Selected Compounds 551
Table J. Number Prefixes 552
Table K. Polyatomic Ions 552
Table L. Flame Test Colors 552
Table M. Aqueous Ion Colors 552
Table N. Colors of Assorted Compounds 552
Table O. Common Acids 553
Table P. pK_a Values for Common Acids 553
Table Q. Common Bases 553
Table R. Common Acid-Base Indicators 553
Figure Z. Bonding Triangle 553
Table S. Symbols Used in Nuclear Chemistry 554
Table U. Selected Radioisotopes 554
Table U. Constants Used in Nuclear Chemistry 554
Figure V. Neutron/Proton Stability Band 554
Table W. Activity Series 555
Table X. Std. Reduction Potentials 555
Table Z. Selected Properties of the Elements 556
Table AA. Bond Dissociation Energies & Bond Lengths 558
Table BB. Thermodynamic Data 559
Table DD. Selected Formulas and Equations 561
Periodic Table of the Elements 563

Table A. Standard Temperature and Pressure

Name	Values		
"Standard" Pressure	1 atm	760 torr	101.3 kPa
Standard Temperature	0 °C	32 °F	273.15 K

atm =	atmosphere
Torr =	millimeter of mercury (mm Hg)
kPa =	kilopascal
°C =	degree Celsius
°F =	degree Fahrenheit
K =	kelvin

Table C. Selected Prefixes

Factor	Number of Units	Prefix	Symbol
10^6	1,000,000	mega-	M
10^3	1,000	kilo-	k
10^{-1}	0.1	deci-	d
10^{-2}	0.01	centi-	c
10^{-3}	0.001	milli-	m
10^{-6}	0.000 001	micro-	? (or u)

Table B. Selected Units

Name	Symbol	Quantity
meter (SI)	m	length
centimeter	cm	
kilogram (SI)	kg	mass
gram	g	
Pascal (SI derived)	Pa	pressure
atmosphere	atm	
mm of mercury	mm Hg	
Torr	Torr	
Kelvin (SI)	K	temperature
degree Celsius	°C	
amt of substance (SI)	mol	mole
Joule (SI derived)	J	energy
kilocalorie	kcal	
second (SI)	s	time
liter	L, ℓ	volume
part per million	ppm	concentration
molarity	$M, \frac{mol}{\ell}$	concentration

Table D. Physical Constants for Water	
Freezing Point @ 1 atm	0 °C = 273.15 K
Boiling Point @ 1 atm	100 °C = 373.15 K
Heat of Fusion	333.6 J/g
Heat of Vaporization	2270 J/g
Specific Heat Capacity (C_p)	4.184 J/g·°C
Freezing Point Depression Constant (K_f)	0.52 °C/m
Boiling Point Elevation Constant (K_b)	1.86 °C/m

Table E. Vapor Pressure and Density of Water

Temp (°C)	P_{vap} (kPa)	density (g/cm^3)
0.01	0.61173	0.99978
1	0.65716	0.99985
4	0.81359	0.99995
5	0.87260	0.99994
10	1.2281	0.99969
15	1.7056	0.99909
20	2.3388	0.99819
25	3.1691	0.99702
30	4.2455	0.99561
35	5.6267	0.99399
40	7.3814	0.99217
45	9.5898	0.99017
50	12.344	0.98799
55	15.752	0.98565
60	19.932	0.98316
65	25.022	0.98053
70	31.176	0.97775
75	38.563	0.97484
80	47.373	0.97179
85	57.815	0.96991
90	70.117	0.96533
95	84.529	0.96192
100	101.32	0.95475
105	120.79	0.95475

Figure F. Phase Diagram for Water

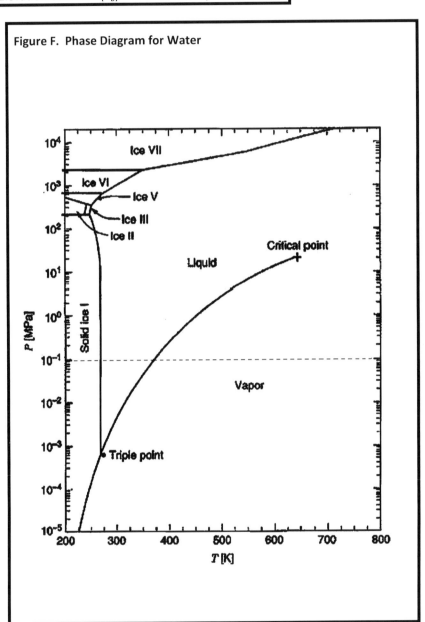

Table G. Solubility Guidelines

Ions That Form SOLUBLE Compounds	EXCEPT with	Ions That Form INSOLUBLE Compounds	EXCEPT with
Group I ions (Li$^+$, Na$^+$, etc.)		carbonate (CO$_3^{2-}$)	Group I ions, ammonium (NH$_4^+$)
ammonium (NH$_4^+$)		chromate (CrO$_4^{2-}$)	
nitrate (NO$_3^-$)		phosphate (PO$_4^{3-}$)	
hydrogen carbonate (HCO$_3^-$)		sulfite (SO$_3^{2-}$)	
chlorate (ClO$_3^-$)		sulfide (S^{2-})	Group I ions, Group II ions, NH$_4^+$
perchlorate (ClO$_4^-$)			
acetate (C$_2$H$_3$O$_2^-$ or CH$_3$COO$^-$)	Ag$^+$	hydroxide (OH$^-$)	Group I ions, NH$_4^+$, Ba^{2+}, Sr^{2+}, Tl$^+$
halides (Cl$^-$, Br$^-$, I$^-$)	Ag$^+$, Cu$^+$, Pb^{2+}, Hg$_2^{2+}$	oxide (O^{2-})	
sulfates (SO$_4^{2-}$)	Ca^{2+}, Sr^{2+}, Ba^{2+}, Ag$^+$, Pb^{2+}		

Table H. K_{sp} Values for Some Insoluble Salts at 25 °C

Compound	K_{sp}
MgCO$_3$	1.0×10^{-5}
PbCl$_2$	1.7×10^{-5}
BaF$_2$	2.0×10^{-6}
CuCl	1.0×10^{-6}
PbI$_2$	1.6×10^{-8}
AgOH	1.0×10^{-8}
BaCO$_3$	8.1×10^{-9}
CaCO$_3$	3.8×10^{-9}
SrCO$_3$	9.4×10^{-10}
AgCl	1.8×10^{-10}
BaSO$_4$	1.1×10^{-10}
CaF$_2$	3.9×10^{-11}
Mg(OH)$_2$	1.0×10^{-11}
Ag$_2$CrO$_4$	9.0×10^{-12}
CuI	5.0×10^{-12}
AgBr	3.3×10^{-13}
PbSO$_4$	2.5×10^{-13}
PbCO$_3$	1.6×10^{-13}
Mn(OH)$_2$	4.0×10^{-14}
PbCrO$_4$	1.8×10^{-14}
Fe(OH)$_2$	1.6×10^{-14}
AgI	1.5×10^{-16}
Zn(OH)$_2$	7.9×10^{-18}
FeS	4.0×10^{-18}
HgCl	2.0×10^{-18}
ZnS	1.0×10^{-23}
PbS	8.4×10^{-28}
CdS	3.6×10^{-29}
Al(OH)$_3$	1.6×10^{-34}
CuS	8.7×10^{-36}
Fe(OH)$_3$	1.3×10^{-36}
Ag$_2$S	2.0×10^{-50}
HgS	3.0×10^{-53}

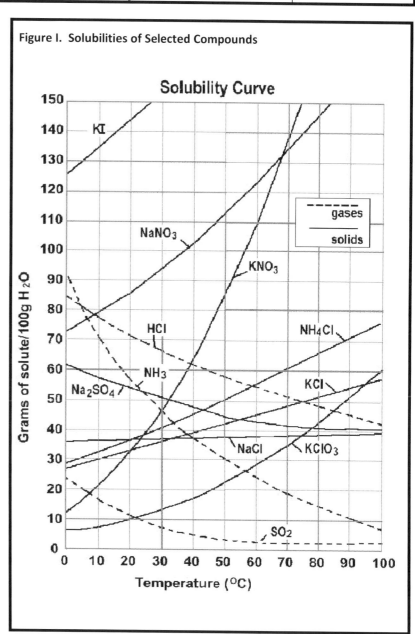

Figure I. Solubilities of Selected Compounds

Chemistry Reference Tables

Table J. Number Prefixes

Number	Inorganic	Organic	Number	Inorganic	Organic
1	mono-	meth-	6	hexa-	hex-
2	di-	eth-	7	hepta-	hept-
3	tri-	prop-	8	octa-	oct-
4	tetra-	but-	9	nona-	non-
5	penta-	pent-	10	deca-	dec-

Table K. Polyatomic Ions

ion	formula	ion	formula	ion	formula	ion	formula
americyl	AmO_2^{2+}	ammonium	NH_4^+	cyanate	OCN^-	dichromate	$Cr_2O_7^{2-}$
carbonyl	CO^{2+}	hydronium	H_3O^+	thiocyanate	SCN^-	imide	NH^{2-}
thiocarbonyl	CS^{2+}	iodyl	IO_2^+	selenocyanate	$SeCN^-$	molybdate	MoO_4^{2-}
chromyl	CrO_2^{2+}	nitrosyl	NO^+	tellurocyanate	$TeCN^-$	peroxide	O_2^{2-}
neptunyl	NpO_2^{2+}	thionitrosyl	NS^+	hydroxide	OH^-	oxalate	$C_2O_4^{2-}$
plutoryl	PuO_2^{2+}	phosphoryl	PO^+	iodate	IO_3^-	phthalate	$C_8H_4O_4^{2-}$
selinyl	SeO^{2+}	thiophosphoryl	PS^+	methanolate	CH_3O^-	selenate	SeO_4^{2-}
selenoyl	SeO_2^{2+}	phospho	PO_2^+	methanethiolate	CH_3S^-	disulfide	S_2^{2-}
thionyl / sulfinyl	SO^{2+}	acetate	CH_3COO^-	ethanolate	$C_2H_5O^-$	sulfate	SO_4^{2-}
sulfonyl / sulfuryl	SO_2^{2+}	amide	NH_2^-	permanganate	MnO_4^-	thiosulfate	$S_2O_3^{2-}$
uranyl	UO^{2+}	hydroxylamide	$NHOH^-$	nitrate	NO_3^-	dithionate	$S_2O_4^{2-}$
vanadyl	VO^{2+}	azide	N_3^-	superoxide	O_2^-	silicate	SiO_3^{2-}
mercury (II)	Hg^{2+}	hydrazide	$N_2H_3^-$	tetraborate	$B_4O_7^{2-}$	borate	BO_3^{3-}
mercury (I)	Hg_2^{2+}	bromate	BrO_3^-	carbide	C_2^{2-}	arsenate	AsO_4^{3-}
		chlorate	ClO_3^-	carbonate	CO_2^{2-}	phosphate	PO_4^{3-}
		cyanide	CN^-	chromate	CrO_4^{2-}	orthosilicate	SiO_4^{4-}

Table L. Flame Test Colors

Element	Color	Element	Color	Element	Color
Ba	yellow-green	K	pink	Pb	blue
Ca	orange-red	Li	fuchsia	Sb	pale green
Cu	blue-green	Mg	bright white	Sr	red
Fe	gold	Na	yellow	Zn	blue-green

Table M. Aqueous Ion Colors

Ion	Color	Ion	Color
Cu^+	green	V^{2+}	violet
Cu^{2+}	blue	V^{3+}	blue-green
Fe^{2+}	yellow-green	CrO_4^{2-}	yellow
Fe^{3+}	orange-red	$Cr_2O_7^{2-}$	orange
Cr^{3+}	violet [$Cr(NO_3)_3$] to green [$CrCl_3$]	$Cu(NH_3)_4^{2+}$	dark blue
Ni^{2+}	green	$FeSCN^{2+}$	red-brown (wine-red to dark orange)
Mn^{2+}	pink	Co^{2+}	pink
Mn^{7+}	purple (e.g., the MnO_4^- ion)	$CoCl_4^{2-}$	blue
Pb^{3+}	blue-green (Pb^{2+} and Pb^{4+} are clear)	$Ti(H_2O)_6^{3+}$	purple

Table N. Colors of Assorted Compounds

Compound	Color	Compound	Color
F_2	pale yellow gas	NO	colorless gas
Cl_2	green-yellow gas	NO_2	brown gas
Br_2	red-brown liquid	metallic sulfides	sulfides of transition metals tend to be black
I_2	dark metallic solid; dark violet vapor		
S_8	yellow odorous solid	metallic oxides	oxides of colored transition metals tend to be colored
PbI_2	bright yellow precipitate		
Fe_2O_3	reddish-brown (rust)		

Chemistry Reference Tables

Table O. Common Acids

Formula	Name
HCl (aq)	hydrochloric acid
HNO_3 (aq)	nitric acid
H_2SO_4 (aq)	sulfuric acid
H_3PO_4 (aq)	phosphoric acid
H_2CO_3 (aq)	carbonic acid
$HC_2H_3O_2$ (aq) or CH_3COOH (aq)	ethanoic acid (acetic acid)

Table Q. Common Bases

Formula	Name
NaOH (aq)	sodium hydroxide
KOH (aq)	potassium hydroxide
$Ca(OH)_2$ (aq)	calcium hydroxide
NH_3 (aq)	aqueous ammonia

Table R. Common Acid-Base Indicators

Indicator	pH Range of Color Change	Color Change
bromophenol blue	3.0 – 4.6	yellow–purple
methyl orange	3.2 – 4.4	red–yellow
bromocresol green	3.8 – 5.4	yellow–blue
methyl red	4.4 – 6.2	red–yellow
litmus	5.5 – 8.2	red–blue
bromothymol blue	6.0 – 7.6	yellow–blue
phenol red	6.8 – 8.4	yellow–red
thymol blue	8.0 – 9.6	yellow–blue
phenolphthalein	8.2 – 10	clear–pink

Table P. pK_a Values for Common Acids

Acid	pK_a	Conj. Base
H_2O	15.7	OH^-
HPO_4^{2-}	12.6	PO_4^{3-}
HCO_3^-	10.2	CO_3^{2-}
NH_4^+	9.2	NH_3
HCN	9.1	CN^-
$H_2PO_4^-$	7.2	HPO_4^{2-}
H_2S	7.0	HS^-
H_2CO_3	6.4	HCO_3^-
CH_3COOH	4.8	CH_3COO^-
HCOOH	3.7	$HCOO^-$
HNO_2	3.3	NO_2^-
HF	3.2	F^-
$C_6H_8O_7$ (citric acid)	3.1	$C_6H_7O_7^-$
H_3PO_4	2.2	$H_2PO_4^-$
HSO_4^-	2.0	SO_4^{2-}
HNO_3	-1.4	NO_3^-
H_3O^+	-1.7	H_2O
HCl	-7.0	Cl^-
HBr	-9.0	Br^-
HI	-10	I^-
$HClO_4$	-10	ClO_4^-
H_2SO_4	-12	HSO_4^-

Any acid with a pK_a value less than 1 is a strong acid; any base whose conjugate acid has a pK_a value greater than 14 is a strong base.

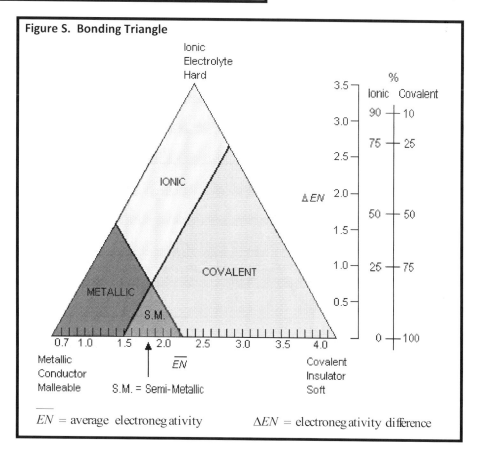

Figure S. Bonding Triangle

$\overline{EN}$ = average electronegativity ΔEN = electronegativity difference

Table T. Symbols Used in Nuclear Chemistry		
Name	Notation	Symbol
alpha particle	$_2^4\text{He}$ or $_2^4\alpha$	α
beta particle (electron)	$_{-1}^{0}e$ or $_{-1}^{0}\beta$	β^-
gamma radiation	$_0^0\gamma$	γ
neutron	$_0^1 n$	n
proton	$_1^1\text{H}$ or $_1^1 p$	p
positron	$_{+1}^{0}e$ or $_{+1}^{0}\beta$	β^+

Table V. Constants Used in Nuclear Chemistry	
Constant	Value
mass of an electron (m_e)	0.00055 amu
mass of a proton (m_p)	1.00728 amu
mass of a neutron (m_n)	1.00867 amu
Becquerel (Bq)	1 disintegration/second
Curie (Ci)	3.7×10^{10} Bq

Figure W. Neutron/Proton Stability Band

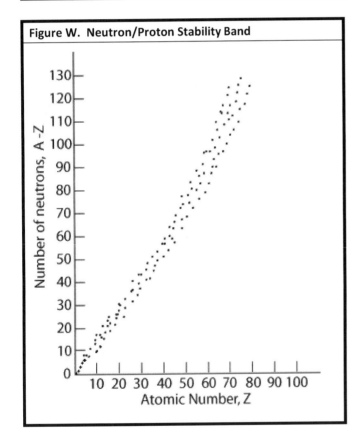

Table U. Selected Radioisotopes		
Nuclide	Half-Life	Decay Mode
^{3}H	12.26 y	β^-
^{14}C	5730 y	β^-
^{16}N	7.2 s	β^-
^{19}Ne	17.2 s	β^+
^{24}Na	15 h	β^-
^{27}Mg	9.5 min	β^-
^{32}P	14.3 d	β^-
^{36}Cl	3.01×10^5 y	β^-
^{37}K	1.23 s	β^+
^{40}K	1.26×10^9 y	β^+
^{42}K	12.4 h	β^-
^{37}Ca	0.175 s	β^-
^{51}Cr	27.7 d	α
^{53}Fe	8.51 min	β^-
^{59}Fe	46.3 d	β^-
^{60}Co	5.26 y	β^-
^{85}Kr	10.76 y	β^-
^{87}Rb	4.8×10^{10} y	β^-
^{90}Sr	28.1 y	β^-
^{99}Tc	2.13×10^5 y	β^-
^{131}I	8.07 d	β^-
^{137}Cs	30.23 y	β^-
^{153}Sm	1.93 d	β^-
^{198}Au	2.69 d	β^-
^{222}Rn	3.82 d	α
^{220}Fr	27.5 s	α
^{226}Ra	1600 y	α
^{232}Th	1.4×10^{10} y	α
^{233}U	1.62×10^5 y	α
^{235}U	7.1×10^8 y	α
^{238}U	4.51×10^9 y	α
^{239}Pu	2.44×10^4 y	α
^{241}Am	432 y	α

Table X. Activity Series

Metals		Non-metals
Reacts with cold H_2O and acids, replacing hydrogen. Reacts with O_2, forming oxides.	Cs ↑ Rb K Na Li Ba Sr Ca	F_2 ↑ Cl_2 Br_2 I_2
Reacts with steam (not cold H_2O) and acids, replacing hydrogen. Reacts with O_2, forming oxides.	Mg Be Al Mn Zn Cr Fe Cd	
Does not react with H_2O. Reacts with acids, replacing hydrogen. Reacts with O_2, forming oxides.	Co Ni Sn Pb H_2	
Reacts with O_2, forming oxides.	Sb Bi Cu	
Fairly unreactive, forming oxides only indirectly.	Ag Hg Au Pt	

Arrows indicate direction from LEAST to MOST active elements. An element can replace any element below itself on the activity series.

Note that the order of elements in the activity series is similar to, though not quite identical with, the order of elements/ions in the table of Std. Reduction Potentials (Table Y).

Table Y. Std. Reduction Potentials

Half-Reaction	E^0 (V)
$Li^+ + e^- \rightleftharpoons Li(s)$	−3.040
$Cs^+ + e^- \rightleftharpoons Cs(s)$	−3.026
$Rb^+ + e^- \rightleftharpoons Rb(s)$	−2.98
$K^+ + e^- \rightleftharpoons K(s)$	−2.931
$Ba^{2+} + 2e^- \rightleftharpoons Ba(s)$	−2.912
$Sr^{2+} + 2e^- \rightleftharpoons Sr(s)$	−2.899
$Ca^{2+} + 2e^- \rightleftharpoons Ca(s)$	−2.868
$Na^+ + e^- \rightleftharpoons Na(s)$	−2.71
$Mg^{2+} + 2e^- \rightleftharpoons Mg(s)$	−2.372
$Be^{2+} + 2e^- \rightleftharpoons Be(s)$	−1.85
$Al^{3+} + 3e^- \rightleftharpoons Al(s)$	−1.66
$Mn^{2+} + 2e^- \rightleftharpoons Mn(s)$	−1.029
$2H_2O + 2e^- \rightleftharpoons H_2(g) + 2OH^-$	−0.828
$Zn^{2+} + 2e^- \rightleftharpoons Zn(s)$	−0.762
$Cr^{3+} + 3e^- \rightleftharpoons Cr(s)$	−0.74
$Fe^{2+} + 2e^- \rightleftharpoons Fe(s)$	−0.44
$Cr^{3+} + e^- \rightleftharpoons Cr^{2+}$	−0.42
$Cd^{2+} + 2e^- \rightleftharpoons Cd(s)$	−0.40
$Co^{2+} + 2e^- \rightleftharpoons Co(s)$	−0.28
$Ni^{2+} + 2e^- \rightleftharpoons Ni(s)$	−0.25
$Sn^{2+} + 2e^- \rightleftharpoons Sn(s)$	−0.13
$Pb^{2+} + 2e^- \rightleftharpoons Pb(s)$	−0.13
$2H^+ + 2e^- \rightleftharpoons H_2(g)$	0.000
$S(s) + 2H^+ + 2e^- \rightleftharpoons H_2S(g)$	+0.14
$Sn^{4+} + 2e^- \rightleftharpoons Sn^{2+}$	+0.15
$Cu^{2+} + e^- \rightleftharpoons Cu^+$	+0.159
$Cu^{2+} + 2e^- \rightleftharpoons Cu(s)$	+0.340
$Cu^+ + e^- \rightleftharpoons Cu(s)$	+0.520
$I_2(s) + 2e^- \rightleftharpoons 2I^-$	+0.54
$Fe^{3+} + e^- \rightleftharpoons Fe^{2+}$	+0.77
$Ag^+ + e^- \rightleftharpoons Ag(s)$	+0.800
$Hg_2^{2+} + 2e^- \rightleftharpoons 2Hg(l)$	+0.80
$Hg^{2+} + 2e^- \rightleftharpoons Hg(l)$	+0.85
$2Hg^{2+} + 2e^- \rightleftharpoons Hg_2^{2+}$	+0.91
$Br_2(l) + 2e^- \rightleftharpoons 2Br^-$	+1.07
$Pt^{2+} + 2e^- \rightleftharpoons Pt(s)$	+1.188
$O_2(g) + 4H^+ + 4e^- \rightleftharpoons 2H_2O$	+1.23
$Cl_2(g) + 2e^- \rightleftharpoons 2Cl^-$	+1.36
$Au^{3+} + 3e^- \rightleftharpoons Au(s)$	+1.52
$Co^{3+} + e^- \rightleftharpoons Co^{2+}$	+1.82
$F_2(g) + 2e^- \rightleftharpoons 2F^-$	+2.87

E^0 values at 1 M concentration and 1 atm.

Table Z. Selected Properties of the Elements

atomic #	atomic symbol	element name	atomic mass (IUPAC 2005)	melting point °C	boiling point °C	specific heat capacity (J/g·K)	density (g/mL)	electro-negativity (Pauling)	1st ionization potential (kJ/mol)	common oxidation states
89	Ac	actinium	227	1050	3200	—	10.1	1.1	499	+3
13	Al	aluminum	26.98	660	2467	0.9	2.7	1.61	578	+3
95	Am	americium	243	994	2607	0.11	13.7	1.3	578	+3,4,5,6
51	Sb	antimony	121.8	631	1950	0.21	6.69	2.05	834	+3,5
18	Ar	argon	39.95	−189.2	−185.7	0.52	0.00178		1521	0
33	As	arsenic	74.92	817	617	0.33	5.73	2.18	947	±3,+5
85	At	astatine	210	302	337	—	—	2.2	917	
56	Ba	barium	137.3	725	1640	0.204	3.5	0.89	503	+2
97	Bk	berkelium	247	986	—	—	14	1.3	601	+3,4
4	Be	beryllium	9.012	1278	2970	1.82	1.85	1.57	899	+2
83	Bi	bismuth	209.0	271	1560	0.12	9.75	2.02	703	+3,5
107	Bh	bohrium	272	—	—	—	—	—	—	—
5	B	boron	10.81	2079	2550	1.02	2.34	2.04	801	+3
35	Br	bromine	79.90	−7.2	58.8	0.473	3.12	2.96	1140	±1,+5
48	Cd	cadmium	112.4	320.9	765	0.23	8.65	1.69	868	+2
20	Ca	calcium	40.08	839	1484	0.63	1.55	1	590	+2
98	Cf	californium	251	—	—	—	—	1.3	608	+3
6	C	carbon	12.01	3367	4827	0.71	2.25	2.55	1086	±4
58	Ce	cerium	140.1	798	3257	0.19	6.66	1.12	534	+3,4
55	Cs	cesium	132.9	28.4	669	0.24	1.87	0.79	376	+1
17	Cl	chlorine	35.45	−101	−34.6	0.48	0.00321	3.16	1251	−1
24	Cr	chromium	52.00	1857	2672	0.45	7.19	1.66	653	+3,2,6
27	Co	cobalt	58.93	1495	2870	0.42	8.9	1.88	760	+2,3
112	Cn	copernicum	285	—	—	—	—	—	—	—
29	Cu	copper	63.55	1083	2567	0.38	8.96	1.9	745	+2,1
96	Cm	curium	247	1340	—	—	13.5	1.3	581	+3
110	Ds	darmstadtium	281	—	—	—	—	—	—	—
105	Db	dubnium	268	—	—	—	—	—	—	—
66	Dy	dysprosium	162.5	1412	2567	0.17	8.55	1.22	573	+3
99	Es	einsteinium	252	—	—	—	—	1.3	619	+3
68	Er	erbium	167.3	1529	2868	0.17	9.07	1.24	589	+3
63	Eu	europium	152.0	822	1529	0.18	5.24		547	+3,2
100	Fm	fermium	257	—	—	—	—	1.3	627	+3
114	Fl	flerovium	289	—	—	—	—	—	—	—
9	F	fluorine	19.00	−219.8	−188.1	0.82	0.0017	3.98	1681	−1
87	Fr	francium	223	27	677			0.7	380	+1
64	Gd	gadolinium	157.3	1313	3273	0.23	7.9	1.2	593	+3
31	Ga	gallium	69.72	29.8	2403	0.37	5.9	1.81	579	+3
32	Ge	germanium	72.63	947.4	2830	0.32	5.32	2.01	762	+4,2
79	Au	gold	197.0	1064	3080	0.128	19.3	2.54	890	+3,1
72	Hf	hafnium	178.5	2227	4600	0.14	13.3	1.3	659	+4
108	Hs	hassium	270	—	—	—	—	—	—	—
2	He	helium	4.003	−272.2	−268.9	5.193	0.000179		2372	0
67	Ho	holmium	164.9	1474	2700	0.16	8.8	1.23	581	+3
1	H	hydrogen	1.008	−259.1	−252.9	14.304	0.0000699	2.2	1312	±1
49	In	indium	114.8	156.6	2080	0.23	7.31	1.78	558	+3
53	I	iodine	126.9	113.5	184	0.214	4.93	2.66	1008	−1,+5,7
77	Ir	iridium	192.2	2410	4130	0.13	22.4	2.2	878	+4,3,6
26	Fe	iron	55.85	1535	2750	0.44	7.86	1.83	762	+3,2
36	Kr	krypton	83.80	−157	−152	0.248	0.00374	3	1351	0
57	La	lanthanum	138.9	920	3454	0.19	6.15	1.1	538	+3
103	Lr	lawrencium	262	—	—	—	—	—	—	+3
82	Pb	lead	207.2	327.5	1740	0.13	11.4	2.33	716	+2,4
3	Li	lithium	6.968	180.5	1342	3.6	0.543	0.98	520	+1
116	Lv	livermorium	293	—	—	—	—	—	—	—
71	Lu	lutetium	175.0	1663	3402	0.15	9.84	1.27	524	+3
12	Mg	magnesium	24.31	649	1090	1.02	1.74	1.31	738	+2
25	Mn	manganese	54.94	1244	1962	0.48	7.43	1.55	717	+2,3,4,6,7

Chemistry 1 — Chemistry Reference Tables — Page: 557

atomic #	atomic symbol	element name	atomic mass (IUPAC 2005)	melting point °C	boiling point °C	specific heat capacity (J/g·K)	density (g/mL)	electro-negativity (Pauling)	1st ionization potential (kJ/mol)	common oxidation states
109	Mt	meitnerium	276	—	—	—	—	—	—	—
101	Md	mendelevium	258	—	—	—	—	1.3	635	+3,2
80	Hg	mercury	200.6	-38.9	357	0.139	13.5	2	1007	+2,1
42	Mo	molybdenum	95.95	2617	4612	0.25	10.2	2.16	684	+6,3,5
115	Mc	moscovium	288	—	—	—	—	—	—	—
60	Nd	neodymium	144.2	1016	3127	0.19	7	1.14	533	+3
10	Ne	neon	20.18	-248	-248.7	0.904	0.0009	—	2081	0
93	Np	neptunium	237	640	3900	0.12	20.2	1.36	605	+5,3,4,6
28	Ni	nickel	58.69	1453	2730	0.44	8.9	1.91	737	+2,3
113	Nh	nihonium	284	—	—	—	—	—	—	—
41	Nb	niobium	92.91	2468	4742	0.26	8.57	1.6	652	+5,3
7	N	nitrogen	14.01	-209.9	-195.8	1.04	0.00125	3.04	1402	-3
102	No	nobelium	259	—	—	—	—	1.3	642	+2,3
118	Og	oganesson	294	—	—	—	—	—	—	—
76	Os	osmium	190.2	3045	5030	0.13	22.6	2.2	839	+4,6,8
8	O	oxygen	16.00	-218.4	-183	0.92	0.00143	3.44	1314	-2
46	Pd	palladium	106.4	1554	3140	0.24	12	2.2	804	+2,4
15	P	phosphorus	30.97	44.1	280	0.77	1.82	2.19	1012	-3
78	Pt	platinum	195.1	1772	3827	0.13	21.4	2.28	868	+4,2
94	Pu	plutonium	244	641	3232	0.13	19.8	1.28	585	+4,3,5,6
84	Po	polonium	209	254	962	0.12	9.32	2	812	+4,2
19	K	potassium	39.10	63.25	760	0.75	0.86	0.82	419	+1
59	Pr	praseodymium	140.9	931	3017	0.19	6.77	1.13	527	+3,4
61	Pm	promethium	145	1042	3000	0.18	7.26	—	535	+3
91	Pa	protactinium	231.0	1570	4000	0.12	15.4	1.5	568	+5,4
88	Ra	radium	226	700	1140	0.12	5	0.9	509	+2
86	Rn	radon	222	-71	-61.8	0.09	0.00973	—	1037	0
75	Re	rhenium	186.2	3180	5600	0.13	21	1.9	760	+7,4,6
45	Rh	rhodium	102.9	1966	3727	0.242	12.4	2.28	720	+3,4,6
111	Rg	roentgentium	280	—	—	—	—	—	—	—
37	Rb	rubidium	85.47	38.9	686	0.363	1.53	0.82	403	+1
44	Ru	ruthenium	101.1	2310	3900	0.238	12.4	2.2	710	+4,3,6,8
104	Rf	rutherfordium	267	—	—	—	—	—	—	—
62	Sm	samarium	150.4	1074	1794	0.2	7.52	1.17	545	+3,2
21	Sc	scandium	44.96	1541	2832	0.6	2.99	1.36	633	+3
106	Sg	seaborgium	271	—	—	—	—	—	—	—
34	Se	selenium	78.97	217	685	0.32	4.79	2.55	941	+4,-2,+6
14	Si	silicon	28.09	1410	2355	0.71	2.33	1.9	787	±4
47	Ag	silver	107.9	962	2212	0.235	10.5	1.93	731	+1
11	Na	sodium	22.99	97.8	883	1.23	0.971	0.93	496	+1
38	Sr	strontium	87.62	769	1384	0.3	2.54	0.95	549	+2
16	S	sulfur	32.07	112.8	444.7	0.71	2.07	2.58	1000	-2
73	Ta	tantalum	180.9	2996	5425	0.14	16.6	1.5	761	+5
43	Tc	technetium	98	2172	4877	0.21	11.5	1.9	702	+7,4,6
52	Te	tellurium	127.6	449.5	989.8	0.2	6.24	2.1	869	+4,6,-2
117	Ts	tennessine	292	—	—	—	—	—	—	—
65	Tb	terbium	158.9	1365	3230	0.18	8.23	—	569	+3,4
81	Tl	thallium	204.4	303	1457	0.13	11.9	1.62	589	+1,3
90	Th	thorium	232.0	1750	4790	0.12	11.7	1.3	587	+4
69	Tm	thulium	168.9	1545	1950	0.16	9.32	1.25	597	+3,2
50	Sn	tin	118.7	232	2270	0.227	7.31	1.96	709	+4,2
22	Ti	titanium	47.87	1660	3287	0.52	4.54	1.54	659	+4,3,2
74	W	tungsten	183.8	3410	5660	0.13	19.3	2.36	770	+6,4
92	U	uranium	238.0	1132	3818	0.12	19	1.38	598	+6,3,4,5
23	V	vanadium	50.94	1890	3380	0.49	6.11	1.63	651	+5,2,3,4
54	Xe	xenon	131.3	-111.8	-107.1	0.158	0.00589	2.6	1170	0
70	Yb	ytterbium	173.1	819	1196	0.15	6.97	—	603	+3,2
39	Y	yttrium	88.91	1523	3337	0.3	4.47	1.22	600	+3
30	Zn	zinc	65.38	419.6	906	0.39	7.13	1.65	906	+2
40	Zr	zirconium	91.22	1852	4377	0.27	6.51	1.33	640	+4

Table AA. Bond Dissociation Energies & Bond Lengths

Values given are *homolytic* bond dissociation energies, meaning that the electrons are divided equally between the two atoms.

Hydrogen Compounds

Bond	D (kJ/mol)	r (pm)
H – H	432	74
H – B	389	119
H – C	411	109
H – Si	318	148
H – Ge	288	153
H – Sn	251	170
H – N	386	101
H – P	322	144
H – As	247	152
H – O	459	96
H – S	363	134
H – Se	276	146
H – Te	238	170
H – F	565	92
H – Cl	428	127
H – Br	362	141
H – I	295	161

Group VIIA Compounds

Bond	D (kJ/mol)	r (pm)
F – F	155	142
Cl – Cl	240	199
Br – Br	190	228
I – I	148	267
At – At	116	?
I – O	201	?
I – F	273	191
I – Cl	208	232
I – Br	175	?

Group IIIA Compounds

Bond	D (kJ/mol)	r (pm)
B – B	293	?
B – O	536	?
B – F	613	?
B – Cl	456	175
B – Br	377	?

Group IVA Compounds

Bond	D (kJ/mol)	r (pm)
C – C	346	154
C = C	602	134
C ≡ C	835	120
C – Si	318	185
C – Ge	238	195
C – Sn	192	216
C – Pb	130	230
C – N	305	147
C = N	615	129
C ≡ N	887	116
C – P	264	184
C – O	358	143
C = O	799	120
C ≡ O	1072	113
C – B	356	?
C – S	272	182
C = S	573	160
C – F	485	135
C – Cl	327	177
C – Br	285	194
C – I	213	214
Si – Si	222	233
Si – N	355	?
Si – O	452	163
Si – S	293	200
Si – F	565	160
Si – Cl	381	202
Si – Br	310	215
Si – I	234	243
Ge – Ge	188	241
Ge – N	257	?
Ge – F	470	168
Ge – Cl	349	210
Ge – Br	276	230
Ge – I	212	?
Sn – F	414	?
Sn – Cl	323	233
Sn – Br	273	250
Sn – I	205	270
Pb – F	331	?
Pb – Cl	243	242
Pb – Br	201	?
Pb – I	142	279

Group VA Compounds

Bond	D (kJ/mol)	r (pm)
N – N	167	145
N = N	418	125
N ≡ N	942	110
N – O	201	140
N = O	607	121
N – F	283	136
N – Cl	313	175
P – P	201	221
P – O	335	163
P = O	544	150
P = S	335	186
P – F	490	154
P – Cl	326	203
P – Br	264	?
P – I	184	?
As – As	146	243
As – O	301	178
As – F	484	171
As – Cl	322	216
As – Br	458	233
As – I	200	254
Sb – Sb	121	?
Sb – F	440	?
Sb – Cl$_{(5)}$	248	?
Sb – Cl$_{(3)}$	315	232

Group VIA Compounds

Bond	D (kJ/mol)	r (pm)
O – O	142	148
O = O	494	121
O – F	190	142
S = O	522	143
S – S$_{(8)}$	226	205
S = S	425	149
S – F	284	156
S – Cl	255	207
Se – Se	172	?
Se = Se	272	215

Table BB. Thermodynamic Data

Standard enthalpy of formation $(\Delta H°_f)$ & standard entropy $(S°)$ for selected compounds. Note that standard enthalpy values are in kilojoules per mole, whereas entropy values are in joules per mole·Kelvin.

Subst.	State	$\Delta H°_f$ ($\frac{kJ}{mol}$)	$S°$ ($\frac{J}{mol \cdot K}$)
Ag	s	0	42.6
Ag^+	aq	105.79	72.7
AgCl	s	−127.01	96.2
AgBr	s	−100.4	107.1
$AgNO_3$	s	−124.4	140.9
Al	s	0	28.3
Al^{+3}	aq	−538.4	−321.7
$AlCl_3$	s	−704	110.7
Al_2O_3	s	−1675.7	50.9
$Al(OH)_3$	s	−1277	
Ba	s	0	62.8
$BaCl_2$	s	−858.6	123.7
$BaCO_3$	s	−1216.3	112.1
$Ba(NO_3)_2$	s	−992	214
BaO	s	−553.5	70.4
$Ba(OH)_2$	s	−998.2	112
$BaSO_4$	s	−1473.2	132.2
Be	s	0	10
BeO	s	−599	14
Br_2	ℓ	0	152.2
Br−	aq	−121	82
C	s	0	5.7
CCl_4	ℓ	−135.4	216.4
$CHCl_3$	ℓ	−134.5	201.7
CH_4	g	−74.8	186.2
C_2H_2	g	+226.7	200.8
C_2H_4	g	+52.3	219.5
C_2H_6	g	−84.7	229.5
C_3H_8	g	−103.8	269.9
CH_3OH	ℓ	−238.7	126.8
C_2H_5OH	ℓ	−277.7	160.7
$C_6H_{12}O_6$	s	−1275	212
CO	g	−110.53	197.6
CO_2	g	−393.51	213.6
CO_3^{-2}	aq	−675.23	−56.9
Ca	s	0	41.4
Ca^{+2}	aq	−543.0	−53.1
$CaCl_2$	s	−795.8	104.6
$CaCO_3$	s	−1206.9	92.9
CaO	s	−634.92	39.8
$Ca(OH)_2$	s	−986.1	83.4
$Ca_3(PO_4)_2$	s	−4126	241
$CaSO_4$	s	−1434.1	106.7
Cd	s	0	51.8
Cd^{+2}	aq	−75.92	−73.2
$CdCl_2$	s	−391.5	115.3
CdO	s	−258.35	54.8
$Cd(OH)_2$	s	−561	96
CdS	s	−162	65
$CdSO_4$	s	−935	123
Cl_2	g	0	223.0
Cl^-	aq	−167.080	56.5
ClO_4^-	aq	−128.10	182.0
Cr	s	0	23.8
Cr_2O_3	g	−1139.7	81.2
Cu	s	0	33.2
Cu^+	aq	+71.7	40.6
Cu^{+2}	aq	+64.8	−99.6
CuO	s	−157.3	42.6
Cu_2O	s	−168.6	93.1
$Cu(OH)_2$	s	−450	108
CuS	s	−53.1	66.5
Cu_2S	s	−79.5	120.9
$CuSO_4$	s	−771.4	107.6
F^-	aq	−335.35	−13.8
F_2	g	0	202.7
Fe	s	0	27.3
$Fe(OH)_3$	s	−823.0	106.7
FeO	s	−272	61
Fe_2O_3	s	−824.2	87.4
Fe_3O_4	s	−1118.4	146.4
$FeSO_4$	s	−929	121
H_2	g	0	130.6
H^+	aq	0	0.0
HBr	g	−36.29	198.6
HCO_3^-	aq	−689.93	91.2
HCl	g	−92.31	186.8
HF	g	−273.30	173.7
HI	g	26.50	206.5
HNO_3	aq	−174.1	155.6
HPO_4^{-2}	aq	−1299.0	−33.5
HSO_4^-	aq	−886.9	131.8
H_2O	ℓ	−285.830	69.9
H_2O	g	−241.826	188.7
$H_2PO_4^-$	aq	−1302.6	90.4
H_2S	g	−20.6	205.7
Hg	ℓ	0	76.0
Hg^{+2}	aq	170.21	−32.2
HgO	cr	−90.79	70.3
I^-	aq	−56.78	111.3
I_2	s	0	116.1
K	s	0	64.2
K^+	aq	−252.14	102.5
KBr	s	−393.8	95.9
KCl	s	−436.7	82.6
$KClO_3$	s	−397.7	143.1
$KClO_4$	s	−432.8	151.0
KNO_3	s	−494.6	133.0
Mg	s	0	32.7
Mg^{+2}	aq	−467.0	−138.1
$MgCl_2$	s	−641.3	89.6
$MgCO_3$	s	−1095.8	65.7
MgO	s	−601.60	26.9
$Mg(OH)_2$	s	−924.5	63.2
$MgSO_4$	s	−1284.9	91.6
Mn	s	0	32.0
Mn^{+2}	aq	−220.8	−73.6
MnO	s	−385.2	59.7
MnO_2	s	−520.0	53.0
N_2	g	0	191.5
NH_3	g	−45.94	192.3
NH_4^+	aq	−133.26	113.4
NO_2^-	aq	−104.6	123.0
NO_3^-	aq	−206.85	146.4
N_2H_4	ℓ	+50.6	121.2
NH_4Cl	s	−314.4	94.6
NH_4NO_3	s	−365.6	151.1
NO	g	+90.2	210.7
NO_2	g	+33.2	240.0
N_2O	g	+82	220
N_2O_4	g	+9.2	304.2
Na	s	0	51.2
Na^+	aq	−240.34	59.0
Na_2CO_3	s	−1131	136
$NaHCO_3$	s	−948	102
NaCl	s	−411.2	72.1
NaF	s	−573.6	51.5
$NaNO_3$	s	−467	116
NaOH	s	425.6	64.5
Ni	s	0	29.9
$NiCl_2$	s	−316	107
NiO	s	−239.7	38.0
OH^-	aq	−230.015	−10.8
O_2	g	0	205.0
P_4	s	0	164.4
PCl_3	g	−287.0	311.7
PCl_5	g	−374.9	364.5
PH_3	g	+5	210
PO_4^{-3}	aq	−1277.4	−222
Pb	s	0	64.8
Pb^{+2}	aq	0.92	10.5
$PbBr_2$	s	−278.7	161.5
$PbCl_2$	s	−359.4	136.0
PbO	s	−219.0	66.5
PbO_2	s	−277.4	68.6
PbS	s	−100	91
$PbSO_4$	s	−920	149
S	s	0	31.8
SO_2	g	−296.81	248.1
SO_3	g	−395.7	256.7
SO_4^{-2}	aq	−909.34	20.1
S	−	2	aq
Si	s	0	18.8
SiO_2	s	−910.7	41.8
Sn	s	0	51.6
Sn^{+2}	aq	−8.9	−17.4
SnO_2	s	−577.63	52.3
Zn	s	0	41.6
Zn^{+2}	aq	−153.39	−112.1
ZnI_2	s	−208.0	161.1
ZnO	s	−350.46	43.6
ZnS	s	−206.0	57.7

Table CC. Some Common & Equivalent Units and Approximate Conversions

Some Common & Equivalent Units					
Length	1 in. (inch)	=	2.54 cm		
	12 in.	=	1 ft. (foot)		
	3 ft.	=	1 yd. (yard)		
	5,280 ft.	=	1 mi. (mile)	=	1,760 yd
Mass	1 lb. (pound)	=	16 oz.	~	454 g
	1 ton	=	2000 lb.		
	1 tonne	=	1000 kg		
Volume	1 pinch	=	≤ 1/8 teaspoon		
	3 teaspoons	=	1 tablespoon (Tbsp)		
	2 tablespoons	=	1 ounce		
	8 oz. (ounces)	=	1 cup		
	2 cups	=	1 pint		
	2 pints	=	1 quart		
	4 quarts	=	1 gallon		

Some APPROXIMATE Conversions					
Length	1 cm	~	width of a small paper clip		
	6 in.	~	length of a (US) dollar bill		
	1 ft.	~	30 cm		
	1 m	~	1 yd.		
	1 mi.	~	1.6 km		
	0.6 km	~	1 mi.		
Volume	1 pinch	~	≤ 1/8 teaspoon		
	1 mL	~	10 drops		
	1 teaspoon (tsp)	~	5 mL	~	60 drops
	1 tablespoon (Tbsp)	=	3 tsp	~	15 mL
	2 Tbsp.	=	1 fl. oz.	~	30 mL
	1 C (cup)	=	8 fl. oz.	~	250 mL
	1 qt. (quart)	~	1 L		
Mass	1 small paper clip	~	1 gram (g)		
	1 nickel (5¢ coin)	~	5 g		
	1 oz.	~	30 g		
	1 pound (lb.)	=	16 oz.	~	0.5 kg
	1 ton	=	2000 lb.	~	1 tonne
Speed	60 mi./h	~	100 km/h	~	30 m/s
Density	air	~	1 g/L		
	fresh water	~	1 g/mL	~	8.3 lb./gal.
		~	1 tonne/m^3	~	1 ton/yd.^{3}

Table DD. Selected Formulas and Equations

Density	$D = \dfrac{M}{V}$	D = density	M = mass	V = volume
Mole Conversions	1 mol = [molar mass] g (molar mass = formula weight = gram formula mass) 1 mol = 22.4 L of gas at 0 °C and 1 atm 1 mol = 6.022×10^{23} molecules, atoms, or particles			
Percent Error	% error $= \dfrac{\text{measured value} - \text{accepted value}}{\text{accepted value}} \times 100\%$			
Percent Composition	% composition $= \dfrac{\text{mass of part}}{\text{mass of whole}} \times 100\%$			
Concentration	molarity $(M) = \dfrac{\text{moles of solute}}{\text{liter of solution}}$ normality $(N) = \dfrac{(\text{moles of solute})(\text{dissociation factor})}{\text{liter of solution}}$ molality $(m) = \dfrac{\text{moles of solute}}{\text{kg of solvent}}$ mole fraction $(\chi_A) = \dfrac{\text{moles of A}}{\text{total moles}}$ parts per million (ppm) $= \dfrac{\text{grams of solute}}{\text{grams of solvent}} \times 1{,}000{,}000$			
Gases	$\dfrac{P_1 V_1}{T_1} = \dfrac{P_2 V_2}{T_2}$ $PV = nRT$ P = pressure V = volume (L) n = moles T = temperature (K) $R = 0.0821 \dfrac{\text{L·atm}}{\text{mol·K}} = 8.31 \dfrac{\text{L·kPa}}{\text{mol·K}} = 62.4 \dfrac{\text{L·torr}}{\text{mol·K}}$			
Pressure	1 atm = 101.3 kPa = 760 torr = 760 mm Hg = 29.92 in. Hg = 1.013 bar = 14.7 psi $P_A = \chi_A P_T$ P_A = partial pressure of A χ_A = mole fraction of A $P_T = P_A + P_B + P_C + \ldots$ P_T = total pressure			
Titration	$N_A V_A = N_B V_B$ N_A = normality of H_3O^+ N_B = normality of OH^- V_A = volume of acid V_B = volume of base			
Colligative Properties	$\Delta T_f = i m K_f$ $\Delta T_b = i m K_b$ $\pi = i M R T = N R T$ ΔT_f = freezing point depression (°C) ΔT_b = boiling point elevation (°C) K_f = freezing point depression constant (H_2O = 0.52 °C/m) K_b = boiling point elevation constant (H_2O = 1.86 °C/m) i = van't Hoff factor (dissociation factor) π = osmotic pressure m = molality M = molarity (mol/L) N = normality (mol/L) R = ideal gas const. T = temperature (K)			
Acid-Base	$\text{pH} = -\log[H_3O^+]$ $\text{pOH} = -\log[OH^-]$ $\text{pH} + \text{pOH} = 14$ $K_w = K_a \cdot K_b = 1 \times 10^{-14}$ $K_a = \dfrac{[H_3O^+][A^-]}{[HA]}$ $K_b = \dfrac{[H^+B][OH^-]}{[B]}$ $\text{p}K_a = -\log(K_a)$ $\text{p}K_b = -\log(K_b)$ $\text{pH} = \text{p}K_a + \log \dfrac{[\text{base}]}{[\text{acid}]}$			
Equilibrium	$K_p = K_c (RT)^{\Delta n}$ K_p = gas press. equil. const. K_c = molar conc. equil. const. Δn = change in # moles			
Heat	$q = m C_p \Delta T$ $q = m \Delta H_f$ $q = m \Delta H_v$ q = heat m = mass C_p = specific heat capacity ΔT = change in temperature ΔH_f = heat of fusion ΔH_v = heat of vaporization			
Thermodynamics	$\Delta G° = \Delta H° - T \Delta S°$ $\Delta G° = -RT \ln K$ $G°$ = standard free energy $H°$ = standard enthalpy $S°$ = standard entropy T = temperature (K)			
Electrochemistry	$I = \dfrac{q}{t}$ $\Delta G° = -nFE°$ I = current (amperes) q = charge (Coulombs) t = time (seconds) F = Faraday's constant = 96,000 Coulomb per mole electrons n = moles of electrons $E°$ = standard reduction potential			
Temperature	K = °C + 273.15 °C = (°F − 32) * 5/9 K = Kelvin °C = degrees Celsius °F = degrees Fahrenheit			
Radioactive Decay	$A = A_0 \left(\dfrac{1}{2}\right)^{t/\tau_{1/2}}$ A = amount left A_0 = original amount t = total elapsed time $\tau_{1/2}$ = half-life number of half-lives $= \dfrac{t}{\tau_{1/2}}$			

Periodic Table of the Elements

Period	1 IA	2 IIA		3 IIIB	4 IVB	5 VB	6 VIB	7 VIIB	8 VIIIB	9 VIIIB	10 VIIIB	11 IB	12 IIB	13 IIIA	14 IVA	15 VA	16 VIA	17 VIIA	18 VIIIA
1	1 ±1 **H** hydrogen 1.008		1s																2 **He** helium 4.003
2	3 +1 **Li** lithium 6.968	4 +2 **Be** beryllium 9.012	2s / 2p											5 +3 **B** boron 10.81	6 ±4 **C** carbon 12.01	7 −3 **N** nitrogen 14.01	8 −2 **O** oxygen 16.00	9 −1 **F** fluorine 19.00	10 **Ne** neon 20.18
3	11 +1 **Na** sodium 22.99	12 +2 **Mg** magnesium 24.31	3s / 3p											13 +3 **Al** aluminum 26.98	14 ±4 **Si** silicon 28.09	15 −3 **P** phosphorus 30.97	16 −2 **S** sulfur 32.07	17 −1 **Cl** chlorine 35.45	18 **Ar** argon 39.95
4	19 +1 **K** potassium 39.10	20 +2 **Ca** calcium 40.08	3d	21 +3 **Sc** scandium 44.96	22 +3,4 **Ti** titanium 47.87	23 +5,2,3,4 **V** vanadium 50.94	24 +3,2,6 **Cr** chromium 52.00	25 +2,3,4,6,7 **Mn** manganese 54.94	26 +3,2 **Fe** iron 55.85	27 +2,3 **Co** cobalt 58.93	28 +2,3 **Ni** nickel 58.69	29 +2,1 **Cu** copper 63.55	30 +2 **Zn** zinc 65.38	31 +3 **Ga** gallium 69.72	32 +4,2 **Ge** germanium 72.63	33 ±3,4,2 **As** arsenic 74.92	34 −2,4,6 **Se** selenium 78.97	35 −1 **Br** bromine 79.90	36 **Kr** krypton 83.80
5	37 +1 **Rb** rubidium 85.47	38 +2 **Sr** strontium 87.62	4d	39 +3 **Y** yttrium 88.91	40 +4 **Zr** zirconium 91.22	41 +5,3 **Nb** niobium 92.91	42 +6,3,5 **Mo** molybdenum 95.95	43 +7,4,6 **Tc** technetium 98	44 +3,4,6,8 **Ru** ruthenium 101.1	45 +3 **Rh** rhodium 102.9	46 +2,4 **Pd** palladium 106.4	47 +1 **Ag** silver 107.9	48 +2 **Cd** cadmium 112.4	49 +3 **In** indium 114.8	50 +4,2 **Sn** tin 118.7	51 ±3,5 **Sb** antimony 121.8	52 −2,4,6 **Te** tellurium 127.6	53 −1 **I** iodine 126.9	54 **Xe** xenon 131.3
6	55 +1 **Cs** cesium 132.9	56 +2 **Ba** barium 137.3	† 5d	71 +3 **Lu** lutetium 175.0	72 +4 **Hf** hafnium 178.5	73 +5 **Ta** tantalum 180.9	74 +6,4 **W** tungsten 183.8	75 +7,4,6 **Re** rhenium 186.2	76 +4,6,8 **Os** osmium 190.2	77 +4,3,6 **Ir** iridium 192.2	78 +4,2 **Pt** platinum 195.1	79 +3,1 **Au** gold 197.0	80 +2,1 **Hg** mercury 200.6	81 +1,3 **Tl** thallium 204.4	82 +2,4 **Pb** lead 207.2	83 +3,5 **Bi** bismuth 209.0	84 **Po** polonium 209	85 **At** astatine 210	86 **Rn** radon 222
7	87 +1 **Fr** francium 223	88 +2 **Ra** radium 226	‡ 6d	## **Lr** lawrencium 262	## **Rf** rutherfordium 267	## **Db** dubnium 268	## **Sg** seaborgium 271	## **Bh** bohrium 272	## **Hs** hassium 270	## **Mt** meitnerium 276	## **Ds** darmstadtium 281	## **Rg** roentgenium 280	## **Cn** copernicium 285	## **Nh** nihonium 284	## **Fl** flerovium 289	## **Mc** moscovium 288	## **Lv** livermorium 293	## **Ts** tennessine 292	## **Og** oganesson 294

Key:
- atomic # → 29
- atomic symbol → **Cu**
- English element name → copper
- +2,1 ← ions commonly formed
- 63.55 ← atomic mass (rounded)

Legend: Gases | Liquids | Metalloids

† lanthanides (rare earth metals) 4f

| 57 +3 **La** lanthanum 138.9 | 58 +3,4 **Ce** cerium 140.1 | 59 +3,4 **Pr** praseodymium 140.9 | 60 +3 **Nd** neodymium 144.2 | 61 +3 **Pm** promethium 145 | 62 +3,2 **Sm** samarium 150.4 | 63 +3,2 **Eu** europium 152.0 | 64 +3 **Gd** gadolinium 157.3 | 65 +3 **Tb** terbium 158.9 | 66 +3,4 **Dy** dysprosium 162.5 | 67 +3 **Ho** holmium 164.9 | 68 +3 **Er** erbium 167.3 | 69 +3 **Tm** thulium 168.9 | 70 +3,2 **Yb** ytterbium 173.1 |

‡ actinides 5f

| 89 +3 **Ac** actinium 227 | 90 +4 **Th** thorium 232.0 | 91 +4 **Pa** protactinium 231.0 | 92 +6,3,4,5 **U** uranium 238.0 | 93 +5,3,4,6 **Np** neptunium 237 | 94 +4,3,5,6 **Pu** plutonium 244 | 95 +3,4,5,6 **Am** americium 243 | 96 +3 **Cm** curium 247 | 97 +3,4 **Bk** berkelium 247 | 98 +3 **Cf** californium 251 | 99 +3 **Es** einsteinium 252 | 100 +3 **Fm** fermium 257 | 101 +3,2 **Md** mendelevium 258 | 102 +2,3 **No** nobelium 259 |

Index

acceleration, 61
accuracy, 36, 38, 72
acid, 7, 124, 174, 297, 438, 536, 537, 540, 543, 548, 555, 563
air, 137, 138, 148, 225, 255, 395, 416, 491, 562
alkali, 263
alkali metal, 263
alkaline, 263
alkaline earth metal, 263
alpha decay, 193
alpha particle, 193, 195, 208
Amonton, 128
amphoteric, 537
amplitude, 223
assumptions, 140
atom, 74, 75, 112, 123, 162, 163, 164, 166, 167, 168, 169, 170, 173, 174, 183, 199, 200, 201, 204, 214, 221, 223, 224, 230, 233, 238, 244, 246, 247, 257, 265, 266, 267, 269, 272, 273, 274, 279, 280, 281, 286, 287, 300, 301, 307, 308, 309, 310, 311, 313, 314, 317, 318, 319, 320, 323, 324, 332, 335, 337, 338, 339, 351, 353, 412, 419, 420, 434, 435, 436
atomic mass, 157, 163, 164, 183, 184, 256, 257, 258, 259, 351, 352, 353, 360
atomic number, 164, 165, 196, 237, 254, 258, 259
atomic radius, 272
band of stability, 192, 193, 194
base, 7, 73, 74, 75, 413, 438, 536, 537, 540, 543, 548, 555, 563
basic, 84, 168, 389, 533, 540, 541
beta decay
    beta minus, 193
    beta plus, 194
beta particle, 195
binding energy, 199, 200
Bohr model, 219
boiling, 7, 108, 110, 111, 113, 119, 122, 129, 151, 205, 262, 263, 264, 330, 343, 372, 392, 393, 394, 563
boiling point, 7, 108, 111, 113, 122, 129, 151, 262, 263, 264, 330, 343, 372, 392, 393, 394, 563
boiling point elevation, 392, 394, 563
bonding, 17, 278, 279, 306, 307, 323, 324
boson, 180
buffer, 533, 548
burning, 111, 113, 433
calorimeter, 30, 31, 43

calorimetry, 478, 479
catalyst, 514, 519
cell, 396, 442, 443
charge, 75, 85, 129, 163, 164, 165, 168, 169, 177, 180, 181, 190, 193, 214, 247, 259, 266, 273, 274, 279, 282, 283, 284, 286, 289, 290, 291, 294, 311, 314, 332, 337, 338, 343, 412, 434, 436, 437, 438, 563
chemical bond, 112, 279, 307, 335, 342, 428, 500
chemical equilibrium, 522
chemical formula, 112, 148, 174, 278, 281, 286, 287, 290, 300, 306, 356, 359, 360, 366, 367, 392, 419
chemical reaction, 7, 17, 18, 30, 43, 125, 130, 163, 167, 229, 244, 246, 352, 381, 402, 403, 407, 408, 411, 419, 427, 428, 435, 446, 447, 452, 453, 456, 496, 497, 502, 514, 517, 518, 519, 523
chemical symbol, 165
circuit, 208
collision
    elastic, 135
color, 178
combination, 10, 19, 74, 145, 174, 223, 402, 497, 548
combined gas law, 139, 140
combustion, 402, 408, 411, 435, 508, 510, 511
composition, 7, 162, 166, 174, 190, 359, 402
compound, 76, 96, 98, 99, 111, 112, 114, 133, 148, 164, 167, 174, 257, 281, 282, 283, 284, 287, 289, 290, 291, 297, 299, 307, 314, 342, 343, 351, 353, 359, 360, 361, 362, 365, 366, 380, 382, 392, 397, 407, 408, 412, 435, 436, 437, 447, 448, 498, 499, 500, 502, 503, 507, 509, 523, 537, 539
concentration, 46, 85, 99, 100, 372, 378, 387, 394, 396, 397, 514, 520, 524, 541, 542, 551, 557
conjugate, 537
Cornell Notes, 9
current, 206
debye, 340
decomposition, 402, 407, 435
density, 7, 45, 85, 86, 87, 88, 90, 97, 98, 100, 108, 111, 113, 262, 330, 552, 563
dipole, 340, 343, 344, 345, 374, 377, 495

dipole moment, 340, 343, 495
direction, 36, 37, 474
dissolve, 7, 28, 46, 112, 264, 345, 372, 373, 374, 375, 377, 378, 380, 384, 387, 388, 392, 393, 394, 395, 404, 416, 438, 520, 536
distance, 23, 176, 192, 195
double displacement, 402, 408
double replacement, 408, 411, 412, 413
effusion, 155, 157
electrolysis, 417
electrolytic cell, 442, 443
electromagnet, 206
electron, 162, 163, 166, 167, 168, 170, 171, 193, 194, 197, 199, 213, 214, 215, 219, 221, 223, 224, 228, 230, 231, 234, 238, 239, 240, 244, 245, 246, 247, 248, 249, 254, 263, 265, 266, 267, 306, 307, 308, 309, 311, 313, 314, 318, 319, 320, 324, 435, 556
electron capture, 194
electronegativity, 254, 269, 270, 280, 281, 306, 330, 331, 332, 333, 334, 335, 343
electrostatic, 7, 108, 224
element, 9, 111, 112, 114, 123, 163, 164, 165, 167, 173, 183, 213, 228, 229, 231, 234, 238, 240, 248, 254, 255, 256, 257, 260, 265, 266, 269, 270, 274, 281, 287, 289, 290, 291, 333, 343, 359, 360, 407, 412, 421, 422, 435, 436, 437, 500, 557
empirical formula, 350, 359, 360, 361, 362
energy, 135, 170, 176, 195, 199, 200, 206, 207, 208, 219, 473, 474, 475, 477, 484, 493
    kinetic, 473, 474, 493
enthalpy, 7, 375, 491, 494, 495, 497, 498, 499, 500, 507, 509, 517, 561, 563
entropy, 7, 375, 472, 491, 494, 496, 497, 561, 563
equilibrium, 18, 85, 493, 514, 522, 523, 524, 529, 530, 540, 542
expand, 477, 483
fermion, 180
fission, 190, 204, 205
fluid, 116, 129, 404
force, 176, 181, 192, 193, 199, 219

formula, 14, 30, 39, 61, 86, 112, 124, 136, 145, 148, 156, 157, 174, 230, 281, 283, 284, 286, 287, 288, 290, 291, 294, 300, 314, 333, 340, 343, 351, 356, 359, 360, 361, 362, 366, 367, 392, 397, 437, 554, 563
free energy, 7, 496, 563
  Gibbs free energy, 496, 497
fuse, 206
fusion, 115, 116, 117, 190, 206, 484, 485, 563
galvanic cell, 442
gamma ray, 195, 207, 208
gas, 4, 7, 89, 90, 101, 108, 109, 110, 115, 116, 118, 120, 122, 124, 125, 128, 129, 130, 131, 132, 133, 134, 135, 136, 137, 138, 139, 140, 141, 144, 145, 146, 148, 150, 152, 155, 156, 167, 240, 244, 247, 266, 267, 343, 351, 356, 378, 381, 394, 395, 397, 404, 477, 483, 484, 486, 495, 500, 554, 563
gas constant, 89, 90, 133, 145, 156, 397, 563
gravity, 176
half-life, 201, 202, 208
heat, 7, 13, 18, 30, 31, 33, 34, 43, 85, 90, 129, 131, 168, 200, 205, 206, 262, 266, 343, 375, 403, 404, 408, 416, 417, 473, 474, 475, 476, 477, 478, 479, 483, 484, 485, 486, 487, 492, 493, 495, 497, 499, 500, 502, 503, 505, 507, 508, 509, 510, 511, 517, 530, 563
heat capacity, 343, 563
heat of fusion, 484, 485, 563
heat of vaporization, 484, 486, 563
heating curve, 485, 486
hydrocarbon, 362, 408, 411
hydrogen bond, 330, 343, 344, 345, 374
hydroxide, 124, 278, 288, 383, 533, 553, 554, 555
ideal gas law, 146
inorganic, 18
intermolecular force, 342, 343, 375, 377, 382
ion, 164, 165, 166, 214, 247, 264, 266, 267, 270, 274, 279, 280, 283, 286, 287, 288, 289, 290, 291, 293, 294, 297, 311, 314, 343, 344, 345, 374, 377, 380, 381, 382, 407, 413, 416, 427, 436, 437, 500, 533, 537, 554
isotope, 183, 184, 190, 196, 208
Kelvin, 140, 146
kinetic energy, 97, 130, 131, 145, 151, 155, 157, 378, 473, 474, 493, 495, 517, 520

kinetic molecular theory, 108, 128
kinetics, 7, 18
limiting reactant, 456, 457, 458, 459
limiting reagent, 456
liquid, 33, 34, 100, 108, 109, 110, 114, 115, 116, 117, 118, 119, 120, 121, 122, 129, 130, 135, 151, 206, 264, 279, 343, 345, 378, 381, 393, 395, 404, 473, 483, 484, 485, 554
mass defect, 199, 200
mass number, 164, 165, 183, 184, 196
matter, 4, 16, 17, 23, 84, 87, 108, 109, 112, 114, 120, 123, 128, 173, 309, 351, 423, 428, 438
melt, 117, 206, 264
metal, 13, 45, 90, 168, 208, 280, 289, 309, 333, 342, 343, 344, 407, 412, 416, 434, 437, 442, 443, 453
metathesis, 408
metric system, 74, 77, 145
mixture, 108, 111, 112, 113, 114, 156, 395
molecule, 120, 122, 123, 151, 157, 164, 167, 256, 260, 299, 300, 309, 313, 317, 318, 319, 320, 332, 337, 338, 339, 340, 342, 343, 353, 359, 380, 394, 404, 419, 420, 422, 447, 495, 499
momentum, 135, 473
negative charge, 163, 168, 229, 274, 282, 283, 284, 289, 311, 332, 337, 338
neutralization, 536
neutralize, 208
neutrino, 180
neutron, 163, 171, 199, 204, 215, 556
noble gas, 240, 244, 247, 264, 266, 267, 270
nonmetal, 280, 289, 333, 342, 344
non-metal, 261, 262, 280
non-metal, 280
non-metal, 307
non-metal, 309
non-metal, 407
non-metal, 412
nonpolar bond, 334
nuclear, 123, 124, 125, 162, 190, 196, 199, 205, 206, 207, 214, 259, 273, 495
nucleus, 162, 163, 164, 166, 168, 169, 170, 171, 199, 204, 214, 221, 223, 230, 231, 266, 272, 273, 274, 319
orbital, 166, 224, 230, 232, 234, 238, 239, 248, 249, 323, 324
organic, 18, 281, 508
osmosis, 396, 397

osmotic pressure, 397, 563
oxidation, 7, 433, 434, 435, 436, 437, 438, 441, 442
oxidation number, 7, 433, 436, 437
oxidation state, 436
particle, 108, 163, 168, 171, 214, 215, 223, 229, 443, 495, 514, 556
percent error, 39, 72
percent yield, 7, 446, 464, 465
period, 74, 76, 229, 231, 260, 270
periodic law, 258
periodic table, 17, 112, 213, 231, 234, 237, 238, 240, 244, 245, 247, 248, 254, 257, 258, 259, 260, 263, 264, 270, 351, 353
periodic trend, 7
phase change, 483
physical, 7, 18, 74, 84, 108, 109, 111, 112, 113, 120, 123, 173, 230, 257, 258, 260, 330, 392, 417, 495
plasma, 206, 483
positive charge, 163, 168, 169, 208, 266, 273, 274, 282, 284, 289, 332, 337
power, 81, 195, 202, 207
precision, 36, 38, 40, 44, 48, 52, 72, 491, 502, 507, 515
pressure, 134, 137, 138, 139, 141, 146, 205, 477
propagation, 204
proton, 163, 171, 199, 204, 214, 215, 533, 556
quantum, 18, 166, 170, 171, 219, 221, 223, 224, 228, 238
quantum mechanics, 18
quantum number, 170, 171, 224
quark, 176, 181, 192, 193, 194
radiation, 74, 75, 195, 207, 556
radioactive, 190, 196, 201, 205, 207, 208
radioactive decay, 191, 192, 193, 194, 195, 196, 197, 201
reactant, 30, 31, 407, 447, 452, 456, 457, 458, 459, 505, 520
reaction, 7, 31, 43, 62, 123, 124, 173, 204, 205, 206, 352, 380, 381, 403, 404, 407, 408, 411, 412, 413, 416, 419, 420, 427, 428, 433, 434, 435, 438, 441, 442, 443, 447, 452, 453, 456, 457, 459, 464, 465, 496, 500, 502, 503, 505, 507, 508, 509, 510, 511, 515, 516,517, 518, 519, 520, 522, 523, 524, 529, 536, 548
reagent, 456
redox, 435, 438, 441
reduction, 7, 433, 434, 435, 438, 441, 442, 563

relative error, 38, 39, 42, 43, 45, 46, 62
salt, 100, 112, 114, 393, 394, 443, 536
scientific notation, 49, 52, 72, 78, 80, 81, 82, 83
significant figures, 14, 47, 48, 49, 52, 53, 54, 57, 72, 84, 94, 96, 99, 109
single displacement, 402, 407
single replacement, 407, 411, 412, 416, 435
solid, 72, 108, 109, 110, 113, 114, 115, 116, 118, 120, 121, 129, 130, 135, 166, 215, 264, 279, 343, 365, 380, 393, 404, 428, 442, 473, 483, 484, 485, 554
solubility, 373, 380, 383, 384
soluble, 263, 264, 342, 373, 382
solute, 342, 373, 374, 375, 377, 378, 380, 382, 384, 387, 388, 389, 392, 393, 394, 395, 396, 397
solution, 11, 14, 46, 100, 112, 372, 373, 374, 375, 378, 380, 381, 384, 387, 388, 389, 392, 393, 394, 395, 397, 404, 420, 427, 442, 443, 453, 540, 541, 542, 548
solvent, 113, 342, 343, 373, 374, 375, 377, 378, 382, 384, 387, 392, 393, 394, 395, 396, 539
specific heat, 13, 30, 43, 90, 343, 563

specific heat capacity, 13, 30, 43, 90, 476, 477, 478, 485, 486, 563
spectrum, 217
speed, 193, 194, 199, 204
standard model, 178
stoichiometry, 7, 18, 356, 372, 446, 447, 448, 449, 452, 453, 456, 458, 464, 465, 502, 505
strong force, 176, 181, 192
surroundings, 484, 492, 493
synthesis, 402, 407, 435
system, 9, 11, 13, 16, 19, 27, 33, 36, 38, 47, 55, 59, 64, 73, 80, 84, 91, 96, 109, 111, 115, 473, 483, 484, 491, 492, 493, 502, 507
temperature, 134, 137, 138, 139, 140, 141, 146, 208, 473, 474, 477, 478, 479, 480, 485, 486, 487, 492, 493
theoretical yield, 464
theory, 7, 11, 19, 22, 23, 130, 162, 166, 167, 168, 169, 174, 306, 317, 318, 397, 419, 515, 524, 533
thermal equilibrium, 493
thermal expansion, 473
thermometer, 41, 57, 473, 493
uncertainty, 38, 39, 40, 41, 42, 43, 44, 45, 46, 47, 48, 50, 52, 53, 57, 58, 61, 62, 63, 66
unit, 4, 39, 72, 73, 74, 76, 77, 82, 85, 88, 91, 92, 93, 94, 96, 97, 98, 99, 145, 163, 190, 226, 340, 387, 452, 458

units, 14, 17, 43, 55, 57, 72, 73, 74, 75, 76, 84, 85, 89, 90, 91, 93, 96, 97, 109, 111, 115, 140, 145, 146, 156, 157, 353, 356, 477, 498, 502, 507
valence, 246, 247, 248, 249, 257, 260, 262, 263, 264, 270, 278, 280, 306, 307, 313, 314
valent, 247, 257, 266, 267, 280, 308, 309, 311, 324, 437
vapor, 7, 115, 120, 122, 151, 392, 395, 404, 554
velocity, 195
VSEPR, 306, 317, 318, 319, 320, 321, 324, 325
water, 17, 28, 30, 31, 43, 45, 74, 99, 112, 113, 114, 117, 119, 124, 138, 151, 174, 205, 225, 255, 263, 264, 297, 330, 339, 343, 345, 365, 366, 367, 377, 380, 381, 382, 384, 387, 388, 389, 392, 393, 394, 395, 396, 404, 412, 416, 417, 438, 442, 493, 533, 536, 537, 539, 540, 542, 562
wave, 171, 217, 223
wavelength, 171
weak force, 176, 192, 193
weight, 57
wire, 206
work, 58, 59, 81, 193
yield, 464, 465